TRACE ANALYSIS OF SPECIALTY AND ELECTRONIC GASES

TRACE ANALYSIS OF SPECIALTY AND ELECTRONIC GASES

Edited by

WILLIAM M. GEIGER
Consolidated Sciences
Houston, Texas

MARK W. RAYNOR
Matheson, Advanced Technology Center
Longmont, Colorado

Published by John Wiley & Sons, Inc., Hoboken, New Jersey.
Published simultaneously in Canada.

For general information on our other products and services please contact our Customer Care Department within the United States at (800) 762-2974, outside the United States at (317) 572-3993 or fax (317) 572-4002.

Wiley also publishes its books in a variety of electronic formats. Some content that appears in print, however, may not be available in electronic formats. For more information about Wiley products, visit our web site at www.wiley.com.

Library of Congress Cataloging-in-Publication Data:

Trace analysis of specialty and electronic gases / edited by William M. Geiger, Consolidated Sciences, Houston, TX, Mark W. Raynor, Matheson, Advanced Technology Center, Longmont, CO.
  pages cm
Includes bibliographical references and index.
ISBN 978-1-118-06566-2 (cloth)
1. Gases—Analysis. 2. Trace elements—Analysis. 3. Gases—Spectra. I. Geiger, William M., 1948–
II. Raynor, Mark W., 1961–
QD121.T73 2013
543—dc23   2012050030

Printed in the United States of America.

10 9 8 7 6 5 4 3 2 1

CONTRIBUTORS

FLORIAN ADLER, JILA, National Institute of Standards and Technology, and University of Colorado, Department of Physics, Boulder, CO

KRIS A. BERTNESS, National Institute of Standards and Technology, Boulder, CO

MARTINE CARRE, Air Liquide, Jouy-en-Josas, France

DANIEL R. CHASE, Matheson, Advanced Technology Center, Longmont, CO

KEVIN C. COSSEL, JILA, National Institute of Standards and Technology, and University of Colorado, Department of Physics, Boulder, CO

DANIEL COWLES, Air Liquide Balazs NanoAnalysis, Dallas, TX

DARON DECKER, Agilent Technologies, Pearland, TX

WILLIAM M. GEIGER, Consolidated Sciences, Pasadena, TX

TRACEY JACKSIER, Air Liquide, Newark, DE

SUHAS N. KETKAR, Air Products and Chemicals, Allentown, PA

ROBERT LASCOLA, Savannah River National Laboratory, Aiken, SC

RAFAL LEWICKI, Rice University, Houston, TX

BARBARA MARSHIK, MKS Instruments, Inc., Methuen, MA

SCOTT MCWHORTER, Savannah River National Laboratory, Aiken, SC

GLENN M. MITCHELL, Matheson, Advanced Technology Center, Longmont, CO

JORGE E. PÉREZ, CIC Photonics, Inc., Albuquerque, NM

MARK W. RAYNOR, Matheson, Advanced Technology Center, Longmont, CO

LEONARD M. SIDISKY, Supelco, Bellefonte, PA

KOHEI TARUTANI, Air Liquide Japan, Ibaraki, Japan

FRANK K. TITTEL, Rice University, Houston, TX

STEPHEN VAUGHAN, Custom Gas Solutions, Durham, NC

JUN YE, JILA, National Institute of Standards and Technology, and University of Colorado, Department of Physics, Boulder, CO

CONTENTS

2 Sample Preparation and ICP–MS Analysis of Gases for Metals 21

Tracey Jacksier, Kohei Tarutani, and Martine Carre

3 Novel Improvements in FTIR Analysis of Specialty Gases 43

Barbara Marshik and Jorge E. Pérez

4 Emerging Infrared Laser Absorption Spectroscopic Techniques for Gas Analysis 71

Frank K. Tittel, Rafal Lewicki, Robert Lascola, and Scott McWhorter

LIST OF FIGURES

LIST OF TABLES

FOREWORD

In December 1996, with the help of a talented group of contributing authors, I put together a book for Wiley entitled *Specialty Gas Analysis: A Practical Guidebook*. This book was directed toward the analysis of ultrapure gases, exotic gases, reactive gases, blended gases, and/or combinations thereof. At that time in its history, the gas industry was moving headlong into the business of producing these types of gases for industries such as semiconductor manufacturing, not because it was a high-volume market, but more because it was a high-margin market. These gas products, by their very nature, were difficult to make, contain, and transport. As such, the prices they demanded were significantly higher than those of more common gases such as bulk gases, inert gases, industrial gases, and even medical-grade gases. In addition, the purity and accuracy requirements of these gases demanded greater analytical scrutiny for QA/QC in production, distribution, and inline analysis. This, in turn, drove the science involved in performing these analyses to new levels, demanding more research, better equipment, and more challenging methods. The book was intended to be a helpful resource for those entering this area of analysis and/or research, as well as to serve as an overview of current practices for those already involved in this field.

Several years prior to putting this book together, I had been tasked with building, equipping, and staffing a new gas analysis laboratory for a major semiconductor manufacturer. My efforts were stifled by the lack of available literature on the subject. I found this confusing, as it was apparent that a substantial group of people were already actively working in this area — so where was the literature? It turns out that the majority of those involved in this type of analysis worked for gas producers and/or distributors, and their analytical methods were considered by their companies to be proprietary. The proprietary designation had inhibited these scientists from sharing what they had learned, which in many cases went against the nature of their background and education, as most had, at one point or another, participated in peer-reviewed research. With the help of the contributing authors, we were able to convince most of their management organizations that their contributions to the publication of this book would serve as a showcase for their companies' talents, and could therefore be a useful tool for sales and marketing. Few could resist this approach, and most consented. As such, *Specialty Gas Analysis* provided an outlet for these researchers and/or practitioners, and gave them the opportunity to share what they had discovered, as well as to benefit from the work of others — much more in line with traditional scientific training. *Specialty Gas Analysis* was well received and has sold many copies since its publication some 16 years ago.

Recently, I was asked by Wiley to review a proposal for a new book that was directed toward updating and expanding *Specialty Gas Analysis*. I was very pleased to see that the proposal had been submitted by a contributor to *Specialty Gas Analysis*. The proposal was thorough and well designed. The book was to be put together by two excellent, seasoned scientists in this particular field: Mark Raynor and Bill Geiger. After careful review, I gave it my highest recommendation.

Having now seen the manuscript, I can honestly say that it delivers even more than I had hoped for initially. This excellent new book will serve as an invaluable resource for those involved in this area of science, and is a much-needed update to, and expansion of the somewhat-dated book *Specialty Gas Analysis*. You will find the pages of *Trace Analysis of Specialty and Electronic Gases* full of useful information, insight, and guidance by an excellent group of scientists from a variety of backgrounds in this field. The list of contributing authors was very well chosen and reads like a "Who's Who" of gas analysts. Their combined experience numbers in the hundreds of years, and their generosity in sharing their knowledge and experience will be a vital resource to anyone working in trace analysis of these challenging matrices. *Trace Analysis of Specialty and Electronic Gases* will increase the knowledge base in this area of analytical chemistry by sharing the combined knowledge of these highly experienced scientists across a broad range of practitioners.

JEREMIAH D. HOGAN

May 2012

ACKNOWLEDGMENTS

The idea for this book had crossed our minds on more than one occasion before we were approached by the publisher, who suggested that we put together a book based on an article we had written in *Spectroscopy*. However, this would not have been practical had we not met the many friends in industry and academia who were willing to make contributions of their personal time — all this while maintaining the responsibilities of their "day jobs." We would also like to recognize the efforts of Cindi Foster, Mark Ripkowski, and John Durr of CONSCI for compiling Appendix A, and Jesus Anguiano, also with CONSCI, for Appendix B.

Most of all we want to thank Sarah Meyer Truswell for the multipurpose role she played in keeping us on task; vetting reference documents; checking grammar, style, and English; performing the detailed electronic typesetting; and doing the artwork, all of which may have been beyond our skill sets.

W. M. G. and M. W. R.

ACRONYMS

AED	Atomic emission detection
AM	Amplitude modulation
AN	Army/Navy
AOM	Acousto-optic modulator
APCI	Atmospheric pressure chemical ionization
APIMS	Atmospheric pressure ionization mass spectrometry
ASB	Aluminum silicon bronze composite
AU	Absorbance unit
BSI	British Standards Institute
CAS	Conventional absorption spectroscopy
CCT	Collision cell technology
CE	Cavity enhanced
CEP	Carrier envelope phase
CFC	Chlorofluorocarbons
CGA	Compressed Gas Association

CIC	Compound-independent calibration
CID	Collision-induced dissociation
CLS	Classical least squares
CP	Cool plasma
CPAS	Conventional photoacoustic spectroscopy
CRDS	Cavity ring-down spectroscopy
CTFE	Chlorotrifluoroethylene
CW	Continuous wave
DC	Direct current
DFB	Distributed feedback
DFCS	Direct frequency-comb spectroscopy
DID	Discharge ionization detection
DIN	Deutsches Institut für Normung
DISS	Diameter-indexed stainless steel
DL	Detection limit
DOAS	Differential optical absorption spectroscopy
EC	External cavity
ECD	Electron capture detection
EFNI	Electroformed nickel
EI	Electron impact
EPA	(U.S.) Environmental Protection Agency
ESG	Electronic specialty gases
ETFE	Ethyltrifluoroethylene
ETV	Electrothermal vaporization
FDA	(U.S.) Food and Drug Administration
FEP	Fluorinated ethylene propylene
FET	Field-effect transistor
FID	Flame ionization detector
FPD	Flame photometric detection
FS	Fused silica
FSR	Fused silica (adapter fitting) with removable liner
FTIR	Fourier transform infrared
FTS	Fourier transform spectrometer
FWHH	Full width at half height

GC	Gas chromatography
GFAA	Graphite furnace atomic absorption
GLC	Gas-liquid chromatography
GSC	Gas-solid chromatography
GSV	Gas sampling valve
HCFC	Hydrochlorofluorocarbons
HEMT	High electron mobility transistor
HID	Helium ionization detection
HITRAN	High-resolution transmission molecular absorption
HNF	Highly nonlinear fiber
HR	High resolution
ICL	Interband cascade laser
ICOS	Integrated cavity output spectroscopy
ICP	Inductively coupled plasma
ID	Internal diameter
ILS	Intracavity laser spectroscopy
IMS	Ion mobility spectrometry
IP	Ionization potential
ITRS	International Technology Roadmap for Semiconductors
KED	Kinetic energy discrimination
LAS	Laser absorption spectroscopy
LD	Laser diode
LDL	Lower detectable limit
LFPG	Low-frost-point generator
LIBS	Laser-induced breakdown spectroscopy
LIDAR	Light-induced detection and ranging
LIF	Laser-induced fluorescence
MCT	Mercury cadmium telluride
MEK	Methyl ethyl ketone
MEMS	Microelectromechanical system
MFC	Mass flow controller
MIP	Microwave-induced plasma
mR	Microresonator
MS	Mass spectrometry

MSD	Mass selective detection
NDIR	Nondispersive infrared
NICE	Noise-immune cavity-enhanced
NIST	National Institute of Standards and Technology
NPL	(U.K.) National Physical Laboratory
NPT	National pipe thread
OD	Outer diameter
OES	Optical emission spectrometry
OFCS	Optical frequency-comb spectroscopy
OHMS	Optical heterodyne molecular spectroscopy
OPO	Optical parametric oscillators
PA	Photoacoustic
PAEK	Poly(aryether ketone)
PAS	Photoacoustic spectroscopy
PC	Polycarbonate
PDA	Photodiode array
PDID	Pulsed discharge ionization detection
PEEK	Poly(ether ether ketone)
PEG	Poly(ethylene glycol)
PEK	Poly ether ketone
PEKEKK	Poly(ether ketone ether ketone ketone)
PEKK	Poly(ether ketone ketone)
PFA	Perfluoroalkoxy
PFC	Perfluorocarbons
PFPD	Pulsed flame photometric detection
PLOT	Porous-layer open tubular (column)
PLS	Partial least squares
PMT	Photo multiplier tube
PPS	Poly(phenylene sulfide)
PTFE	Poly(tetrafluoroethylene)
PVDF	Poly(vinyldene fluoride)
QCL	Quantum cascade laser
QCM	Quartz crystal microbalance
QEPAS	Quartz-enhanced photoacoustic spectroscopy

QMS	Quadrupole mass spectrometry
QTF	Quartz tuning fork
RDC	Ring-down cavity
RF	Radio frequency
RRF	Relative response factor
SAW	Surface acoustic waves
SIM	Selected ion monitoring
SNR	Signal-to-noise ratio
SRM	Standard reference materials
SS	Stainless steel
TDLAS	Tunable diode laser spectroscopy
THF	Tetrahydrofuran
TIC	Total ion current
TOF	Time of flight
VCR	Vacuum coupling radiation
VIPA	Virtually imaged phased array
VSL	Van Swinden Laboratory or Dutch Metrology Institute
WCOT	Wall-coated open tubular (column)
WM	Wavelength modulation
ZDV	Zero dead volume

CHAPTER 1

INTRODUCTION TO GAS ANALYSIS: PAST AND FUTURE

SUHAS N. KETKAR

Air Products and Chemicals, Allentown, Pennsylvania

It is useful to look at the historical developments that contributed to the growth of analytical techniques, with a particular emphasis on the analysis of gases, not with the intent to provide a comprehensively detailed history of gas analysis, but to offer a peek into the development of the rudimentary techniques that began in the eighteenth century and how they evolved to the sophisticated, sensitive analyzers of the twenty-first century. This development has been influenced by discoveries and developments in all branches of science and engineering. Many of the analytical techniques discussed here are detailed in later chapters, especially those that have found application in the analysis of specialty and electronic gases.

The beginnings of gas analysis can be traced back to the mid-eighteenth century during the period when the constituents of air were discovered. At that time air was considered to be indivisible. Johann Beecher and George Stahl postulated the presence of phlogiston to explain how air initially supports combustion but later gets saturated with phlogiston and does not support combustion. In 1750, Joseph Black discovered carbon dioxide during a reaction of magnesium carbonate and sulfuric acid. He demonstrated that the gas produced during this reaction was the same as that

Trace Analysis of Specialty and Electronic Gases.
By William M. Geiger and Mark W. Raynor. Copyright © 2013 John Wiley & Sons, Inc.

produced from combustion and that this gas was not atmospheric air. The experiments that Black carried out are thought to have introduced the art of gas analysis [1].

In 1774, Joseph Priestley discovered oxygen. He indicated that he had discovered an air that was five or six times as good as common air, and called it "dephlogisticated air." Although he had concluded that this air constituted about 20 % of atmospheric air, he was not willing to give up his belief in the phlogiston theory. The following year Antoine Lavoisier made quantitative measurements of Priestley's experiment to propose the laws of conservation of mass as well as the theory of combustion. Lavoisier's careful measurements of the weights of the reactants and products in a chemical reaction are regarded as one of the first truly quantitative measurements of a chemical reaction. Lavoisier also investigated the composition of water and air. Using volumetric measurements during experiments by heating mercury, he concluded that air was one-fifth oxygen. This was the beginning of the use of the volumetric method for gas analysis and the first quantitative measurements made on gases.

1.1 The Beginning

The beginning of the industrial revolution in the nineteenth century created a need for the analysis of gas composition in mining and, specifically, the iron industry, which relied on the combustion of charcoal, or coke. Sampling apparatus was developed to obtain samples, which were shipped to central laboratories for analysis. One of the earliest examples of such apparatus is from 1885 (see Figure 1.1) [2].

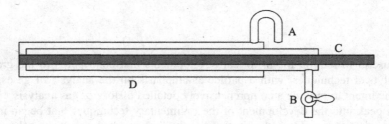

Figure 1.1   Sampling apparatus used for sampling from a furnace [2].

The sampling apparatus was made of a metal alloy, usually copper or iron. The gas sample flows through tube A, which is kept cool by flowing water through concentric tube C. Sample tube A is connected to a sampling tube and an aspirator. The sampling tube can be removed and sent to a laboratory for analysis. For gas sampling in mines, small sampling tubes were used (Figure 1.2).

These tubes were made of fusible glass and were evacuated in a laboratory using mercury pumps and sealed. Once inside the mine, the seals were opened to obtain the atmospheric sample inside the mine and then resealed. Similar sampling tubes and apparatus were developed to capture gases dissolved in liquids as well as gases formed during chemical reactions.

a   b

Figure 1.2   Small sampling tube for gas sampling in mines [2].

Analytical methods based on selective absorption and combustion were used to quantitatively measure the composition of the gases in the sample. In the late nineteenth and early twentieth centuries, it was recognized that shipping samples to a laboratory for analysis was a time-consuming process and that many industries needed faster analyses. Early in the twentieth century, Alexander Wright and Company of London, England, introduced the Simmance Abady Precision Combustion Recorder, which was capable of automatically measuring the percentage of carbon dioxide in flue gases as fast as every 3 minutes [3]. This is one of the first commercial instruments capable of making quantitative gas measurements automatically and displaying the results on a recorder in real time. One of these early instruments, along with the output measured is shown in Figure 1.3.

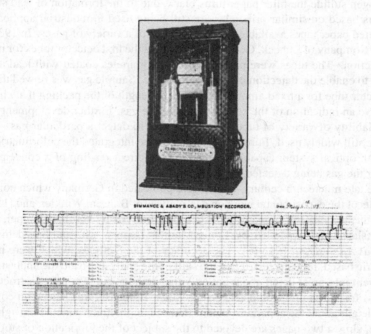

Figure 1.3   Simmance Abady Combustion Recorder with output reel [3].

Although the Simmance Abady CO_2 recorder was able to provide measurements every 3 minutes, there was a need to obtain measurements at even faster rates. In 1907, Strache et al. described the Autolysator, which was able to provide carbon dioxide concentration at any instant of time [4]. In 1906, Haber described the gas refractometer, which used light refraction, for measurement of the composition of gases. Carl Zeiss of Jena, Germany, introduced a commercial gas refractometer [2].

By 1910, volumetric methods had been established for the measurement of gases, which included oxygen, hydrogen, nitrogen, ammonia, hydrogen chloride, chlorine, carbon monoxide, nitric oxide, silicon tetrafluoride, and acetylene. These gases are used routinely in the semiconductor fabrication industry today. Although the methods developed in the early twentieth century were not as sophisticated or sensitive as the techniques available today, they nonetheless laid the foundation for rapid development in the field of gas analysis. Gas analysis methods, developed to meet the needs of the industrial revolution, also found applications in other fields. The method for the analysis of carbon dioxide found application in the field of medicine for the analysis of expired breath. In 1920, August Krogh described an apparatus to measure carbon dioxide in expired breath to an accuracy of 0.001 % [5].

In the early twentieth century, reactive methods relying on color-changing reactions were being used for the detection of inorganic compounds. An early application of this to gas detection was the use of lead acetate to detect hydrogen sulfide gas. This method used a filter paper dipped in a solution of lead acetate. Upon exposure to hydrogen sulfide the filter paper turns black, due to the formation of lead sulfide. Detectors based on similar principles are still being used in industrial applications, with coated paper tapes available for the detection of a variety of gases. In 1937, the Draeger Company of Lubeck, Germany, introduced the first detector tubes for mobile gas detection. The tubes were packed with solid granules coated with lead acetate solution to enable the detection of hydrogen sulfide. Sample gas was flowed through the detector tube for a fixed amount of time. The length of the packing that changed color gave an indication of the concentration of the gas. Further development led to the availability of variety of tubes, each designed to detect a particular gas. These tubes are still widely used. Further refinements have integrated the colorimetric concept to an optical system capable of enabling the direct reading of a concentration value for the gas being detected.

In the late nineteenth century, books were published in Germany which collected the results of the experimental work on gas analysis of Bunsen, Winkler, and Hempel. In 1902 the Macmillan Company published *Gas Analysis* by Walther Hempel, which was translated and expanded by Louis Monroe Dennis, who published his own book, *Gas Analysis*, in 1913 [6]. The accuracy of the methods, which were mainly volumetric based, relied primarily on the skill of the analyst. To aid the analyst, these books went into great detail about not only the methods but also the construction of apparatus for the sampling and analysis of gases. Emphasis was placed on covering minute details about the apparatus, details that one might find trivial in hindsight. For example, almost two pages are devoted to the subject of the lubrication of stopcocks used in gas analysis apparatus. Detailed recipes are given for making lubricants using commonly available ingredients such as Vaseline and paraffin. These books were almost certainly used as reference guides by scientists working in laboratories performing gas analysis.

In 1920, H. A. Daynes reported the development of the Katharometer, an instrument to give an automatic indication of the presence of small quantities of hydrogen in air [7]. This instrument was based on measuring the heat loss from two heating coils surrounded by the gas. The much higher thermal conductivity of hydrogen as compared to air resulted in the Katharometer being very sensitive to small amounts of hydrogen in air. In 1935, GOW–MAC Instruments of Newark, New Jersey, introduced an automotive engine analyzer [8]. This instrument used a thermal conductivity detector to measure carbon dioxide in engine exhaust. This led to the development of binary gas analyzers which were used by the U.S. Navy during World War II. Thermal conductivity detectors played a key role in the development of gas chromatography and its application to gas analysis. Thermal conductivity binary gas analyzers are still used for the measurement and control of binary gas mixtures used in the semiconductor industry (e.g., phosphine in hydrogen, arsine in hydrogen). In 1989, Jonathan Stag disclosed the use of acoustic measurements to determine the composition of a binary fluid mixture. Commercial analyzers based on this are available and are finding use in real-time control of binary mixtures produced using dynamic blending.

1.2 Gas Chromatography

The technique of chromatography owes its existence to the pioneering work of Ramsay, who used adsorbents to separate mixtures of gases and vapors. Tswett used adsorption to separate plant pigments and has been credited for coining the word *chromatography*, meaning writing in color. Chromatography as a technique to analyze gases was developed in the mid-1940s by Erika Cremer and her student, Fritz Prior. Prior's doctoral dissertation from 1947 describes the first chromatograph used to analyze for gaseous samples [9]. Although rudimentary by today's standards, it had all the components of the modern gas chromatograph (GC). The gas sample was introduced using a gas burette, hydrogen was used as the carrier gas, and silica gel and activated carbon were used in the column, which was kept at a constant temperature by submerging it in a liquid bath. A thermal conductivity cell was used as a detector, and was coupled to a recorder, which served as the output device. The original setup and a sample chromatogram are shown in Figures 1.4 and 1.5 [10,11].

The need for more sensitive analysis, not just for gases, spurred innovation in all components of the simple gas chromatograph developed by Cremer and Prior. Gas burettes gave way to gastight syringes, which were replaced by metallic injection valves with elastomeric seals. Multiport rotary valves were introduced which could accommodate column switching and other methods that were developed for gas chromatography. To meet the need for lower impurity levels in semiconductor gases, better more compatible elastomers were used. Diaphragm valves, which rely on metal-to-metal seals, became available for use. These valves have exceptionally long lifetimes, which is an added advantage, and they are used routinely in process monitoring. In 1957, Marcel Golay proposed using very long (>90 m) tubes of narrow diameter (0.25 μm) lined with a very thin liquid film which would significantly improve the separating power of a chromatographic column. These capillary columns, or Golay columns, drastically improved the speed of chromatographic analysis by enabling the separation of multiple impurities in a short time [12]. Molecular sieves

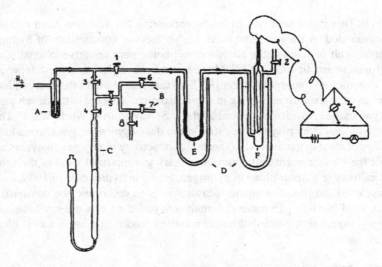

Figure 1.4   GC system used in Prior's work in 1945–1947. A, Adsorbent for purification of the carrier gas (hydrogen); B, sample inlet system; C, buret containing mercury with niveau glass for sample introduction; D, Dewar flasks; E, separation column (1 cm OD containing in a 20 cm length, silica gel or activated charcoal); F, thermal conductivity detector; 1 to 8 are glass stopcocks. A vacuum pump was connected to the system at stopcock 8. With kind permission from Springer Science+Business Media: [10], fig. 1.

(or zeolites) of different pore sizes began to be used for the analysis of gases. The 1980s saw the development of porous polymer packings for columns such as Porapak (Millipore Corporation), Chromosorb (Johns Manville Corporation), and HayeSep (Hayes Separation, Inc.) which found widespread acceptance. The HayeSep A column found application in the air separation industry since it could separate nitrogen, oxygen, argon, and carbon monoxide at room temperature. The analysis of reactive and corrosive gases was facilitated by the use of precolumns and back-flush techniques, which prevented the sample gases from attacking the downstream components of the gas chromatograph [13–15]. The needs of the space program led to the development of micro-GCs to monitor for contaminants in spacecraft air.

Due to their ruggedness, capacity, and simple separation requirements, packed columns continue to be used for a wide variety of applications in the gas industry and other industrial and process environments, despite the advancement of capillary columns. The capillary column proposed by Golay in 1957 drastically improved the speed of chromatographic analysis, but perhaps more important, it provided higher plate counts, making it a more efficient column capable of separating close-boiling analytes in complex mixtures. These capillary columns are also known as *open tubular columns*, and with the traditional use of a liquid film coating as the stationary phase, are sometimes referred to as *wall-coated open tubular* (WCOT) *columns*. A solid stationary phase can also be achieved in these columns by applying a thin coating of a solid material. These types of columns, known as *porous layer*

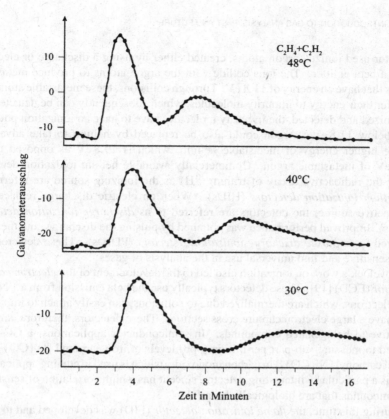

Figure 1.5 Separation of ethylene and acetylene. *x*-axis, time (min); *y*-axis, galvanometer deflection. With kind permission from Springer Science+Business Media: [11], fig. 3.

open tubular (PLOT) *columns*, have found widespread applications in gas analysis. Molecular sieve–coated PLOT columns are useful in the separation of permanent gases, whereas alumina- or divinylbenzene-coated PLOT columns have found use in the separation of C_1 to C_{10} hydrocarbons in specialty gases. In some applications for the analysis of electronic specialty gases the GC is outfitted with two columns, one with a molecular sieve stationary phase and the other with a divinylbenzene stationary phase to analyze for all impurities that are of interest.In 1979, S. C. Terry proposed a gas chromatograph integrated on a 2-inch silicon substrate in 1979 [16,17]. This miniaturization resulted in not only in a very small footprint, but also in a faster analysis time and reduction in the amount of gas needed for analysis, all of which were requirements of the space program. In 1995, Microsensor Technologies, Inc. introduced a portable GC based on this concept.

Along with advances in column technology was the development of specialized ionization detectors in the late 1950s. Work by J. E. Lovelock demonstrated the use of argon metastable atoms for detecting impurities in the effluent of a GC [18]. This detector used ionized argon atoms, created either by using a discharge or electrons from a beta emitter. The ions collide with the argon atoms to produce metastable atoms that have an energy of 11.8 eV. Through collisions, these metastable atoms can transfer their energy to impurity molecules, which subsequently can be detected. To be ionized and detected, the impurity molecules have to have an ionization potential (IP) below 11.8 eV. Argon could also be replaced by helium to take advantage of the higher energy of metastable helium, which is 19.8 eV as opposed to the 11.2 eV of metastable argon. Commercially available helium ionization detectors using the radioactive decay of tritium (^{3}H) as the ionizing source are referred to as *helium ionization detectors* (HIDs). When an electric discharge replaces the radioactive source, the detectors are referred to as *discharge ionization detectors* (DIDs). Improved performance was obtained by pulsing the discharge, and these are referred to as *pulsed discharge ionization detectors* (PDIDs). These detectors are very sensitive and find universal use in the analysis of gases.

Lovelock's work on ionization also led to the development of the *electron capture detector* (ECD) [19]. These detectors typically use the beta emission from a ^{63}Ni foil. The electrons, which are thermalized due to collisions, can easily attach to impurities that have a large electron capture cross sections. These detectors, therefore, are very sensitive to halogenated compounds. In semiconductor applications, a GC–ECD is used to measure sub-part per billion (ppb) levels of iron carbonyl [$Fe(CO)_5$] and nickel carbonyl [$Ni(CO)_4$] in carbon monoxide, which is used in etching applications. This is a particularly interesting detector, since it has a high correlation of sensitivity to compounds that are biologically harmful.

During this time, the *flame ionization detector* (FID) was developed and today is the most widely used gas chromatographic detector [20]. An FID uses a hydrogen/air flame to ionize the analyte molecules. The analyte molecules are ionized due to the thermal emission from microscopic carbon particles that are generated during the combustion process. FIDs are sensitive to molecules containing C–H bonds. Just like the use of thermocouple detectors for continuous gas analysis, FIDs are also widely used for continuous monitoring of hydrocarbons in gas streams.

An unusual but surprisingly useful method of detection developed in the early 1950s has become a staple in the high-purity bulk gas business. This detector is based on the reduction of mercury oxide to elemental mercury by the reducing analyte. The first application, for the determination of ethylene, involved detection of the reduced mercury based on the extent of blackening of a sensitized strip [21]. Quantification was also possible using the strong mercury ultraviolet (UV) absorption at 253.7 nm. In the late 1980s, Trace Analytical introduced a GC system utilizing a reduction gas detector to achieve sub-ppb detection of hydrogen and carbon monoxide in bulk inert gases of interest to the semiconductor fabrication industry.

In the late 1950s, the *flame photometric detector* (FPD) (or the *emissivity detector*, as it was called then) was developed [22]. It was not as sensitive as the ionization detectors that were introduced around the same time. However, the FPD was very sensitive to compounds containing phosphorus or sulfur, which chemiluminesce in a flame. Pulsing of the flame in the *pulsed flame photometric detector* (PFPD) provides much higher sensitivities and selectivities for the detection of phosphorus and sulfur.

The PFPD is also used for the detection of compounds containing nitrogen, arsenic, tin, selenium, germanium, tellurium, and others.

In the late 1980s, Hewlett-Packard introduced a GC with an *atomic emission detector* (AED). The AED utilizes elemental emission lines generated in a microwave induced plasma to characterize the constituents in the plasma. With GC/AED the components separated by the GC column, in sequence enter the helium plasma, where they are atomized and excited. The atomic emission lines specific to the elements present in the molecule are detected with a spectrophotometer and used to identify and quantify the components. GC/AED provides a selective approach to impurity detection and is used most commonly for the analysis of impurities in hydride gases such as phosphine, arsine, and silane. Meanwhile, over a number of years from the 1960s on, what had earlier been stand-alone instruments were being interfaced with gas chromatographs to become powerful hyphenated technologies, such as GC/MS and GC–ICP–MS.

1.3 Ion Chromatography

As the name implies, ion chromatography is used to separate ions in an aqueous sample. Ions in a solution have different affinities to the resins in a chromatographic column, which leads to the separation of these ions. Ion chromatography as a general analytical technique was established after Small and others at Dow Chemical Company demonstrated the use of stripper columns to suppress the background due to the chromatographic eluent [23]. This suppression of the eluent ions enabled the use of conductivity detection to detect ions of interest. Ion chromatography is used to analyze for anions in the hydrolysis solutions of specialty gases. The anions detected can then be quantified and related back to the impurities present in the specialty gas. This technique has found use in the semiconductor industry, where there are strict requirements for acidic and basic contaminants such as hydrogen chloride, nitric acid, sulfur dioxide, and ammonia in clean room air. Impingers are used to sample clean room air, and the solution is analyzed using ion chromatography [24]. A similar technique is used to determine acidic and basic contaminants in purge gases used for 157-nm lithographic tools.

1.4 Mass Spectrometry

Mass spectrometry had its origin in the pioneering work of J. J. Thomson. During his experimental work using cathode ray tubes in 1910, he discovered that under the influence of a magnetic and an electrostatic field, charged particles followed different parabolic paths, which are dependent on their charge-to-mass ratio. The apparatus he used was referred to as Thomson's parabola spectrograph. By 1912, this apparatus had demonstrated the existence of stable isotopes by measuring ^{20}Ne and ^{22}Ne. F. W. Aston, who was working in the Cavendish Laboratory along with Thomson, is credited with coining the term *mass spectrograph*. During his career Aston discovered 212 of the 287 naturally occurring isotopes.

Advances in vacuum technology and detectors led to the commercial introduction of mass spectrometers. The first commercial mass spectrometers used a combination of electrostatic and magnetic fields to disperse ions of different masses. The sample was ionized using electron impact ionization, and either electric or magnetic scanning was used to separate ions of different mass-to-charge ratios. These mass spectrometers were very large in size and were typically installed in central laboratories where the samples could be sent for analysis. Gas analysis was limited to relatively high concentrations, and complex mixtures required solving simultaneous equations for measurements. Mass spectrometers found increased use in the field of gas analysis with the advent of quadrupole mass spectrometers. In the early 1950s, Wolfgang Paul and his co-workers developed a mass spectrometer based on the motion of charged particles in a quadrupolar electric field [25]. The field was generated by utilizing a combination of radio-frequency (RF) and direct-current (DC) fields applied to four cylindrical rods. The opposite rods were connected together and applied with a voltage of

$$U + V\cos(wt) \text{ and } U - V\cos(wt) \tag{1.1}$$

The placement of cylindrical rods at the corners of a square was chosen such that the radius of the rods was 1.144 times the radius of the circle inscribed between the rods. This placement resulted in the generation of a quadrupolar field, and the trajectory of charged particles in such a field is described by the Mathieu equation. The ratio of the RF and DC voltages and the peak amplitude of the RF voltage determine the mass-to-charge ratio of the ions that would be on stable trajectories and detectable by the detector. In a strict sense, these are not spectrometers but, rather, filters. Nonetheless, these devices are described routinely as quadrupole mass spectrometers. Compared to magnetic mass spectrometers, quadrupole mass spectrometers are small in size and enjoy widespread use in physics and chemistry laboratories. Mass spectrometers typically use electron impact ionization in ionizing a sample. In this form of ionization, typically, 70 eV electrons are used for the ionization and only about 1.0×10^{-6} of the sample molecules are ionized. In gas analysis applications, even with such low ionization efficiency, quadrupole mass spectrometers have found use for detecting impurities in the low-part per million ppm) and even the ppb range. Some cryogenic air separation plants use these types of mass spectrometers to monitor and control the purity of the gases that are produced. As indicated previously, the use of gas chromatography with electron impact ionization mass spectrometry (EI–MS) has also become widespread for trace impurity analysis.

Atmospheric pressure ionization sources were developed many years later than were electron impact ionization sources. The first reported coupling of an ionizer operated at atmospheric pressure to a mass spectrometer was demonstrated by Horning et al. from the Baylor College of Medicine, who observed that significant sensitivity improvement over other sources could be achieved [26]. This discovery was a milestone in the development of mass spectrographic application for detecting impurities in the sub-ppb range in both air monitoring and high-purity gas analysis applications. In 1976, Kambara and Kanomata published work demonstrating the use of atmospheric pressure ionization mass spectrometry (APIMS) for the analysis of impurities in nitrogen at the sub-ppb level. They used a novel approach of low-pressure collisions to overcome the issues with cluster formation that are present in high-pressure ionization sources [27]. Hitachi introduced a commercial APIMS instrument, based on the work of Kambara and Kanomata, to analyze for sub-ppb impurities in gases. Accurate determination of impurities in gases at the sub-ppb level required gas standards at these levels. This was accomplished by using dynamic dilution of ppm-level gas standards, which could easily be made using gravimetric techniques. Air Liquide patented a dynamic dilution system based on double dilution, and Praxair patented a single-dilution system capable of achieving high dilution ratios [28,29]. Air Liquide teamed up with VG Gas Analysis, and Praxair paired with Extrel Corporation, to develop commercial APIMS instrumentation tailored for sub-ppb analysis of inert gases. APIMS instrumentation is used widely in the semiconductor fabrication industry to certify bulk inert gases to meet specifications at part per trillion ppt) levels. Early APIMS applications of gas analysis were limited to nitrogen, argon, helium, and gases with a higher IP than the typical impurities of significance to the industry. Novel approaches were used to overcome some of these limitations. Clustering reactions were utilized to detect low levels of nitrogen in argon and water in oxygen [30,31]. Another novel solution to utilize proton transfer reactions was developed to detect impurities in oxygen [32].

1.5 Ion Mobility Spectrometry

Drift cells were used to study gas-phase ion molecule reactions in the 1950s, and by the late 1960s, a Plasma Chromatograph, an instrument based on the drift cells, was introduced. This instrument used an atmospheric pressure ionizer which was coupled to a drift cell operating at atmospheric pressure. Ions moving under the influence of a field gradient in the drift cells were separated according to their mobilities. This technique is very sensitive, and ion mobility spectrometers based on this have found widespread use in environmental monitoring, and in chemical warfare agent and explosives detection. Most airports are equipped with ion mobility–based instruments for explosive detection. Research at Air Products and Chemicals, Inc. led to the development of ion mobility spectrometers for detecting trace levels of impurities in gases [33,34]. In the late 1990s, SAES of Milan, Italy, using this licensed technology, introduced an ion mobility spectrometer to detect sub-ppb impurities in bulk inert gases for the semiconductor industry.

1.6 Optical Spectroscopy

The origins of optical spectroscopy can be traced back to Isaac Newton, who in the late seventeenth century, demonstrated that light from the sun could be separated into a series of colors and is thought to have coined the word *spectrum* [35]. His original apparatus, which used a prism as a dispersing unit, could be thought of as the first spectrometer. The eighteenth and nineteenth centuries heralded significant developments. In 1729, Pierre Bougeur observed that the amount of light that could pass through a liquid decreased with increasing thickness; In 1760, Johann Heinrich Lambert published the law of absorption, which was based partially on Bougeur's observations; and in 1776, Alessandro Volta demonstrated that sparking different materials produced different colors. Volta used this to identify some gases by the color of their emissions, which can be thought of as the beginning of emission spectroscopy. In 1851, M. A. Masson produced the first spark emission spectroscope. In the following year, August Beer published a paper extending the absorption law of Lambert to solutions. The law of absorption, known as the Beer–Lambert law, is the basis for determining the concentration of different constituents in a sample in absorption spectroscopy. The first commercially available quartz prism spectrograph was introduced by Adam Hilger, Ltd., of London.

Nondispersive infrared (NDIR) spectroscopy is used widely in gas analysis. In 1937–1938, Luft and Lehrer, working in the laboratories of the German chemical company I. G. Farbenindustrie, developed a NDIR-based process analyzer that was used widely by the German chemical industry. Use of optical filters, tuned to the impurity of interest, resulted in instruments that were targeted the detection of a single gas species. Miniaturization has resulted in compact analyzers that have the ability to detect multiple impurities and have found widespread use in industry. With the advent of minicomputers, Fourier transform infrared spectroscopy (FTIR)–based instrumentation became available for gas analysis. In this instrument, a broadband light source is used in a Michelson interferometer. The sample cell is in one arm of the spectrometer. The Fourier transform of the recorded interferogram is related to the absorption characteristics of the sample. The advantage of the FTIR instrument is its ability to record the entire spectrum of a sample without the need for a dispersing element, thereby allowing determination of multiple impurities using a single instrument. Advances in this technology have led to the development of instruments that can detect impurities at levels in the low-ppb range [36].

The development of tunable diode lasers led to the introduction of instruments based on traditional absorption spectroscopy as well as cavity ring-down spectroscopy (CRDS) for the analysis of gases. The development of quantum cascade lasers (QCLs) led to the availability of light sources covering the range of 4 to 12 µm [37]. This has provided the sources needed for the detection of a broad range of gases. A variety of manufacturers have targeted different applications spanning the range from traditional gas analysis to environmental analysis to expired breath analysis.

1.7 Metals Analysis

Atomic emission spectroscopy using inductively coupled plasma (ICP) was developed for multielemental analysis of liquid samples in the 1960s. The Applied Research Laboratories in the United States introduced the first commercial instrument in 1974. Issues with optical interferences in emission spectroscopy, especially for samples with high matrix concentrations, led to the development of ICP sources suitable to be coupled to mass spectrometers [38,39]. Both Sciex, Inc. in Canada and VG Isotopes in the U.K. introduced ICP–MS instruments for the analysis of elemental impurities in liquid samples in the early 1980s. Hutton et. al. reported the first gas-phase determination of elemental impurities in 1990 using an ICP–MS instrument [40]. In 1992, Jahl and Barnes reported the use of a sealed ICP source coupled to an emission spectrometer for determining elemental impurities in specialty gases [41]. To meet the needs of the semiconductor industry, methods were developed for trapping the elemental impurities for subsequent analysis on an ICP–MS instrument. Hydrolyzable gases were dissolved in water to obtain aqueous samples, and nonhydrolyzable gases were bubbled through acids or bases to trap the elemental impurities. Inert gases, where the elemental impurities were particulate in nature, were filtered and the filter digested in acid. Coupling of GCs to an ICP–MS enabled the determination of gas-phase metallic impurities.

1.8   Species-Specific Analyzers

The reactivity of oxygen and the corrosivity of water vapor (moisture) have detrimental effects in many applications. This is particularly the case for the application of gases in the semiconductor industry. This has led to tight specifications for oxygen and moisture in all the gases used in semiconductor fabrication. The specifications vary by the application and range from ppt to ppm levels. Analytical instruments have been developed which are targeted specifically to the determination of oxygen or water.

1.8.1   Oxygen Analyzers

Electrochemical Analyzers   The use of electrochemical cells to study the oxidation and reduction of a material under the influence of an electrical stimulation dates back to the 1880s, when Kohlrausch described a two-electrode cell with platinum electrodes [42]. In 1952, Hersch developed an electrochemical cell with a platinum cathode and a lead anode in a potassium hydroxide solution acting as an electrolyte [43]. The depolarization of the cathode by the oxygen in the sample gas passing through the solution gives rise to a galvanic current, which is proportional to the concentration of oxygen in the gas in contact with the cathode. Researchers at Monsanto further refined this concept to make a rugged instrument for the monitoring of trace oxygen in petrochemical plant streams [44].

Electrochemistry, the study of the chemical response of a system to an electrical stimulation, is used routinely to study the oxidation and reduction that a material undergoes during this electrical stimulation. Many analytical techniques based on this have been developed for the study of solutions and surfaces. Two techniques based on an application of electrochemistry have been used for the analysis of oxygen and water in gases. In 1960, Keidel described the development of a coulometric analyzer for determining trace quantities of oxygen [45]. This analyzer relied on the quantitative electrochemical conversion of oxygen to electron current. Due to the quantitative conversion, Faraday's law could be used to convert to oxygen concentration the current measured. Moreover, since no current was generated in the absence of oxygen, there was very little background.

In the 1970s, Teledyne introduced an analyzer based on a sealed electrochemical cell, and Delta F Corporation introduced a coulometric analyzer for the measurement of trace oxygen. Refinements to these are still in use. The coulometric analyzer is currently capable of measuring oxygen in inert gases at levels as low as 100 ppt.

Analyzers based on the use of solid-state electrolytes were introduced in the 1960s. It had been recognized since the late nineteenth century that some solids begin to conduct electricity at elevated temperatures. It was also recognized that doped zirconia conducts oxygen ions at elevated temperatures. Panson and Ruka of Westinghouse Electric Corporation patented an "oxygen gauge" which used a solid-state oxygen sensor [46]. The sensor was based on mixed valence oxides that exhibited a resistance change at temperatures as low as 350 °C when the oxygen partial pressure changed. Sensors based on these are used widely in the automotive industry. Applications in the gas industry are limited since coexisting impurities such as hydrogen, carbon monoxide, and methane interfere with the determination of oxygen.

1.8.2 Paramagnetic Analyzers

The presence of two unpaired electrons gives rise to the paramagnetic properties of oxygen. Oxygen's large paramagnetic susceptibility has been exploited to determine low oxygen levels. Linus Pauling used the large paramagnetic susceptibility of oxygen to build an "oxygen meter" in response to the needs of the U.S. military. The meter consisted of a torsion balance with two hollow glass spheres. In the 1940s, under a contract from Cal Tech, A. O. Beckman, a staff member there, started manufacturing these instruments for the U.S. military. A variety of instruments based on the paramagnetic susceptibility of oxygen are available commercially. The coexistence of other paramagnetic gases, such as nitrous oxide, carbon dioxide, and water, in the sample gas exhibit small interference effects.

The paramagnetic analyzer is not very sensitive, and is typically not used for oxygen measurements below 0.1 %. Nonetheless, it is the analyzer used most commonly for analyzing oxygen impurity in fluorine-containing laser gas mixes that are used for photolithography.

1.8.3 Moisture Analyzers

In 1958, F. A. Keidel of DuPont patented an "apparatus for water determination" [47]. This apparatus utilized a film of a hygroscopic material, phosphorus pentoxide (P_2O_5), which became electrically conductive when wet. On the application of an electrical voltage, the current measured gave an indication of the amount of moisture absorbed according to Faraday's law. Meeco of Warrington, Pennsylvania, licensed this technology from Dupont and introduced a moisture analyzer for gases using a phosphorus pentoxide cell. Initially, the analyzers were used to measure moisture in fluorocarbons; however, the approach soon found applications in the natural gas industry, where corrosion in pipelines was a major issue. With the growth of the semiconductor industry in the 1980s, it found itself the leader in the analysis of moisture in semiconductor gases. As the specifications for moisture for semiconductor gases approached the low- and sub-ppb levels, analyzers based on spectroscopic techniques gained prominence.

Laser Spectroscopy–Based Moisture Analyzers   A number of technological advances have led to the development of laser spectroscopy–based moisture analyzers. The main advance occurred due to the needs of the long-haul telecommunications industry, which resulted in the development of low-cost tunable diode lasers in the region 1.3 to 2 μm. The wavelength of these diodes [typically, gallium arsenide (GaAs)–based devices] can be varied by varying the temperature and/or the diode current. Since water has an absorption line in the near infrared at 1.39253 μm, and quartz has regions of high transparency in the near infrared near 1.33 and 1.55 μm, these GaAs laser diodes could be employed for moisture detection in gases. Two approaches have been developed for moisture analyzers using diode lasers. The first uses traditional absorption spectroscopy. Analyzers based on this use a low-pressure absorption cell, typically at approximately 100 torr, to avoid issues with pressure broadening [48,49]. Measurements have been performed at sub-ppb levels in both inert and corrosive gases used in the semiconductor industry utilizing this approach. Delta F Corporation introduced an instrument based on tunable diode laser spectroscopy (TDLAS) to detect moisture in inert gases. The second approach used is cavity ring-down spectroscopy (CRDS) [50–53]. In this approach, the absorption cell is a high-finesse optical cavity. A pulse of light is injected into this cavity and the time decay of the transmitted light is measured. The exponential time decay depends on the concentration of the absorbing species, which in this case is water vapor, via the relation

$$\tau(\nu) = \frac{d}{c(1 - R + \sigma(\nu)Nd)} \tag{1.2}$$

where c is the speed of light, d is the cell length, R is the mirror reflectivity, N is the molecular density (concentration), σ is the absorption cross section, τ is the ring-down time, and ν is the frequency.

Thus, a time measurement can be used to measure the concentration. This approach has some advantages over absorption spectroscopy. The cell can be operated at atmospheric pressure since pressure broadening has minimal impact on the measurement. Changes in the intensity of the diode laser have no effect since the concentration measured is based on the decay time. However, the degradation of the cavity mirrors will be detrimental to the measurement sensitivity. Tiger Optics introduced a CRDS-based instrument to measure sub-ppb levels of moisture in inert gases. Both the TDLAS- and the CRDS-based instruments have been developed further to be able to determine moisture in corrosive gases. Excellent review articles are available which have compared the performance of moisture analyzers based on different techniques [54–57].

1.9 Sensors

The history of gas sensors can be traced back to the observation by Wagner and Hauffe in 1938 that atoms and molecules can interact with semiconductor surfaces and influence the conductivity. In 1962, Seiyama et al. described the first semiconductor-based gas sensor which relied on changes in the resistivity. In 1968, the Taguchi gas sensor, based on the change in conductivity of a tin oxide film, was introduced in Japan for the detection of combustible and reducing gases [58]. Advances in microelectromechanical system (MEMS) technology have enabled the miniaturization of sensors based on this and other technologies. Sensors based on changes in conductivity, mass, optical properties, and work function of the sensing element are widely available. In the post-9/11 world, a lot of focus has been placed on the detection of biological and chemical weapons. Funding has been available to develop instrumentation to detect chemical and biological agents in ambient air. Two types of piezoelectric sensors, one based on surface acoustic waves (SAWs) and the other based on quartz crystal microbalance (QCM), have been developed for this application. By changing the active adsorbing film on these sensors, it is anticipated that they will find widespread use in the specialty gas industry.

1.10 The Future

The successful demonstration of a solid-state transistor at Bell Labs in 1947, followed by the development of the integrated circuit by Kilby at Texas Instruments, started the electronic revolution. The path of the semiconductor industry following Moore's law has resulted in ever-shrinking dimensions, increasing computation prowess, and decreasing costs of electronic devices. Sensors, often referred to as *electronic noses*, have greatly benefited by this development [59,60]. Computational power available in a small package has led to the use of arrays of MEMS-based SAW sensors as well as QCM sensors to detect multiple coexisting impurities. This has been accomplished by developing and characterizing the adsorption properties of coatings and using fast computation to deconvolute the response of the sensor arrays to multiple impurities. This trend is likely to continue and enable the detection of virtually any molecule in trace quantities.

Development of laser-based spectroscopic methods continues, including cavity-enhanced techniques such as frequency-comb spectroscopy, a broadband approach allowing for detection of multiple species in a matrix with sensitivities matching or exceeding FTIR.

The need for field-portable analytical instruments for environmental monitoring has also led to the miniaturization of instruments that would have been unfathomable considering the size of original devices. Miniature FTIR instruments as well as miniature quadrupole mass spectrometers have been developed [61,62].

The analysis of specialty and electronic gases is a niche market for sensor and analytical instrument developers. New developments in sensor technology to meet the growing needs of homeland security and environmental monitoring will eventually lead to the availability of these systems to meet the needs of the specialty gas industry. One can envision the value (price/performance) of these devices to reach a point where they are ubiquitous in the specialty gas industry. It is possible to imagine, in a not too distant future, a sensor integrated into every container of a specialty gas that will provide the user with real-time information on the quality (purity). It is also possible to imagine a widespread array of sensors in a semiconductor fabrication facility that monitors the quality of the gas in the entire gas distribution system. With the availability of real-time purity information of the gases used in semiconductor processing, one can envision the use of feedback loops to fine-tune the operating conditions of processing instruments based on the quality of the gases.

REFERENCES

1. Szabádvary, F. (1966). *History of Analytical Chemistry*. New York: Pergamon Press.

2. Dennis, L. M. (1913). *Gas Analysis*. New York: Macmillan.

3. Low, F. R. (1911). The Simmance-Abady CO_2 Recorder. In *Power and the Engineer*. New York: McGraw-Hill, pp. 1446–1448.

4. Strache, M., Genzken, Z. & Jahoda, R. (1906). Gasanalyse. *Zeitschrift für Analytische Chemie*, *45*, 632–638

5. Krogh, A. (1920). A gas analysis apparatus accurate to 0.001 % mainly designed for respiratory exchange work. *Biochemical Journal*, *14*, 3–4.

6. Hempel, W. (1902). *Methods of Gas Analysis* (L. M. Dennis, Trans.). New York: Macmillan. (original work published 1889).

7. Daynes, H. A. (1920). The process of diffusion through a rubber membrane. *Proceedings of the Royal Society A: Mathematical, Physical and Engineering Sciences*, *97*, 685, 286–307.

8. GOW–MAC Instrument Company. *About us*. Retrieved from http://www.gow-mac.com/about.html.

9. Prior, F. (1947). *Über die bestimmung der adsorptionswärmen von gasen und dämpfen, unter anwendung der chromatographischen methode auf die gasphase*. Unpublished doctoral dissertation. University of Innsbruck, Austria.

10. Bobleter, O. (1996). Exhibition of the first gas chromatographic work of Erika Cremer and Fritz Prior. *Chromatographia*, *43*, 7/8, 444–446.

11. Cremer, E., & Müller, R. (1951). Trennung und quantitative Bestimmung kleiner Gasmengen durch Chromatographie. *Mikrochemie/Mikrochimica Acta*, *36/37*, 553–560.

12. Golay, M. J. E. (1958). In V. J. Coates, H. J. Noebels, and I. S. Fagerson (Eds.), *Gas Chromatography (1957 East Lansing Symposium)*. Academic Press: New York, pp. 1–13.

13. Andre, C. E., & Mosier, A. R. (1973). Direct gas chromatographic analysis of aqueous solutions of aliphatic *N*-nitrosamines. *Analytical Chemistry*, *45*, 2, 372–373.

14. Fett, E. R. (1963). Backflush applied to capillary column-flame ionization detector gas chromatography systems. *Analytical Chemistry*, *35*, 3, 419–420.

15. Laurensa, J. B., Petrus de Coning, J., & Swinley, J. M. (2001). Gas chromatographic analysis of trace gas impurities in tungsten hexafluoride. *Journal of Chromatography A, 911*, 107–112.

16. Terry, S. C. (1975). *A gas chromatography system fabricated on a silicon wafer using integrated circuit technology.* Doctoral dissertation. Stanford University. Retrieved from ProQuest Dissertations and Theses (7525618).

17. Terry, S. C., Jerman, J. H., & Angell, J. B. (1979). A gas chromatographic air analyzer fabricated on a silicon wafer. *IEEE Transactions on Electron Devices, 26*, 12, 1880–1886.

18. Lovelock, J. E. (1958). A sensitive detector for gas chromatography. *Journal of Chromatography A, 1*, 35–46.

19. Lovelock, J. E., & Lipsky, S. R. (1960). Electron affinity spectroscopy: a new method for the identification of functional groups in chemical compounds separated by gas chromatography. *Journal of the American Chemical Society, 82*, 2, 431–433.

20. Harley, J., Nel, W., & Pretorius, V. (1958). Flame ionization detector for gas chromatography. *Nature, 181*, 4603, 177–178.

21. Stitt, F., Tomimatsu, Y., & Tjensvold, A. (1953). *Process for determination of ethylene in gases.* U.S. Patent No. 2,648,598. Washington, DC: U.S. Patent and Trademark Office.

22. Institute of Petroleum (Great Britain), & Desty, D. H. (1958). Gas chromatography, 1958. *Proceedings of the Second Symposium Organized by the Gas Chromatography Discussion Group Under the Auspices of the Hydrocarbon Research Group of the Institute of Petroleum and the Koninklijke Nederlandse Chemische Vereniging Held at the Royal Tropical Institute, Amsterdam, May 19–23, 1958.*

23. Small, H., Stevens, T. S., & Bauman, W. C. (1975). Novel ion exchange chromatographic method using conductimetric detection. *Analytical Chemistry, 47*, 11, 1801–1809.

24. Komazaki, Y., Hamada, Y., Hashimoto, S., Fujita, T., & Tanaka, S. (1999). Development of an automated, simultaneous and continuous measurement system by using a diffusion scrubber coupled to ion chromatography for monitoring trace acidic and basic gases (HCl, HNO_3, SO_2, and H_3) in the atmosphere. *Analyst, 124*, 1151–1157.

25. Dawson, P. H. (1976). *Quadrupole Mass Spectrometry and Its Applications.* Amsterdam: Elsevier Scientific.

26. Horning, E. C., Horning, M. G., Carroll, D. I., Dzidic, I., & Stillwell, R. N. (1973). New picogram detection system based on a mass spectrometer with an external ionization source at atmospheric pressure. *Analytical Chemistry, 45*, 6, 936–943.

27. Kambara, H., & Kanomata, I. (1977). Chemical ionization by a needle electron source. *International Journal of Mass Spectrometry and Ion Physics, 24*, 4, 453–463.

28. Mettes, J., Kimura, T., & Schack, M. (1991). *Process for producing low-concentration gas mixtures, and apparatus for producing the same.* U.S. Patent No. 5,054,309. Washington, DC: U.S. Patent and Trademark Office.

29. Leggett, G., & Sonricker, M. (1993). *Calibration for ultra high purity gas analysis.* U.S. Patent No. 5,214,952. Washington, DC: U.S. Patent and Trademark Office.

30. Hunter, E. J., Homyak, A. R., & Ketkar, S. N. (1998). Detection of trace nitrogen in bulk argon using proton transfer reactions. *Journal of Vacuum Science and Technology A: Vacuum, Surfaces, and Films, 16*, 5, 3127–3130.

31. Scott, A. D., Hunter, E. J., & Ketkar, S. N. (1998). Use of a clustering reaction to detect low levels of moisture in bulk oxygen using an atmospheric pressure ionization mass spectrometer. *Analytical Chemistry, 70*, 9, 1802–1804.

32. Ketkar, S. N., Scott, A. D., & Hunter, E. J. (2000). The use of proton-transfer reactions to detect low levels of impurities in bulk oxygen using an atmospheric pressure ionization mass spectrometer. *International Journal of Mass Spectrometry, 206*, 1/2, 7–12.

33. Ketkar, S. N., & Dheandhanoo, S. (2001). Use of ion mobility spectrometry to determine low levels of impurities in gases. *Analytical Chemistry, 73*, 11, 2554–2557.

34. Ketkar, S. N., & Dheandhanoo, S. (2003). *Method of improving the performance of an ion mobility spectrometer used to detect trace atmospheric impurities in gases*. U.S. Patent No. 6,639,214. Washington, DC: U.S. Patent and Trademark Office.

35. MIT Spectroscopy Lab. *History. The history of spectroscopy*. Retrieved from http://web.mit.edu/spectroscopy/history/index.html.

36. Rosenthal, P., & Spartz, M. L. (2003). Instrumentation for real-time trace impurity detection of bulk ammonia in production environments. *Gases and Technology, 25*, 8–12.

37. Faist, J., Capasso, F., Sivco, D. L., Sirtori, C., Hutchinson, A. L., & Cho, A. Y. (1994). Quantum cascade laser. *Science, 264*, 5158, 553–556.

38. Date, A. R., & Gray, A. L. (1981). Plasma source mass spectrometry using an inductively coupled plasma and a high resolution quadrupole mass filter. *Analyst, 106*, 1255–1267.

39. Douglas, D. J., & French, J. B. (1981). Elemental analysis with a microwave-induced plasma/quadrupole mass spectrometer system. *Analytical Chemistry, 53*, 1, 37–41.

40. Hutton, R. C., Bridenne, M., Coffre, E., Marot, Y., & Simondet, F. (1990). Investigations into the direct analysis of semiconductor grade gases by inductively coupled plasma mass spectrometry. *Journal of Analytical Atomic Spectrometry, 5*, 6, 463–466.

41. Jahl, M. J., & Barnes, R. M. (1992). Analysis of silane with a sealed inductively coupled plasma discharge. *Journal of Analytical Atomic Spectrometry, 7*, 6, 833–838.

42. Göpel, W., Jones, T. A., Göpel, W., Zemel, J. N., & Seiyama, T. (2008). Historical remarks. In W. Göpel, J. Hesse, and J. N. Zemel (Eds.), *Sensors: Chemical and Biochemical Sensors*, Wiley-VCH. Weinheim, Germany: Part I, Vol. 2.

43. Hersch, P. (1952). Galvanic determination of traces of oxygen in gases. *Nature, 169*, 4306, 792–793.

44. Baker, W. J., Combs, J. F., Zinn, T. L., Wotring, A. W., & Wall, R. F. (1959). The galvanic cell oxygen analyzer. *Industrial and Engineering Chemistry, 51*, 6, 727–730.

45. Keidel, F. A. (1960). Coulometric analyzer for trace quantities of oxygen. *Industrial and Engineering Chemistry, 52*, 6, 490–493.

46. Panson, A. J., & Ruka, R. J. (1971). *Solid state oxygen gauge*. U.S. Patent No. 3,558,280. Washington, DC: U.S. Patent and Trademark Office.

47. Keidel, F. A. (1958). *Apparatus for water determination*. U.S. Patent No. 2,830,945. Washington, DC: U.S. Patent and Trademark Office.

48. Inman, R. S., & McAndrew, J. J. F. (1994). Application of tunable diode laser absorption spectroscopy to trace moisture measurement in gases. *Analytical Chemistry, 66*, 15, 2471–2479.

49. Hovde, D. C., Hodges, J. T., Scace, G. E., & Silver, J. A. (2001). Wavelength-modulation laser hygrometer for ultrasensitive detection of water vapor in semiconductor gases. *Applied Optics, 40*, 829–839.

50. Lehman, S. Y., Bertness, K. A., & Hodges, J. T. (2003). Detection of trace water in phosphine with cavity ring-down spectroscopy. *Journal of Crystal Growth, 250*, 1, 262–268.

51. Lehman, S. Y., Bertness, K. A., & Hodges, J. T. (2004). Optimal spectral region for real-time monitoring of sub-ppm levels of water in phosphine by cavity ring-down spectroscopy. *Journal of Crystal Growth, 261*, 2, 225–230.

52. Leeuwen, N. J., Diettrich, J. C., & Wilson, A. C. (2003). Spectroscopy: periodically locked continuous-wave cavity ringdown spectroscopy. *Applied Optics, 42*, 18, 3670–3677.

53. Dudek, J. B., Tarsa, P. B., Velasquez, A., Wladyslawski, M., Rabinowitz, P., & Lehmann, K. K. (2003). Trace moisture detection using continuous-wave cavity ring-down spectroscopy. *Analytical Chemistry, 75*, 17, 4599–4605.

54. Edwards, C. S., Barwood, G. P., Bell, S. A., Gill, P., & Stevens, M. (2001). A tunable diode laser absorption spectrometer for moisture measurements in the low parts in 10^9 range. *Measurement Science and Technology, 12*, 1214–1218.

55. Bell, S., Gardiner, T., Stevens, M., & Waterfield, K. (2002). *Evaluation of trace moisture sensors*. Teddington, UK: National Physical Laboratory. Retrieved from http://publications.npl.co.uk/npl_web/pdf/cbtlm19.pdf.

56. Abe, H., & Yamada, K. M. T. (2011). Performance evaluation of a trace-moisture analyzer based on cavity ring-down spectroscopy: direct comparison with the NMIJ trace-moisture standard. *Sensors and Actuators A: Physical, 165*, 2, 230–238.

57. Werle, P. (1998). Review of recent advances in laser based gas monitors. *Spectrochimica Acta A, 54*, 197–236.

58. Seiyama, T., Kato, A., Fujiishi, K., & Nagatani, M. (1962). A new detector for gaseous components using semiconductive thin films. *Analytical Chemistry, 34*, 11, 1502–1503.

59. Freidhoff, C. B., Young, R. M., Sriram, S., Braggins, T. T., O'Keefe, T. W., Adam, J. D., et al. (1999). Micro-electro-mechanical systems/vacuum technology: vacuum MEMS and microanalysis–chemical sensing using nonoptical microelectromechanical systems. *Journal of Vacuum Science and Technology A: Vacuum, Surfaces, and Films, 17*, 4, 2300.

60. Briand, D., Wingbrant, H., Sundgren, H., van der Schoot, S. B., Ekedahl, L.-G., Lundström, I., & de Rooij, N. F. (2003). Modulated operating temperature for MOSFET gas sensors: hydrogen recovery time reduction and gas discrimination. *Sensors and Actuators B, 93*, 276–285.

61. Briand, D., Manzardo, O., Hildenbrand, J., Wöllenstein, J., & de Rooij, N. F. (2007). Gas detection using a micromachined FTIR spectrometer. *Proceedings of the IEEE Sensors Conference, Atlanta, GA, October 28–31*, pp. 1364–1367.

62. Malcolm, A., Wright, S., Dash, N., Schwab, M. A., Finlay, A., & Syms, R. R. A. (2010). Miniature mass spectrometer systems based on a microengineered quadrupole filter. *Analytical Chemistry, 82*, 5, 1751–1758.

CHAPTER 2

SAMPLE PREPARATION AND ICP–MS ANALYSIS OF GASES FOR METALS

Tracey Jacksier,[1] Kohei Tarutani,[2] and Martine Carré[3]

[1]Air Liquide, Newark, Delaware
[2]Air Liquide Japan, Ibaraki, Japan
[3]Air Liquide, Jouy-en-Josas, France

2.1 Introduction

The presence of trace metal impurities in electronic specialty gases (ESGs) can modify the electrical properties of the insulating or conducting layers in a semiconductor device. The importance of controlling these impurities in the etching and deposition processes has increased as the feature sizes have decreased and become more complex. Typical metal levels today are around 10 parts per billion (ppb), but can reach below 10 parts per trillion (ppt) for specific metals in certain critical gases.

Some impurities originate from the product source itself, others from the reaction of the gas with the cylinder package. Storage of corrosive ESGs poses special challenges, owing to their corrosive and reactive nature. Therefore, the concentration of the metals in an ESG is also a measure of the quality of cylinder preparation [1]. For example, halogen-containing gases such as boron trichloride can hydrolyze in the presence of moisture and react with metal container surfaces, forming particles that compromise gas purity [2]. Carbon monoxide forms gas-phase metal carbonyls

Trace Analysis of Specialty and Electronic Gases.
By William M. Geiger and Mark W. Raynor. Copyright © 2013 John Wiley & Sons, Inc.

in contact with nickel- or iron-rich metal surfaces [3]. Further, if such impurity-laden gas streams reach the point of use, the wafer surface can become contaminated.

The ability to analyze quantitatively the concentration of metallic impurities at ppb levels and below has emerged as one of the most significant challenges for the quality assurance of ESGs. The ESGs that are used in the fabrication of semiconductor devices range from highly corrosive gases such as hydrogen chloride, chlorine, hydrogen bromide, and boron trichloride, to pyrophoric gases such as silane, to relatively inert gases such as hexafluoroethane, tetrafluoromethane, and nitrous oxide. Two principal approaches can be taken to measure impurities in ESGs: (1) extraction of impurities from the gas before analysis, and (2) direct analysis of the ESG.

Three different sampling methods can be used to extract the metallic impurities from a gas prior to analysis: (1) filtration on a membrane and analysis of the particles retained on the membrane; (2) hydrolysis in an aqueous solution; and (3) trapping a known volume of liquefied gas and analysis of the residue after slow evaporation. Owing to the different chemical and physical properties of the various ESGs, it is difficult to use only one sampling method. Subsequent analysis after sampling is highly dependent on the sampling method chosen as well as on the ESG sampled. For example, for a solution containing a high concentration of chloride (from hydrolysis of hydrogen chloride), it is not possible to analyze arsenic at ppt levels directly due to interferences. Indeed, with traditional nebulization into a quadrupole inductively coupled plasma–mass spectrometer (ICP–MS) there is an isobaric interference of $^{40}Ar^{35}Cl$ on the only isotope of arsenic (^{75}As). Analyzing without correction or with compensation can cause a significant bias in the measurement.

Direct introduction of ESGs into an analyzer, such as an ICP–MS, provides a significant advantage over indirect methods, in that sample preparation is not necessary. However, direct introduction requires modification of commercial equipment. There are also significant issues that must be overcome. Issues include a significant increase in isobaric interferences as well as attack or deposition on the sample cone. To decrease interferences, corrosion, and deposition, the sample gas volume is typically limited to below 5 mL/min, which limits the detection limits achievable. Perhaps the most significant issue to overcome using direct introduction is calibration. Standards for metals in gases are not readily available, making quantification difficult.

2.2   Extraction of Impurities Before Analysis

2.2.1   Filtration Method

The filtration method is generally used for metal analysis in carrier gases and noncorrosive gases where metallic impurities, if present, would be in particulate form. In this method, the suspended particles in the gas stream are retained on a membrane filter installed in a filter housing (Figure 2.1). The impurities captured on the filter are then dissolved in an acid solution, and the resulting solution can be analyzed by ICP–MS. A major advantage of the filtration method is that large quantities of gas can be filtered to concentrate any suspended particles and thereby obtain very low detection limits.

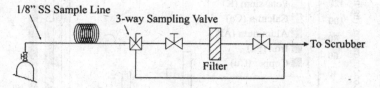

Figure 2.1   Filtration system setup.

To minimize the volume of sample gas necessary to reach ppb/ppt detection limits, it is important to decrease the background impurity concentration of the sampling system. Contamination can originate from three sources: (1) collection of the sample; (2) the purity of the reagents used; and (3) the presence of contaminants in the filter or filter housing. High-purity reagents and appropriate sampling methodologies have been used successfully to reduce the background. However, two of the most important aspects of this method are the selection of the filter housing and the filter. The filter housing must be corrosion resistant and not exhibit catalytic effects in the presence of the gas sampled. The selection of the filter is dependent on the material. In the following sections we discuss studies done to determine appropriate (a) filter and housing materials and (b) preparation procedures to minimize metal contamination from these components. Examples of metal analysis of trifluoromethane and silane by the filtration method are then presented.

Filter Selection   Four 0.2-μm commercial hydrophilic filter materials were evaluated to select the most appropriate filter for trace metal analysis: poly(tetrafluoroethylene) (PTFE), cellulose, polycarbonate (PC), and poly(vinylidenefluoride) (PVF) [4]. Each filter was washed with 16 mL of deionized water containing 800 μL of 30 % hydrochloric acid, and 200 μL of 70 % nitric acid (Tamapure, Tama Chemicals, Japan). The washing solution was subsequently analyzed by ICP–MS, and the results are shown in Figure 2.2.

PVF had the largest impurity levels of any of the filters evaluated. PC showed the lowest blank levels, except for sodium, which was easily removed by washing. Therefore, PC filters were subsequently used (0.2-μm polycarbonate filters, 47 mm, GTTP 04700 m, Millipore Corp., Billerica, MA).

Filter Housing   Contamination can occur from components in the filter housing as well. Therefore, the filter support screen and O-rings were changed to those of higher purity. A Millipore 47-mm high-pressure (700 bar) stainless steel filter housing (XX–4504700, Millipore Corp., Billerica, MA) was fitted with a Teflon support screen (XX440–4702, Millipore Corp., Billerica, MA). The Viton O-rings were replaced with Teflon O-rings (XX45047–10, Millipore Corp., Billerica, MA).

The filter housing, O-rings, and support screen were cleaned with ultrapure water and dried in a clean environment (class 100) to avoid entrainment of particles. The cleaned PC filter was placed in the filter housing using nonmetallic forceps.

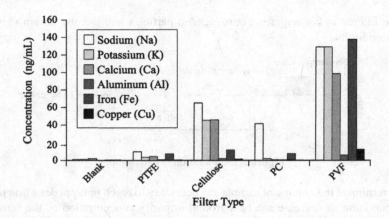

Figure 2.2   Metal impurity content in filters prior to filter washing.

The filter assembly was connected to the sampling system (Figure 2.1). After the requisite amount of gas was passed through the filter, the filter housing was removed from the sampling system. Under a clean hood (class 100) the housing was opened carefully. The filter was removed and placed into a Teflon container containing 20 g of 1 % aqua regia. After 20 minutes, the filter was discarded and the resulting acid solution was analyzed by ICP–MS.

It was necessary to run another filter as an experimental blank, as it is impossible to run a dynamic blank for filtration. It is important to ensure that the "blank" filter is of the same batch as the sample filter because the background impurities in the filters have been observed to change as a function of filter lot number.

To ensure that the sampling system did not contribute contamination, which may occur from valve switching, a system blank was obtained by purging nitrogen through the system and analyzing the "filter blank." The nitrogen gas supplied to the sampling system was prefiltered with an inline PTFE membrane filter (Wafer Guard, WGTF00MRU, Millipore Corp., Billerica, MA). The metal concentrations detected were independent of the nitrogen gas flow time and were the same as those of the filter blanks. Therefore, the metal contamination from the system was considered negligible.

Table 2.1 details the approximate volumes of gases that must be sampled to reach 1 ppt and 1 ppq detection limits. These are the minimum recommended volumes considering uniform detection limits of 100 ppt in 1 % aqua regia for the following elements: aluminum, arsenic, boron, calcium, cadmium, cobalt, copper, iron, potassium, magnesium, manganese, sodium, nickel, phosphorus, lead, antimony, silicon, and zinc. Care must be taken when sampling under conditions where there is not a controlled atmosphere, as larger samples maybe necessary to minimize environmental contributions.

Table 2.1   Minimum quantity of gas that must be sampled to reach the required detection limit (DL) using the filtration method

Gas Sampled	1 ppq DL (m³)	1 ppt DL (m³)
Argon	560	0.56
Helium	11,200	11.20
Hydrogen	11,200	11.20
Nitrogen	800	0.80
Oxygen	700	0.70

Analytical Results   Tables 2.2 and 2.3 illustrate the levels of metals in a high-purity trifluoromethane (CHF_3) cylinder for which 100 g was sampled using the filtration method and two cylinders of high-purity silane for which 2700 g was sampled.

Table 2.2   Metal levels in trifluoromethane obtained by filtration

Element	ppb w/w	Element	ppb w/w
Aluminum	0.25	Manganese	<0.02
Cadmium	0.02	Nickel	<0.02
Chromium	<0.02	Potassium	<0.02
Copper	<0.02	Silicon	<0.02
Iron	<0.02	Sodium	<0.02
Lead	<0.02	Zinc	<0.02
Magnesium	0.09		

2.2.2 Hydrolysis Method

Utilizing the hydrolysis method, the liquid or gas phase of the ESG can be hydrolyzed in an appropriate hydrolysis solution. In this method, there is a reaction between the hydrolysis solution (typically, water) and the ESG (hydrochloric acid, for example). Accordingly, the hydrochloric acid is solubilized. Impurities present in the hydrochloric acid dissolve in the acidic medium. This method is most appropriate for gases that dissolve or have a limited solubility in the hydrolysis solution. The procedures utilized are similar to those described elsewhere for gaseous and liquid-phase sampling [5] and are outlined below. Results for the metals analysis of hydrogen chloride and hydrogen bromide via hydrolysis are cited later in the section to exemplify the method.

Experimental   The hydrolysis sampling system is illustrated in Figure 2.3. As with the filtration method, contamination is a major concern. Every attempt was

Table 2.3   Metal levels in silane obtained by filtration

Element	Cylinder 1 (ppt w/w)	Cylinder 2 (ppt w/w)
Cadmium	1.50	<0.20
Chromium	6.00	<4.00
Copper	15.00	16.00
Iron	88.00	<12.00
Lead	<0.10	<0.10
Lithium	0.07	0.01
Magnesium	9.70	<1.00
Manganese	3.40	3.90
Molybdenum	0.70	0.20
Nickel	17.00	<0.70
Potassium	<1.00	<1.00
Sodium	13.00	<2.00
Zinc	32.00	<2.00

made to decrease the possibility of contamination during sampling. The collection vessels (500 mL), made from a fluoropolymer, were precleaned prior to use (15017–500–25–2S, Berghof America, Pompano Beach, FL).

Unless the cylinder was supplied with a dip tube, it was necessary to invert the cylinder to sample the liquid phase effectively. To ensure that the sampling lines were clean and conditioned, liquid phase was initially passed through the lines. The

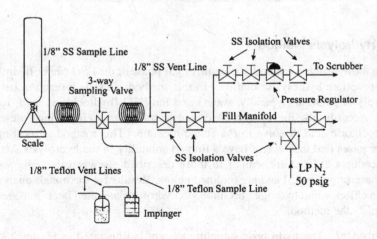

Figure 2.3   Liquid-phase hydrolysis system setup.

flow of liquid through the sampling line was metered with a cross-pattern sampling needle valve, contained in the line. The tubing from the cylinder to the cross-pattern valve (V7) was stainless steel, and the tubing from the valve to the impinger was Teflon. Once the sampling line was purged with liquid-phase, the liquid was directed into an impinger. Liquid was allowed to flow into the impinger until approximately 50 g had been collected. The reaction with the water was extremely exothermic. The amount of liquid sampled was monitored gravimetrically; both the cylinder and the impinger were monitored. It took approximately 15 seconds to collect the sample. It is difficult to control accurately the withdrawal rate of the liquefied gas into the impinger. Once the sample was obtained, the cylinder valve was closed and nitrogen flowed through the system. The sampling line was then removed. The sample was transferred to a 125-mL precleaned Teflon bottle. The sampling device was removed from the cylinder and the cylinder was reinverted.

Analytical Results   Tables 2.4 and 2.5 contain data obtained from replicate liquid-phase hydrolysis sampling of two cylinders of high-purity anhydrous hydrochloric acid and hydrogen bromide, respectively. The replicate samples have good reproducibility, indicating good capture efficiency.

Table 2.4   Liquid-phase hydrochloric acid (50-g sample)

| | Cylinder 1 (ppb w/w) | | Cylinder 2 (ppb w/w) | |
Element	Sample 1	Sample 2	Sample 1	Sample 2
Chromium	<1.00	1.28	<1.0	<1.0
Copper	<1.00	<1.00	<1.0	<1.0
Iron	<1.00	<1.00	1.4	1.8
Molybdenum	<1.00	<1.00	<1.0	<1.0
Nickel	1.69	2.53	<1.0	<1.0

Table 2.5   Liquid-phase hydrogen bromide (50-g sample)

| | Cylinder 1 (ppb w/w) | | Cylinder 2 (ppb w/w) | |
Element	Sample 1	Sample 2	Sample 1	Sample 2
Chromium	<0.46	<0.46	<1.0	<1.0
Copper	<1.16	<1.16	<1.0	<1.0
Iron	<1.16	<1.16	<2.0	<2.0
Nickel	<1.16	<1.16	<1.0	<1.0

2.2.3 Residue Method

The third common method for the analysis of metals in ESGs is the residue method. For ESGs with a liquid phase, the sample is collected after inversion of the cylinder. A sample of liquid phase is sampled into an impinger. The experimental setup is similar to that shown in Figure 2.3, except that the impinger does not contain a hydrolysis solution. After sampling, the liquid is evaporated completely with a low flow rate of nitrogen. The residue remaining in the container after evaporation is subsequently dissolved in an acidic solution. The resulting solution can then be analyzed by ICP–MS. The procedure used for residue sampling and the metal analysis results obtained for boron trichloride using this approach are outlined below.

Experimental   To ensure that the sampling lines were clean and conditioned, initially liquid phase was passed through the lines. The flow of liquid through the sampling line was metered with a cross-pattern sampling needle valve contained in the line. The tubing from the cylinder to the cross-pattern valve was stainless steel, and the tubing from the valve to the impinger was Teflon. Once the sampling line was purged with liquid-phase, the liquid was directed into an impinger. Liquid was allowed to flow into the impinger until approximately 200 g had been collected. The amount of liquid was monitored gravimetrically; both the cylinder and the impinger were monitored. Once the sample had been collected, a low flow of nitrogen was passed over the liquid to facilitate evaporation. Once evaporation was complete, the residue was reconstituted with approximately 20 mL of 1 % nitric acid in a clean hood (class 100).

For boron trichloride, the residue method was chosen (as opposed to the liquid-phase hydrolysis method), owing to the insolubility of the reaction products when boron trichloride contacts water. Upon hydrolysis of boron trichloride, boric acid, an insoluble oxide, is formed, which makes analysis extremely difficult.

Analytical Results   Table 2.6 contains data obtained from liquid-phase sampling of two cylinders of high-purity anhydrous boron trichloride using the residue method. The residue was reconstituted with 20 mL of 1 % nitric acid and subsequently analyzed by ICP–MS.

Table 2.6   Liquid-phase boron trichloride (200-g sample)

Element	Cylinder 1 (ppb w/w)	Cylinder 2 (ppb w/w)
Chromium	0.90	0.67
Copper	8.10	0.47
Iron	0.13	4.63
Nickel	0.09	0.67

2.2.4 Choice of Sampling Method

The choice of sampling method is highly dependent on the ESG. Different chemical and physical properties of the ESGs require different sampling methods, as it is unrealistic to expect that multiple methods can be used routinely in production environments. Thus, the most appropriate sampling method for the metallic impurities of interest and for each specific ESG must be determined (Table 2.7).

Table 2.7 Common gases measured and recommended sampling procedures

Type	Gas Matrix		Filtration	Hydrolysis	Residue
Air Gases	Argon	Ar	×		
	Helium	He	×		
	Hydrogen	H$_2$	×		
	Nitrogen	N$_2$	×		
	Oxygen	O$_2$	×		
Reactive Gases	Ammonia	NH$_3$		×	×
	Boron trichloride	BCl$_3$			×
	Carbon dioxide	CO$_2$		×	×
	Carbon monoxide	CO		×	
	Chlorine	Cl$_2$		×	
	Hydrogen bromide	HBr		×	×
	Hydrogen chloride	HCl		×	
	Nitric oxide	NO		×	
	Silane	SiH$_4$	×		
	Silicon tetrafluoride	SiF$_4$		×	
	Tungsten hexafluoride	WF$_6$		×	×
Nonreactive Gases	Difluoromethane	CH$_2$F$_2$	×		
	Fluoromethane	CH$_3$F	×		
	Hexafluoroethane	C$_2$F$_6$	×		
	Nitrous oxide	N$_2$O	×		
	Octafluoropropane	C$_3$F$_8$	×		
	Silicon hexafluoride	SF$_6$	×		
	Tetrafluoromethane	CF$_4$	×		
	Trifluoromethane	CHF$_3$	×		

If metals are present in a nonreactive gas, either compressed or liquefied, they are likely to be present as particles. These particles are a direct reflection of the cylinder preparation process itself. While the total particle concentration can be obtained by particle counting, the metallic composition must be obtained by an alternative methodology. Filtration is generally utilized for this classification of gases.

The hydrolysis method has a substantial advantage over filtration for reactive gases in that both volatile and nonvolatile species can be retained in the hydrolysis solution. However, the efficiency of this method decreases with increasing flow rate. It is possible using the residue method to obtain lower levels of detection, but volatile species such as arsenic and boron can also evaporate, so that only nonvolatile species will remain. For example, in the case of tungsten hexafluoride (WF_6), volatile compounds such as uranium hexafluoride (UF_6), chromyl fluoride (CrO_2F_2), molybdenum hexafluoride (MoF_6), and thorium tetrafluoride (ThF_4) will be lost utilizing the residue method (Table 2.8).

Table 2.8   Comparison of liquid-phase residue and hydrolysis sampling methods for tungsten hexafluoride

Element	Liquid-Phase Hydrolysis (ppb w/w)	Liquid-Phase Residue (ppb w/w)
Chromium	25	0.2
Copper	<12	0.2
Iron	12	0.4
Magnesium	<10	0.1
Molybdenum	14	0.15
Nickel	<10	0.7
Thorium	30	0.005
Uranium	20	0.02
Zinc	<30	0.3

Although the ESGs used in the manufacture of integrated circuits are used in the gas-phase, it is often preferred to analyze the liquid phase for gases that have a liquid phase. Since it is accepted that the liquid phase contains higher levels of metallic impurities, it is felt that analysis of the liquid phase presents a worst-case scenario [6]. However, in many corrosive gases it is possible to have stable metallic impurities in the liquid phase be present as stable vapor-phase species. Vapor pressures for various metal compounds are known for a limited range. Given that there is a linear relationship between $\log P$ and $1/T$, where P is vapor pressure in torr, and T is the absolute temperature in kelvin (K), one can fit the literature data and extrapolate the fitted results to obtain the vapor pressure at room temperature. The extrapolated vapor pressure data can then be used to obtain the maximum vapor concentration for various metal compounds [7]. The challenge then becomes which method to use when sampling.

As an example, for carbon dioxide contained in a pressurized vessel, it is expected that metallic impurities in the liquid phase will be present as metal carbonates or carbonyls. Volatile carbonyl species may also be present in the gaseous phase. The carbonate species may be present as either particulate or dissolved species. Owing to the ease of the formation of nickel carbonyl [$Ni(CO)_4$] it is important to assess its possible presence [8]. Additionally, iron, chromium, and tungsten carbonyls [$Fe(CO)_5$, $Cr(CO)_6$, $W(CO)_6$] may also be present, but at significantly reduced concentrations with respect to nickel.

To arrive at the most appropriate sampling method, metal impurity data from the three methods were compared. However, prior to this comparison, the proper hydrolysis solution had to be determined. If the primary forms of the metallic impurities were carbonates, then the proper hydrolysis solution would be hydrochloric acid. If the predominant forms were carbonyls, concentrated nitric acid must be used. To determine if the metallic impurities were in dissolved or particulate form, a PC filter was used upstream of the hydrolysis impinger. In this manner the total impurity concentration could be determined by summing the concentration on the filter with that found in the hydrolysis solution. In three separate experiments the hydrolysis solution was varied: 10 % w/w hydrogen chloride, 10 % w/w nitric acid, and 10 % w/w hydrogen chloride/nitric acid.

Referring to Table 2.9, the filter consistently gave higher total metal impurity levels compared to any of the hydrolysis solutions, indicating that most of the impurities were present as particles. Although the total metal content in the hydrochloric acid hydrolysis solution was less than 1 ppb, the concentrations in the nitric acid and hydrochloric acid/nitric acid mixed acid solutions were similar. These sampling methods can have variations of $\pm 20\%$ [9]. The results indicate that the contributions from dissolved carbonates or carbonyls were minimal and that the most appropriate hydrolysis solution is either 10 % nitric acid or 10 % hydrochloric acid/nitric acid. It should also be noted that samples were taken in the order presented in Table 2.9, and that some of the increase in metallic impurities observed in the filtration experiments may be due to the concentration of impurities as the liquid level was depleted.

Method Comparison   To assess the most appropriate sampling method, samples for filtration, hydrolysis, and residue were taken simultaneously from three parallel flowing streams. The cylinder was rolled for 45 minutes to ensure homogenization and suspension of any particulate matter in the cylinder. The cylinder was heated to increase the pressure to 1200 psia to maintain the carbon dioxide in the liquid phase throughout the sampling system. The results, illustrated in Table 2.10, show that the residue method provided the highest level of metallic impurities.

Most of the metallic impurities were determined to be present as particles. The concentration of metals is strongly dependent on the flow and homogeneity of the particles within the cylinder. Sampling for metals (as particulates) should be done at reasonably high flow rates within 1 hour of the homogenization of the source liquid [10].

Higher sample flow rates cause higher concentrations of particles to be sampled, for two reasons. First, the liquefied carbon dioxide in the source is agitated due

Table 2.9 Metallic impurity data for different sampling methods of carbon dioxide (ppb w/w)

Element	10 % HCl/HNO$_3$	Filter	10 % HNO$_3$	Filter	10 % HCl	Filter
Chromium	0.19	0.41	2.25	0.41	<0.1	0.42
Cobalt	<0.04	<0.05	<0.05	0.15	<0.02	0.08
Copper	0.91	0.31	<0.05	0.49	0.29	0.33
Iron	1.32	4.23	0.78	7.54	0.48	21.64
Nickel	0.37	0.51	<0.05	0.40	0.06	0.47
Tungsten	0.19	<0.05	<0.05	<0.05	<0.02	<0.05
Sum	3.02	5.56	3.56	9.04	0.97	22.99
Total	8.85		12.60		23.96	

Table 2.10 Comparison of simultaneous sampling of carbon dioxide

Element	Hydrolysis (ppb w/w)	Filtration (ppb w/w)	Residue (ppb w/w)
Chromium	3.83	3.59	2.60
Iron	1.59	19.98	66.49
Nickel	0.28	2.80	1.55
Total	5.70	23.37	70.64

to more violent vaporization of the liquid at high withdrawal flow rates, causing available particles to be dispersed within the liquid more effectively. Second, higher sampling velocities help overcome gravity and settling of particles for more efficient transport of the particles through the sampling system.

2.2.5 ICP–MS Analysis

ICP–MS is used routinely because of its high sensitivity and multielement capability for trace elemental analysis. Analysis of the solutions from hydrolysis or residue sampling can be rather challenging, owing to high concentrations of acids, which can severely suppress signal intensities, and the complex matrices, which interfere with the analytes of concern. For example, the presence of a chlorine-containing matrix increases the background for ^{51}V and ^{75}As, due to the formation of ^{35}Cl^{16}O$^+$ and ^{40}Ar^{35}Cl$^+$, respectively, severely degrading the detection limits achievable. It is necessary to separate the matrix from the analyte to eliminate potential analytical interferences. By removing the chloride from the sample, arsenic and vanadium can easily be analyzed. This can be done either by evaporation of the samples and then reconstitution in a suitable acid, or by using electrothermal vaporization–ICP–MS (ETV–ICP–MS). ETV is a sample introduction technique for ICP–mass spectrometers. It utilizes a modified graphite furnace to pretreat small volumes of sample and produce a dry aerosol that is transported to the ICP–MS. Boiling off and sweeping the volatile matrix components out of the graphite tube significantly reduces or eliminates the polyatomic interferences caused by these acids. The residual is then vaporized into the ICP–MS, where it is ionized and analyzed. The transient signal for measurement is <3 seconds. This limits to no more than five the number of isotopes that can be analyzed successfully per injection. However, since the matrix is boiled off prior to introduction into the ICP–MS, variations in the acid concentration are not an issue. ETV introduces the sample into the plasma as a "dry aerosol," eliminating the need for precise matrix matching.

Although other analytical tools may be employed to analyze trace metal levels in high bromide, chloride, and fluoride solutions, ETV–ICP–MS is capable of analyzing all elements at 10 to 50 times lower levels than can pneumatic nebulization alone. A modifier is necessary to obtain linear calibration plots below 1 ppb. Palladium acetate is typically used because it is not an element of interest within the ESG metal specifications. A modifier is used to reduce analyte transport loss that occurs with vaporization. A major advantage of using ETV in conjunction with the ICP–MS is that very little sample is required (ca. 20 μL). As a result, it is possible to analyze highly concentrated acid samples without dilution.

Experimental   An Elan 6000 ICP–MS in conjunction with a HGA–600MS electrothermal vaporizer (PerkinElmer Sciex, Ontario, Canada) equipped with platinum interface cones was operated using the conditions in Table 2.11. The ETV unit was controlled by the Elan software. A PerkinElmer AS-60 autosampler was used for the sample introduction.

Table 2.11   Operating and measuring conditions for ICP–MS and ETV

RF power (W)	1100
Plasma argon flow (L/min)	15
Auxiliary flow (L/min)	1.2
Sampler and skimmer cones	Platinum
Sweeps/reading	1
Readings/replicates	40
Number of replicates	3
Dwell time (ms)	10
Integration time (ms)	400
Scan mode	Peak hopping
Sample volume (μL)	20
Injection temperature (°C)	20
Modifier	10 ppm palladium acetate

Spex CertiPrep Claritas PPT grade standards and palladium modifier (Spex, Metuchen, NJ) were used for all calibration solutions. Standards were typically prepared daily spanning the range 0.5 to 100 ppb in the matrix to be analyzed (hydrogen bromide, hydrochloric acid, etc.). Samples were used undiluted. Samples that had concentrations that fell outside a factor of 2 from the highest concentration standard (>200 ppb) were diluted and reanalyzed. Concentrations below 0.25 ppb (depending on the background equivalent concentration of the element) were reported as less-than values.

The ETV temperature program for the determination of ^{57}Fe, ^{51}V, ^{24}Mg, and ^{60}Ni is described in Table 2.12. The optimum vaporization temperatures for iron, vanadium, and nickel are 2000, 2400 and 2200 °C respectively [11]. The modifier, 10 ppm palladium acetate in 1 % nitric acid, was injected into the graphite ETV tube at 20 °C. It was warmed gently to remove the nitric acid matrix. The temperature was then raised to 1500 °C, just below the vaporization temperature for palladium, to remove traces of impurities in the modifier. The system was then allowed to cool before the sample was introduced. Once the sample was introduced, it was warmed gently to remove the matrix. The system was then quickly heated to 1900 °C, the temperature below the optimum vaporization of the analytes. Maximum power heating was necessary to achieve maximum sensitivity. It was necessary to optimize the thermal program for each set of analytes as well as various matrices.

2.3 Direct Analysis of ESGs

Direct analysis requires that the sample be introduced directly into the analytical instrument without any sample preparation step. There have been several attempts to inject ESGs directly into a plasma: ICP–MS [12], ICP–AES [13,14]. However, material compatibility, deposition, plasma stability, or unsuitable detection limits were common issues that limited the applicability of these techniques. Additionally, no gaseous metal standards were readily available, which made calibration difficult.

Even though it is possible to construct a gas-phase generation system for some metals that have relatively high vapor pressure, such as iron carbonyl [$Fe(CO)_5$], the availability of other high-vapor-pressure metal compounds is quite limited. Therefore, a desolvation nebulizer, which allows more efficient aerosol generation, transport, and solvent vapor removal than traditional nebulizers allow, has been evaluated for its ability as a metal standard generator in gases [15]. Calibration/sampling procedures using these approaches and results for direct metals analysis in carbon monoxide are discussed later in this section.

2.3.1 Calibration

Iron Carbonyl Generation System An impinger, constructed from perfluoroalkoxy (PFA) and PTFE, with an inlet and outlet was used to hold the liquid iron carbonyl. The impinger was heated to 30.3 °C, yielding a vapor pressure of 40 mmHg iron carbonyl. A constant flow of filtered argon (Wafer Guard, WGTF00MRU, Milli-

Table 2.12   ETV thermal program for iron, magnesium, vanadium, and nickel in hydrochloric acid

Step	Cell Temp. (°C)	Ramp (s)	Hold (s)	Internal Flow[a] (mL/min)	Gas to
1	110	30	30	300	Vent
2	400	5	15	300	Vent
3	1500	2	6	300	Vent
4	20	1	10	300	Vent
5	110	30	30	300	Vent
6	320	5	15	300	Vent
7	1900	0	6	0	ICP
8	2600	1	7	0	ICP
9	20	1	30	0	ICP

[a] Argon carrier used for introducing the sample into the ICP–MS from the ETV.

Pipette Sequence	Solution	From	To
1	Modifier		
2		Step 1	Step 4
3	Sample		
4		Step 5	End

pore Corp., Billerica, MA) flowed through iron carbonyl, to create a saturated vapor. An additional argon gas stream was used to further dilute the stream in order to obtain various concentrations (Figure 2.4). All lines were heat-traced to prevent condensation. The system was capable of generating iron concentrations of 0 to 50 ng/min.

Desolvation Nebulizer   A Mistral desolvation nebulizer (VG Elemental, Winsford, U.K.) was used to generate particles that were introduced into the ICP–MS. The uptake rate of the nebulizer was 0.2 to 0.4 mL/min. The aerosol from the nebulizer was transported to the spray chamber, which was heated to 162 °C. The vapor and particulates within the aerosol were then transported to the condenser, which was operated at 1 °C. The dry aerosol was then transported into the torch of the ICP–MS.

Metal standards between 50 to 10,000 ppt aqueous iron chloride were nebulized and subsequently evaporated in the heating chamber, where the metal particles were nebulized into small particles. At the condenser, moisture was condensed. The argon gas stream contained a moisture concentration of a 1 °C dew point (6500 ppmv) moisture. The stable moisture concentration in the carrier gas maintained the plasma conditions.

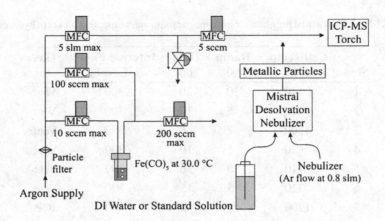

Figure 2.4   Experimental setup for calibration of direct introduction into an ICP–MS using iron carbonyl and a desolvation particle generator.

ICP–MS   The ICP–MS used in this study was a double-focusing sector field type (VG Plasma Trace 1, VG/Fisons, Winsford, U.K.). It was capable of resolution from 400 to 8000. For this study, it was tuned for dry plasma conditions using a resolution of 3000 to 3500. The argon stream from either the iron carbonyl generator or the desolvation nebulizer was selected just prior to the plasma torch (Figure 2.4).

Results   Both the iron carbonyl generation system and the desolvation nebulizer showed a linear response in the range of iron concentrations from 0 to 3 ng/min (Figure 2.5). Both calibrations had the same slopes, indicating that the ionization ratio of iron from iron carbonyl and that of iron as particles generated from the desolvation nebulizer were the same. In addition, the desolvation nebulizer showed nearly 100 % introduction efficiency of iron. It was assumed that other elements gave the same ionization process as that of iron; therefore, calibration of the ICP–MS for the other elements would be possible utilizing the desolvation nebulizer method [16].

Moisture concentration in the carrier gas was a critical parameter for the ICP temperature. The Mistral system generated a stable moisture concentration in the carrier gas from the desolvation nebulizer, so the same ionization condition could be maintained throughout the analysis. The detection limit obtained was 5 ppt w/w iron in argon.

2.3.2   Analysis of Carbon Monoxide

Carbon Monoxide Introduction and Calibration   For the calibration of metals in carbon monoxide, both an iron carbonyl–saturated vapor standard generator and a dissolution nebulizer were used. A carbon monoxide introduction line was attached to the standard generator setup (Figure 2.6). A cylinder of carbon monoxide was connected with a cross-purge unit, after which there was a pressure regulator to

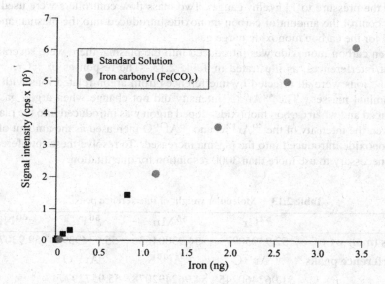

Figure 2.5   Iron calibration curve generated by use of a desolvation nebulizer and iron carbonyl generation system.

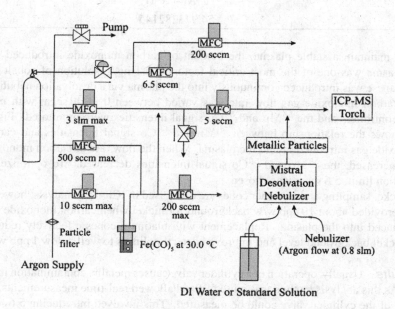

Figure 2.6   Experimental setup for analysis of carbon monoxide by direct introduction into an ICP–MS.

reduce the pressure to $1 \, kg/cm^2$ gauge. Two mass flow controllers were used: the first to control the amount of carbon monoxide introduced into the plasma, and the second for the carbon monoxide purge gas.

When carbon monoxide was introduced into the plasma, there were several significant interferences, as illustrated in Table 2.13. The responses of ^{52}Cr, ^{55}Mn, and ^{56}Fe ions were all affected by interferences from argon-related ions with similar nominal masses. The $^{40}Ar^{15}N$ intensity did not change when argon gas was introduced and when carbon monoxide–doped argon was introduced into the plasma. However, the intensity of the $^{40}Ar^{12}C$ and $^{36}Ar^{16}O$ increased as the amount of carbon monoxide introduced into the plasma increased. To resolve these interferences, it was necessary to use more than 3000 resolution for quantitation.

Table 2.13   Molecular weight of interference peaks

	^{52}Cr	^{55}Mn	^{56}Fe	^{60}Ni
Mass (mol. wt.)	52.9405097	54.9380463	55.9349393	59.930789
Interference peaks	$^{40}Ar^{12}C$	$^{40}Ar^{15}N$	$^{40}Ar^{16}O$	
	51.962460245	54.962492078	55.95729774	
	$^{36}Ar^{16}O$	$^{36}Ar^{16}O^1H$		
	51.962460245	54.965471877		
		$^{40}Ar^{14}N^1H$		
		54.973282145		

To maintain a stable plasma, the amount of carbon monoxide introduced into the plasma was one of the most critical factors. A 1-ng/g solution of cobalt and manganese was introduced continuously into the plasma via a desolvation nebulizer. The carbon monoxide gas flow rate was varied between 0 to 40 sccm with mass flow controllers and the ^{55}Mn and ^{59}Co signal intensities were monitored. Figure 2.7 shows the relationship between ^{55}Mn and ^{59}Co signal intensities and carbon monoxide gas introduction into the plasma. When the flow rate of carbon monoxide was increased, the ^{55}Mn and ^{59}Co signal intensities decreased. To optimize the detection limit, 5.6 sccm was chosen.

Nickel sampling and skimmer cones are often used in ICP–MS analysis; however, they provided about 10 ppb w/w background for nickel when carbon monoxide was introduced into the plasma. Replacement with platinum cones effectively reduced the nickel background level and improved detection limits to well below 1 ppb w/w.

Results   Usually, operation of a cylinder valve causes metallic contamination in the gas. As this analysis method was quite fast and allowed real-time measurements, the effect of the cylinder valve could be measured. This involved introducing 5.6 sccm carbon monoxide into the plasma and 200 sccm for the purge during the analysis. The relationship between metal concentration in carbon monoxide gas and purge time after the cylinder valve had been opened is illustrated in Figure 2.8. Just after

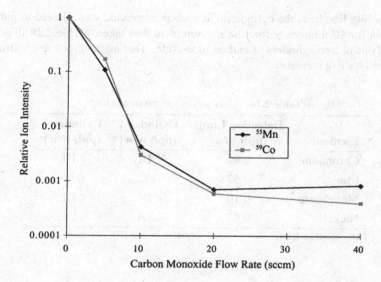

Figure 2.7 Relationship between the signal intensities of ^{55}Mn and ^{59}Co and the carbon monoxide flow rate.

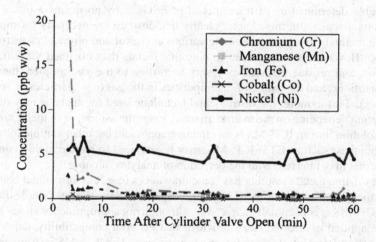

Figure 2.8 Time dependence of metal impurity concentration after a single cylinder valve manipulation.

opening the cylinder valve, chromium concentration increased and then decreased with time. Manganese, iron, cobalt, and nickel showed no time dependence.

For multiple-cylinder analysis, the carbon monoxide cylinder was connected to the sample introduction line. The cylinder valve was purged 10 times with a vacuum, alternating with 3-bar argon purge cycles before introducing carbon monoxide into the sampling line from the cylinder. The carbon monoxide was allowed to purge at 200 sccm for 10 minutes before the measurement was taken. Table 2.14 illustrates the analysis of two cylinders of carbon monoxide. This method took approximately 5 minutes for five elements.

Table 2.14   Analysis of carbon monoxide[a]

Element	Detection Limit (ppb w/w)	Cylinder 1 (ppb w/w)	Cylinder 2 (ppb w/w)
Chromium	0.92	< DL	< DL
Iron	0.57	0.72	0.95
Manganese	0.10	< DL	< DL
Nickel	0.62	< DL	< DL

[a] DL, detection limit.

2.4 Conclusions

The reliable determination of trace metals in ESGs is by no means a trivial task. The results discussed in this chapter clearly demonstrate the necessity for multiple sampling methods to accommodate the various chemical and physical properties of each gas. However, it is important to note that metals data obtained via filtration, hydrolysis, and residue methods may vary according to the gas sampled, the sampling conditions, and the nature of the impurities in the gas (e.g., particles or volatile impurities). Furthermore, the instrumental technique used for analysis may require optimization depending on the matrix. In many cases the aspiration of metal samples in acid solution into an ICP–MS is an effective approach, but other sample introduction techniques, such as ETV–ICP–MS, may be required to handle highly complex matrices that may interfere with the detection of analytes of concern.

Since off-line metal sampling has some drawbacks (e.g., environmental contamination), the quantification of metal impurities by introducing the gas sample directly into an ICP–MS is conceptually attractive. Although not appropriate for all gas types, it can be applied in selected cases provided that material compatibility, deposition, plasma stability, and interference issues are overcome. The ICP–MS instrument may require some hardware modifications and must also be calibrated for the analytes of interest in the presence of the gas matrix. In this work, the analysis of metal impurities in carbon monoxide has been demonstrated by using direct introduction of the gas sample into HR–ICP–MS. Calibration of metallic elements has been accomplished due to the development of a standard gas generator, and the method has provided real-time analysis for the metals of interest with a detection limit of less than 1 ppb w/w. This approach is likely to be adopted for other selected gases in the future.

REFERENCES

1. Zdunek, A., Jacksier, T., Borzio, J., & Rufin, D. (1999). Ultrapure materials–gases: using a nondestructive test to qualify corrosive specialty gas cylinders. *Micro Santa Monica, 17*, 2, 33–39.

2. Ohmi, T., Nakagawa, Y., Nakamura, M., Ohki, A., & Koyama, T. (1996). Formation of chromium oxide on 316L austenitic stainless steel. *Journal of Vacuum Science and Technology A: Vacuums, Surfaces, and Films, 14*, 4, 2505–2510.

3. Tepe, R. K., Vassallo, D., Jacksier, T., & Barnes, R. M. (1999). Iron pentacarbonyl determination in carbon monoxide. *Spectrochimica Acta B: Atomic Spectroscopy, 54*, 13, 1861–1868.

4. Air Liquide Laboratories, Japan (1995). Unpublished results.

5. Schleisman, A. J., & Bollinger, D. (1997). Ultrapure sampling of corrosive-gas cylinders. In J. D. Hogan (Ed.), *Specialty Gas Analysis: A Practical Guidebook*. New York: Wiley-VCH, pp. 81–92.

6. Jacksier, T., Udischas, R., Wang, H.-C., & Barnes, R. M. (1994). Particle- and vapor-phase contributions to metallic impurities in electronic-grade chlorine. *Analytical Chemistry, 66*, 14, 2279–2284.

7. Wang, H.-C., Udischas, R., Doddi, G., & Jacksier, T. (1994). Evaluation of metallic impurities in electronic process gases by a particle counter. *Journal of Aerosol Science, 25*, Supplement 1, 581–582.

8. Tepe, R. K., Vassallo, D., & Jacksier, T. (2000). The synthesis of iron and nickel carbonyl as calibration materials for spectroscopic systems. *Spectrochimica Acta B: Atomic Spectroscopy, 55*, 2, 165–175.

9. Jacksier, T., & Borzio, J. (1997). Field implementation of sampling techniques for metals analysis in electronic specialty gases. *Annual Technical Meeting Institute of Environmental Sciences, 43*, 2, 255–261.

10. Udischas, R., Jacksier, T., Rignon, M., & Graehling, J. (2004). The analysis of high and low K materials by ICP–MS. *Proceedings of the 2004 Semiconductor Pure Water and Chemicals Conference.*

11. Ediger, R. D., & Beres, S. A. (1992). The role of chemical modifiers in analyte transport loss interferences with electrothermal vaporization ICP–mass spectrometry. *Spectrochimica Acta B: Atomic Spectroscopy, 47*, 7, 907–922.

12. Hutton, R. C., Bridenne, M., Coffre, E., Marot, Y., & Simondet, F. (1990). Investigations into the direct analysis of semiconductor grade gases by inductively coupled plasma mass spectrometry. *Journal of Analytical Atomic Spectrometry, 5*, 6, 463–466.

13. Carre, M., Coffre, E., Morin-Adam, A., & Desbonnet, J. (1997). Calibration of high resolution ICP/MS for direct analysis of speciality gases. *European Winter Conference on Plasma Spectrochemistry, Cambridge, UK.*

14. Jacksier, T., & Barnes, R. M. (1994). Quantitative analysis of electronic-grade anhydrous hydrogen chloride by sealed inductively coupled plasma atomic emission spectroscopy. *Journal of Analytical Atomic Spectrometry, 9*, 11, 1299.

15. Montaser, A., Minnich, M. G., McLean, J. A., Liu, H., Caruso, J. A., & McLeod, C. W. (1998). Sample introduction in ICPMS. In A. Montaser (Ed.), *Inductively Coupled Plasma Mass Spectrometry*. New York: Wiley-VCH, pp. 83–264.

16. Suzuki, I., & Tarutani, K. (1997). Comparison of sampling techniques for the analysis of particulate metal impurities in gases. *Analytical Sciences: The International Journal of the Japan Society for Analytical Chemistry, 13,* 5, 833.

CHAPTER 3

NOVEL IMPROVEMENTS IN FTIR ANALYSIS OF SPECIALTY GASES

BARBARA MARSHIK[1] AND JORGE E. PÉREZ[2]

[1]MKS Instruments, Inc., Methuen, Massachusetts
[2]CIC Photonics, Inc., Albuquerque, New Mexico

3.1 Gas-Phase Analysis Using FTIR Spectroscopy

Classical least squares (CLS) is the standard method of determining unknown component concentrations in gas matrices [1]. The spectral peaks of the component of interest are used to calculate the unknown concentration based on a calibration training data set that pairs spectral peaks of the component of interest with validated concentration information. In general, the method uses a spectrum of known concentration and scales it until it matches the unknown spectrum. The amount by which the known spectrum was scaled is then applied to the known concentration, providing a new value predicted for the unknown concentration. This method has been applied to very complex mixtures such as vehicle tailpipe emissions as well as low-ppb-level analysis of water vapor in 100 % ammonia samples. As long as all of the constituents are identified within the region in which the analysis is being made, this method works extremely well. For more complicated gas streams, partial least squares (PLS) and other principal components analysis methods have been used [2,3] which are

Trace Analysis of Specialty and Electronic Gases.
By William M. Geiger and Mark W. Raynor. Copyright © 2013 John Wiley & Sons, Inc.

more suited to compensating for temperature and pressure mismatches. These methods require, however, much more extensive method development and calibrations, which are sometimes not available. Once the PLS method has been created, it can also be more difficult for the end user to modify in case that concentrations or other parameters change. In this chapter the focus is on the use of CLS as the method of analysis, but all spectral changes that are mentioned below are relevant for both CLS and PLS, as they are both based on spectral information associated with known concentration values.

The CLS method is applied to data that are acquired using a Fourier transform infrared (FTIR) spectrometer. FTIR is used in many areas of study and is well suited for the analysis of impurities in the gas phase. The concentration of a particular gas molecule is directly proportional to the absorbance spectra of that molecule.[*]

In this chapter we review the major components and their effects on the line shape of gas molecules, such as temperature, pressure, and pathlength. We also provide information on internal molecular interactions and isotope effects and if not accounted for, how they can distort quantitative results within the method. In the final section we introduce two novel methods for creating a background spectrum facilitating the use of FTIR as a process analyzer and provide an automated robust analytical method.

3.2 Gas-Phase Effects on Spectral Line Shape

The spectral line shape can change depending on a number of factors, including pressure, temperature, pathlength, and matrix gas composition. There are two main effects: Doppler broadening due to the variation of the speed of the molecules, and collisional broadening due to intermolecular collisions of gas molecules. As the pressure in the system increases above 1 torr, collisional broadening dominates the spectral line shape. Detailed information and background are available in a number of articles in the literature.[†] In this chapter we address these effects in more general terms only.

When a gas molecule is at equilibrium with the environment, the number of gas molecules in the excited state follows the Maxwell–Boltzmann distribution. This distribution (which affects the final line shape) is Gaussian in nature and is the most dominant at room temperature and low pressures (<1 torr). It is referred to as *Doppler broadening*. As the temperature of the gas is increased, some of the molecules will speed up, which results in the line shape shifting and broadening out.

---

[*]Peak height, width, and shape greatly affect the ability of the CLS method to provide accurate and repeatable results. For more details on FTIR, see Griffiths, P. A., & de Haseth, J. A. (2007) *Fourier Transform Infrared Spectrometry*, 2nd ed. New York: Wiley-VCH.

[†]See Pope, R. S. (1999). Collisional effects in the absorption spectra of the oxygen A band and nitric oxide fundamental band. Doctoral dissertation. Retrieved from ProQuest Dissertations and Theses (9921302), or Gamache, R. R., Lynch, R., & Brown, L. R. (1996). Theoretical calculations of pressure broadening coefficients for H_2O perturbed by hydrogen or helium gas. *Journal of Quantitative Spectroscopy and Radiative Transfer, 56*, 4, 471–487.

Collisional or pressure broadening occurs when the pressure in a system is increased and there is an exchange of energy that is either elastic or inelastic between the gas molecules as they collide with each other, broadening the line in a Lorentzian manner. Elastic collisions will change the phase of the rotation of the molecule, whereas inelastic collisions change the actual energy transition state of the rotation or vibration of the molecule. Both collisions result in a broadening of the final line shape. This change of state is proportional to the number of molecules that are present in the volume and well as the type. If one molecule is larger than the other, the smaller of the two will transfer more energy to the larger molecule. This broadening has a number of different terms within the literature, including pressure, self, interferent, and collisional broadening, but in this chapter it is addressed as collisional broadening [4–6].

At pressures above 1 torr, the gas-phase line shape is a convolution of both Doppler and collisional broadening, but above 100 torr, collisional broadening begins to dominate the spectral line shape. Although collisional broadening and Doppler broadening occur simultaneously, they are independent of each other, creating a convolution of the two line shapes (Gaussian and Lorentzian), which is represented by the Voigt line shape. In the next few sections we discuss the effects that different elements have on molecules that directly affect the ability to resolve peaks within a gas spectrum. The ability to resolve peaks between different components is the key to quantitative analysis using infrared spectra, and the changes in line shape of the gas molecules greatly affect this process of peak separation.

3.2.1 External Effects on Line Shapes

Both quantitative and qualitative information can be determined from the infrared spectrum, and it is greatly dependent on the ability to distinguish one peak from the next within a gas mixture. In the case of gas-phase FTIR, the peak resolving power is calculated using the full width of the peak at half the total peak height (FWHH). The higher the resolution of the spectral peaks, the greater the ability to separate multiple components having overlapping peaks, provided that the inherent linewidth is $4\,cm^{-1}$ or less. However, higher resolution can also increase the overall noise level within the spectrum, forcing a trade-off between a higher signal-to-noise ratio and the peak resolving power. Peak resolving techniques can be influenced by external or physical means, which include creating an FTIR system that is able to provide high resolution using acquisition hardware, reducing the overall pressure within the gas cell (in particular to less then $100\,mtorr$), lowering the gas sample temperature as well as the choice of effective pathlength of the system.

Data acquisition rates directly affect the final resolution of the spectral peak shapes. In stack and vehicle emission monitoring systems, $0.5\,cm^{-1}$ resolution is generally considered the minimum resolution needed to obtain the detection limits required for sulfur dioxide, nitrogen oxides, and ammonia species. This is due to the severe spectral overlap that these species have with moisture peaks, which can be present in up to 40 % by volume of the gas stream. In the case of specialty gases where moisture levels are much lower, $1\,cm^{-1}$ resolution works well provided that the bulk gas spectral peaks do not interfere greatly with the impurity/analyte of interest.

Figures 3.1 and 3.2 show the effect that the higher peak resolving power has on the ability to distinguish more peaks of a particular component (in this case, nitric oxide at 100 ppm) when in the presence of another component (in this case, water at 15 % by volume). The first figure shows a spectrum of nitric oxide and water from a catalyst at a resolving power of $2\,cm^{-1}$; the second figure shows the same sample but at a higher peak resolving power of $0.5\,cm^{-1}$. It is clear that in the second figure there are more peaks associated only with nitric oxide without overlap with water, providing the ability to detect nitric oxide to much lower limits.

This same effect can be obtained by lowering the gas cell pressure from 760 torr (1 atm) down to the range 50 to 100 mtorr where the collisional broadening no longer dominates, due to many fewer molecular collisions — producing an overall narrowing of the molecular line shape. This allows instruments that are unable to provide high-resolution spectra, such as those capable of only 2 or $4\,cm^{-1}$ resolution, the ability to provide good spectral peak separation.

The caveat to this, however, is that by going to such low pressures to improve peak separation, you are also lowering the overall signal strength, due to the fact that there are far fewer gas molecules available to absorb light at lower pressure. In this case, longer-path gas cells and much stronger infrared light sources must be used to increase the overall signal strength. As the pathlength increases, more light is lost with each reflection from the mirror surface. Therefore, extremely high-reflectance mirror surfaces are required in order to get as much light to the detector as possible. This makes it imperative that the gas stream be particulate-free, as anything that impedes the light reflectivity will quickly degrade the signal strength.

Temperature can also affect the line shape of the gas molecules. In general, the line shape of narrow absorbing molecules is more affected by changes in temperature and pressure than are molecules with broad peaks (e.g., carbon monoxide versus propane). Temperature not only affects the overall signal strength of the component but changes the spectral shape as well, due to different populations in the various energy levels that are temperature dependent. As the temperature of the gas is increased, the line shapes will broaden and the overall signal will decrease. Figure 3.3 shows how the signal strength of a carbon monoxide band changes at 35, 60 and 191 °C. The highest-intensity peak belongs to the lowest temperature, and the lowest intensity peak belongs to the highest temperature. The final peak shape is defined by the Maxwell–Boltzmann distribution as the temperature changes, growing weaker and broadening as the temperature increases.

Additionally, as the temperature is increased, there is a change in the fine structure of the narrow absorbers as well. There will be a decrease in the relative height of the largest peaks and an increase in the smaller peaks at the wings for the same concentration. This reflects the ability for the higher-energy states to be populated as the temperature increases. Figure 3.4 shows the result of 100 ppm carbon dioxide at 35 °C subtracted from 100 ppm carbon dioxide at 191 °C. It is clear from the lower plot that the wings of the 191 °C spectrum are more populated than the spectrum at 35 °C. This phenomenon can be exploited to some extent by including the Q-band

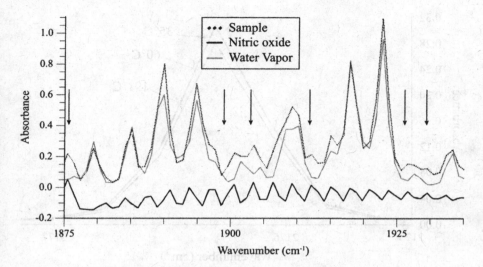

Figure 3.1 Overlaid IR spectral plots of a sample (dashed line) measured at a resolution of $2\,cm^{-1}$, water vapor, and nitric oxide. Arrows indicate nitric oxide peaks with minimal interference from water vapor.

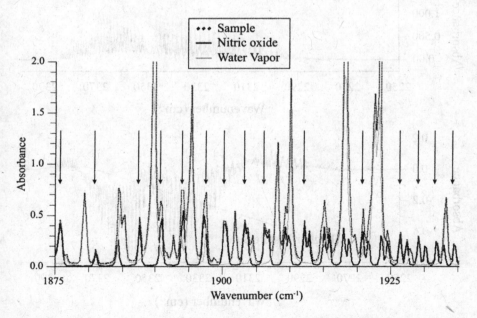

Figure 3.2 IR spectral plots of a sample (dashed line) measured at a resolution of $0.5\,cm^{-1}$, water vapor, and nitric oxide. Arrows indicate nitric oxide, showing a greater number of peaks now available with minimal interference from water vapor.

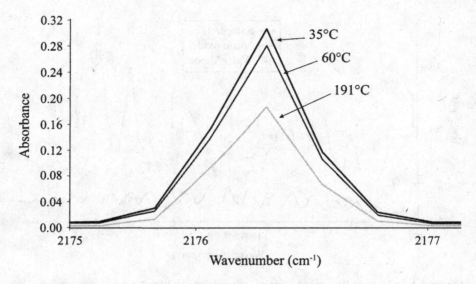

Figure 3.3   Plot showing the effect on the line shape of 100 ppmv carbon monoxide in nitrogen as the temperature is increased from 35 to 191 °C.

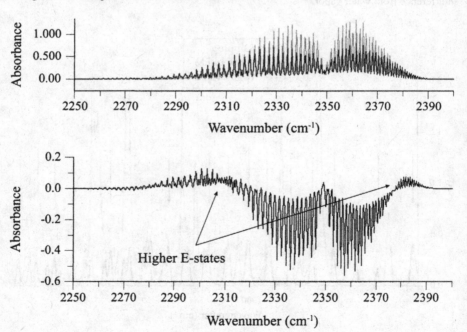

Figure 3.4   Top plot: Spectra of 100 ppm carbon dioxide at 35 °C (gray) and 191 °C (black). Bottom plot: Residue obtained by subtracting the 35 °C spectrum from the 191 °C spectrum. Arrows indicate the 191 °C higher-energy states, which are more populated (positive peaks).

(if it exists) in the analysis region or choosing a large region that spans all the peaks, which can help mitigate the effect when the temperature changes.

When analyzing gas components that are present at high concentrations, many of the peaks throughout the full spectral range may be above 2 AU, which make it impossible to resolve the peaks or even find a region that enables detection. Reducing the overall system pressure to the mtorr range or using a shorter gas cell pathlength, which also results in better peak resolution by reducing the absorbance peaks for the component of interest, may help, but in both cases the lower detection limit will be affected. The trade-off on the choice of gas cell directly affects the minimum level of detection needed for the analysis when changing from a long-path gas cell to a shorter-path gas cell. Figure 3.5 shows the effect of change in pathlength for 40 vol% of carbon dioxide between an effective pathlength of 5.11 m (black spectrum) and a 2-cm pathlength (gray spectrum). It is clear that by changing from the 5.11-m to the 2-cm gas cell, a large number of spectral peaks are now available for carbon dioxide quantification.

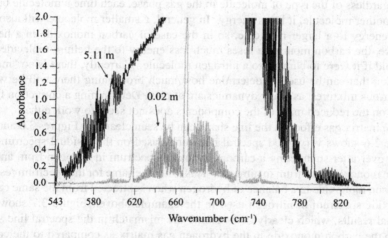

Figure 3.5   IR spectral plots of 40 vol% carbon dioxide using 5.11-m and 2-cm (0.02-m) effective pathlengths.

Pressure, temperature, and pathlength must be corrected for in the final quantitative result so that differences from the calibration spectra that were used in the model are adjusted for the pressure and temperature, and the gas cell pathlength of the sample. While the analysis software can correct for the value of the difference between the calibration spectrum and the unknown spectrum temperature and pressure, it cannot account for the differences in the actual line shape. That is why it is necessary to develop calibrations at the same pressure (or at least within 5 % for carbon monoxide and carbon dioxide and 10 % for other components) and temperature (within approximately ±2 °C) in order to minimize analysis errors due to spectral mismatch. The pathlength affects only the absorbance peak height, not the basic line shape.

3.2.2   Matrix Gas Effects on Line Shapes

This section expands on the effects of the peak line shape due to different types of collisional broadening, including foreign body and self-broadening. If we look at a sample of gas of 10 ppm carbon monoxide in a balance nitrogen at fixed pressure, a narrowing of the FWHH line shape of the carbon monoxide molecule is observed when all the nitrogen molecules are replaced with helium. While the number of total molecules remained the same, the composition of the matrix gas changed. Since there are very few carbon monoxide molecules within the gas mixture (10 ppm total), the carbon monoxide molecules will encounter more of the matrix molecules than will another carbon monoxide molecule. This is referred to as *foreign body broadening*. As the concentration of the analyte (carbon monoxide in this example) is increased, the probability that the carbon monoxide molecule will bump into other carbon monoxide molecules increases, affecting the overall line shape differently from that based only on matrix gas molecule broadening. This type of broadening is termed *self-broadening*.

Regardless of the type of molecule in the gas phase, each time a molecule bumps into another molecule, it loses energy. In general, a smaller molecule will also lose more energy to a larger molecule, so in the case of carbon monoxide in a helium balance, the carbon monoxide loses much less energy to the helium molecule than it would if it were to bump into a nitrogen molecule. Currently, there is no "magic" equation that can be used to determine how much broadening there will be within the various mixtures, as many dynamics are at play. Determining a correction factor based on the reduced mass of the components does not seem to work either.

The matrix gas effect on the line shape can be seen clearly in Figures 3.6 and 3.7. Figure 3.6 shows very good spectral matching based on the residual spectrum that is left over after subtracting a carbon monoxide spectrum in nitrogen from another carbon monoxide spectrum in nitrogen. This is not the same for the spectrum residual of carbon monoxide in a balance of hydrogen when subtracted from the same carbon monoxide spectrum in nitrogen used in the example above. Figure 3.7 shows the residual results, which clearly demonstrate the mismatch in the spectral line shape due to the carbon monoxide in the hydrogen gas matrix as compared to the carbon monoxide in a nitrogen matrix gas.

The best way to reduce the effects of the matrix gas is to create calibration spectra that are based on the same matrix gas as that of the unknown sample. However, it is not always possible to create calibrations in the matrix gas (e.g., creating water calibrations in silane or other reactive gases would be very difficult). It would be much easier and cheaper to create calibrations using nitrogen, helium, or air, then to apply a correction factor that would account for the effect of the collisional broadening.

Although the bias is accounted for by the application of a correction factor between the two different matrix gases, it does not correct for the line shape mismatch. This mismatch will affect the detection limits of the method since the correct line shape was not used during the spectral matching routine. There are some extended versions of CLS that can be used to help minimize this effect once the residual spectral information is obtained [7].

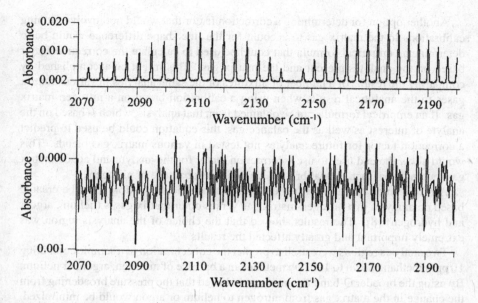

Figure 3.6 Plot of the residual results (below) obtained by subtracting a spectrum of 5 ppm carbon monoxide in nitrogen from another 5 ppm carbon monoxide in nitrogen spectrum (above).

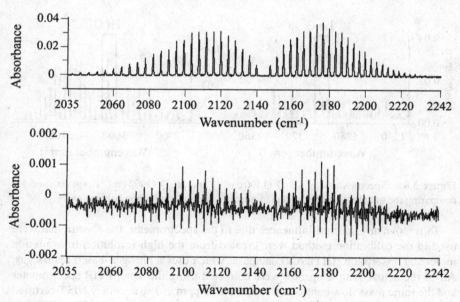

Figure 3.7 Plot of the residual results (below) obtained when subtracting 10.3 ppm carbon monoxide in hydrogen spectrum from a 10.0 ppm carbon monoxide in nitrogen spectrum (above).

Another option for determining a correction factor that would not involve running calibrations in the matrix gas to account for the line shape difference would be to determine an empirical formula that could be used to calculate the correction factor based on the analytes present and the matrix gas. This might be accomplished by comparing the results of a number of mixtures of the same analytes in various matrix gases to the analytical results when using a calibration based on a nitrogen matrix gas. If an empirical formula can be obtained from that analysis, which is based on the analyte of interest as well as the balance gas, this equation could be used to predict a correction factor for future analytes not tested in various matrix gas blends. This would allow the end user to use a correction factor for the analyte and actual matrix gas, but in actual fact, the calibration would be in a balance of nitrogen.

A recent study examined whether or not an empirical equation could be created based on analytes in a nitrogen matrix gas versus other matrix gases (helium, argon, and hydrogen) [8]. The results showed that the choice of the analysis region was extremely important and greatly affected the results.

Two analysis regions were used to predict the concentration of a sample containing 10 ppm methane down to 100 ppb methane in a balance of nitrogen, argon, or helium. By using the broader Q-band region it was believed that the pressure broadening from the change in the matrix gas from nitrogen to helium or argon would be minimized. The methane molecule actually has two regions where the Q-band is fairly strong (see Figure 3.8, peaks labeled Hi Q for the band near 3000 cm^{-1} and Lo Q for the band near 1300 cm^{-1}), so it was used for comparison of the effects of region selection.

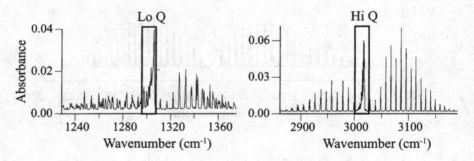

Figure 3.8   Spectra showing Lo Q (1300 cm^{-1}) and Hi Q (3000 cm^{-1}) regions used for comparing the quantitation of methane in nitrogen, helium, and argon balance gases.

To remove any spectral influences due to the spectrometer, the spectra that were used in the calibration method were created from the high-resolution transmission molecular absorption (HITRAN) database, which uses a nitrogen-based matrix gas. All of the actual sample spectra were collected using the same FTIR spectrometer and the same mass flow controllers to dilute a 25 ppm ± 1 % accuracy NIST-certified calibration gas of methane in a balance of nitrogen, helium, or argon. The FTIR was "zeroed" prior to data collection using a purified stream of nitrogen, helium, or argon, and this stream was used for dilution of the 25 ppm methane calibration gas. Figure 3.9 shows the results of comparing the predictions using the HITRAN nitrogen-based method, with $1 \, cm^{-1}$ resolution using either the Lo Q or Hi Q analysis region, to the actual spectra for methane in a balance of nitrogen, helium, or argon. It is clear from the plots shown in Figure 3.9 that there is a difference both in the region where the analysis was done and in the bulk gas that was used.

From the data just presented, the results of the values predicted using the HITRAN method for the 10 ppm standard in nitrogen were 4 % higher than the nominal values, while for argon they were about 6 % higher and for helium about 8 % higher. From this plot it is clear that if an empirical equation could be determined for predicting the effects of the matrix gas upon an analyte of interest, it would be complicated not only by the region that was chosen for the analysis but also by the resolution. Due to this complexity, an empirical equation was not found; however, this study showed that there was a simple correction that could be used to counteract the complication due to the matrix effects, resolution choice, and the spectral mismatch of using a nitrogen-based method for prediction of a completely different matrix gas. This involved comparing the result of the HITRAN 10 ppm calibration spectrum to the 10 ppm spectrum of the standard in nitrogen, argon, or helium and correcting the HITRAN calibration for that difference.

Figure 3.10 shows a plot of the results using the same spectra as those used in Figure 3.9, but in this case the calibration method was modified by taking the result of the 10 ppm methane spectrum and correcting it so that it will read 10 ppm using the HITRAN calibration for each of the three matrix gases tested. This correction factor or span was then used to correct the rest of the HITRAN calibration. Spanning a calibration is used routinely in flame ionization detectors (FIDs) as well as in nondispersive infrared (NDIR) analyzers to remove any drift or changes from the last time the analysis was performed. By applying this technique here to correct for the biases due to the complex nature of the analysis, one can clearly see that a simple span factor greatly reduces the differences between all of the matrix gases (nitrogen, argon, helium) as well as the differences in the regions used for that analysis (Hi Q versus Lo Q). The spread in the values near 2 to 3 ppm is due to the accuracy of the mass flow controller (MFC) used at this dilution level. From 1 ppm down, a different MFC was used, showing a much lower error in the prediction.

For a more dramatic demonstration of this effect, plots of 2, 5 and 10 ppm of methane in helium, argon, hydrogen, and nitrogen balance gases are show in Figure 3.11. The upper plot shows HITRAN calibration based on nitrogen without span correction; the lower plot shows the results after spanning using the 10 ppm standard. In the case of hydrogen the uncorrected results were 35 % higher, and by applying a simple span factor the results were corrected to less than a 1 % error. It is clear that by spanning the HITRAN method, one calibration spectrum in the actual bulk gas can be used to greatly reduce the effect of matrix gas broadening without having to create a full calibration method using the bulk gas.

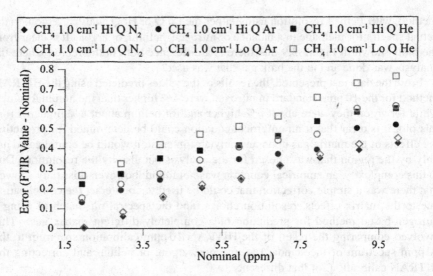

Figure 3.9 Plot showing the results of the difference between the FTIR value reported for methane minus the nominal value along the y-axis and the nominal value along the x-axis. The open symbols represent the results reported using the Lo Q region, and the solid symbols represent the Hi Q region for nitrogen (diamond), argon (circle), and helium (square). Spectral measurements were made at a resolution of $1.0\,\text{cm}^{-1}$.

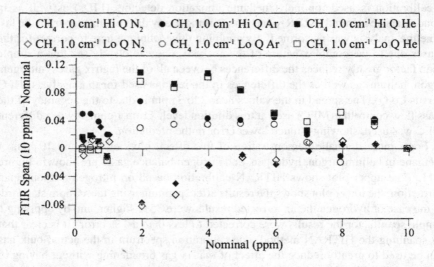

Figure 3.10 Plot showing the results of the difference between the FTIR value reported for methane after the calibration was spanned using 10 ppm spectrum results minus the nominal value along the y-axis against the nominal value along the x-axis. The open symbols represent the results reported using the Lo Q region, and the solid symbols represent the Hi Q region for nitrogen (diamond), argon (circle), and helium (square). Spectral measurements were made at a resolution of $1.0\,\text{cm}^{-1}$.

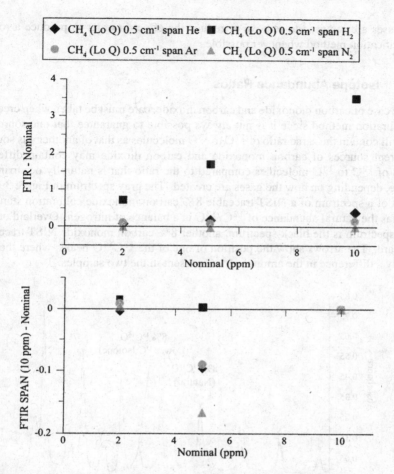

Figure 3.11   Upper plot showing the results of the difference between the FTIR reported value for methane minus the nominal value along the y-axis against the nominal value along the x-axis. The lower plot shows the results after spanning using the 10 ppm results. Only the Lo Q region results are shown for helium (diamond), argon (circle), hydrogen (square), and nitrogen (triangle). Spectral measurements were made at a resolution of $0.5\,cm^{-1}$.

3.3   Factors That Greatly Affect Quantification

Quantification issues can arise from a number of unexpected sources. In some cases the calibration spectra that are used may contain an isotope ratio that is not naturally occurring in nature, which can directly affect the final analysis results. Hydrogen bonding can also produce errors in quantification as well. Other influences include poor background spectra as well as using analysis regions that are inappropriate for either low-ppb or high-percent-level quantification. In all instances, different processes are required to remove or reduce the effect in order to produce a robust quantification method whenever possible.

3.3.1   Isotope Abundance Ratios

In the case of carbon monoxide and carbon dioxide, care must be taken when creating a calibration method since it is not always possible to guarantee that one source of gas will contain the same ratio of ^{12}C to ^{13}C molecules as that of another gas source. Different sources of carbon monoxide and carbon dioxide may contain differing ratios of ^{12}C to ^{13}C molecules compared to the ratio that is naturally occurring in nature, depending on how the gases are created. The gray spectrum in Figure 3.12 is a plot of a spectrum of a NIST-traceable 8 % carbon monoxide calibration standard that has the natural abundance of $^{12}C/^{13}C$ in a balance of nitrogen. Overlaid on the gray spectrum is the black spectrum, another 8 % carbon monoxide NIST-traceable standard. The arrows show the position of one of the $^{13}C^{16}O$ peaks, where there is clearly a difference in the amount of ^{13}C present in the two samples.

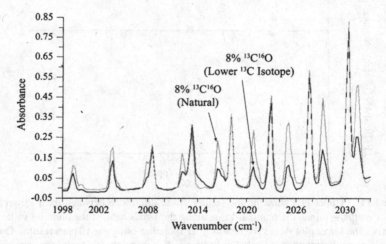

Figure 3.12   Two spectral plots of 8 % carbon monoxide. Arrows indicate the $^{13}C^{16}O$ peak. The carbon monoxide with natural distribution of $^{12}C/^{13}C$ isotopes is shown in gray, and the black spectrum shows a significantly lower ^{13}C isotope content.

When carbon dioxide is present in the gas stream at high concentrations, the particular region shown in Figure 3.12 is one that would be the best choice for the FTIR analysis of carbon monoxide since the carbon dioxide peak obscures the signal of carbon monoxide in other regions. The results of several CLS predictions using this region are shown in Table 3.1. If all of the peaks shown in Figure 3.12 were used in the analysis, then for the case of the natural isotope content, the prediction of the concentration was 8.29 % while that for the lower ^{13}C isotope content was 5.92 % (see entries for A in Table 3.1). If the CLS method is then modified to mask out all other peaks except those for the major isotope $^{12}C^{16}O$, the predictions are less then 2 % off from the certified value of 8.0 %, as shown by the entries for B in Table 3.1.

Table 3.1 Comparison of results when using all carbon monoxide peaks for the analysis versus using single isotopes

Certified CO Standard Concentration	FTIR CLS Quantification	
	A[a]	B[b]
8.01 % CO with natural $^{12}C/^{13}C$ isotope abundance	8.29 %	8.12 %
8.0 % CO with lower ^{13}C isotope abundance	5.92 %	8.02 %

[a] Analysis done using all of the isotopes present in the analysis region in Figure 3.12.
[b] Analysis done with only the $^{12}C^{16}O$ isotope present in the analysis region in Figure 3.12.

It is clear that careful selection of the quantification peaks due to the major isotope only (in this example those associated with $^{12}C^{16}O$) results in the actual concentration value being predicted. Care must be taken when an isotope (other than the natural isotope) can be present in the region of analysis (especially at high concentrations) to ensure that the calculation is based on the major isotope only. The calibration error can be in either the actual calibration gas spectrum that was used to build the calibration method or in the sample that is being analyzed. Any method that involves spectral peak matching, such as CLS, will result in quantification errors if the calibration spectra differ greatly from the unknown spectrum, as was shown clearly in the example above.

3.3.2 Hydrogen Bonding

Just as spectral peak mismatch from different isotope ratios affects the quantification value, hydrogen bonding between hydrogen-containing gases may also affect the quantitative analysis of impurities. This issue is proportional to the strength of the hydrogen bonding that the molecule exhibits, along with its concentration, pressure, and temperature. This effect is widely evident during the impurity analysis of anhydrous hydrogen fluoride.

At low concentrations, hydrogen fluoride exhibits a well-defined absorption feature between 4300 and 3500 cm^{-1} (Figure 3.13). From this absorption spectrum it is easy to assume that the detection of impurities below 3500 cm^{-1} should be without any interference of spectral features from hydrogen fluoride. However, at higher hydrogen fluoride concentrations the situation is completely different. Observing the spectrum of the mid-infrared portion of the spectrum (Figure 3.14) indicates that there are no longer sharply separated hydrogen fluoride peaks. This strong broad absorption feature is due to the hydrogen-bonding interaction between the hydrogen fluoride molecules. As a result, the hydrogen-bonding effect prevents detection of the major impurities normally associated with the production of hydrogen fluoride.

There are three different strategies to overcome the hydrogen-bonding effect [9]. All of them involve moving the conditions inside the sampling cell farther away

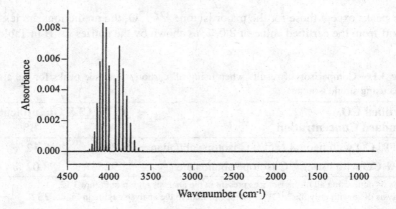

Figure 3.13 Infrared spectrum of hydrogen fluoride at 1 ppm meter measured with a resolution of $2\,cm^{-1}$.

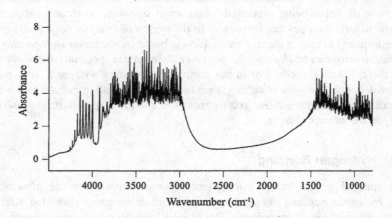

Figure 3.14 Infrared spectrum of anhydrous hydrogen fluoride at room temperature.

from the condensation point of the sample. These strategies include decreasing the pressure, diluting the sample, or increasing the sample temperature.

Due to its simplicity, material compatibility, and lower reduction of sensitivity, increasing the temperature is the strategy recommended. Figure 3.15 shows the same sample of anhydrous hydrogen fluoride as that analyzed in Figure 3.14 but at an elevated temperature. It is evident now that most of the effects of hydrogen bonding have been removed from the spectrum, and the impurities can be measured easily.

3.3.3 Alternative Background Removal Strategies

To obtain good-quality FTIR spectra for quantitation, periodic removal of the spectral interferences caused by the spectrometer components, purge gas quality, as well as

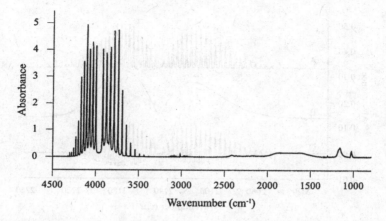

Figure 3.15 Infrared spectrum of anhydrous hydrogen fluoride at 110 °C.

leaks and general changes due to system aging is recommended. Background removal should be carried out at least once every 24 hours, or more frequently, if possible, for best results. As the semiconductor industry pushes for ever-lower detection of impurities in bulk gases used in chip manufacturing and other processes, it may not be feasible to evacuate out the gas cell or flush it with a non-infrared-absorbing gas, which is the general practice for background spectral removal. Alternative methods for FTIR spectral background removal have been developed that do not require the sample gas to be removed from the gas cell, which avoids the introduction of impurities into the gas stream (in particular moisture) as well as facilitates the ability to achieve lower method detection limits.

Although other methods are available, two are presented below that allow for ppb to ppt levels of detection, in particular for water vapor contamination. The analysis of water vapor in a bulk gas stream is most difficult at low levels, due to the fact that it is difficult to remove from the background purge of the FTIR system and must be compensated for in order to analyze the actual gas stream. Discussed below are two methods that allow low-level water vapor as well as other contaminants to be analyzed at ppb to ppt levels.

Auto- or Self-Reference Method In the case of alternative background removal schemes, there is no requirement that a blank nitrogen spectrum ever be collected, as it will not be used in the analysis. In this case once the sample spectrum has been collected, a secondary spectrum will be created from that data file at a much lower resolution, which will then be used as background for the sample file collected [10]. This method has been called *auto-referencing* or *self-referencing*, due to the fact that only one interferogram (or raw spectrum) is collected per sample point. From this raw interferogram two single-beam spectra are generated, one at the required final resolution (in general, this is $1\,cm^{-1}$) and the other by de-resolving the same sample interferogram to $8\,cm^{-1}$ resolution. Unlike the standard FTIR method, where one collects a blank or zero spectrum (run at the same resolution as the sample

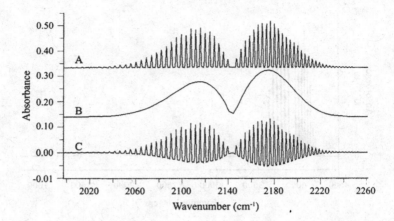

Figure 3.16 Auto-reference spectrum of a narrow-line absorbing molecule: (A) carbon monoxide at $1\,cm^{-1}$ resolution, (B) carbon monoxide at $8\,cm^{-1}$ resolution, and (C) carbon monoxide at the final result of $1\,cm^{-1}$ ratioed to $8\,cm^{-1}$ resolution.

spectrum) and this single-beam spectrum is then stored and ratioed against each sample spectrum created after that point, the auto-reference method generates a new self-referenced background ($8\,cm^{-1}$ resolution) single-beam spectrum every time a sample single-beam spectrum is collected and created. In this manner any change that happens from one scan to the next is removed, making possible a much more stable baseline which never contains shifts due to drift or other changes.

Because the sample is being ratioed to itself at a broader resolution, only those peaks that have a resolution narrower than $8\,cm^{-1}$ will survive the transformation. This method produces a derivative-like spectrum that has a baseline centered at zero, removing from the spectrum all transient and long-term baseline artifacts. Figure 3.16 shows an example of carbon monoxide, which has an absorption spectrum made up of very narrow line shapes. Figure 3.16A is a spectrum of carbon monoxide measured at $1\,cm^{-1}$ resolution, Figure 3.16B is a spectrum of carbon monoxide at a resolution of $8\,cm^{-1}$, and Figure 3.16C shows the final transformation spectrum based upon the ratio of carbon monoxide at $1\,cm^{-1}$ to carbon monoxide at $8\,cm^{-1}$ resolution, resulting in a derivative-like spectrum for carbon monoxide.

In the second example it is clear what becomes of the broad absorbance when the auto-reference method is applied to propane (Figure 3.17). Because the propane spectrum is composed mainly of a single broad feature which has a FWHH greater than $8\,cm^{-1}$, those spectral features do not change when it is converted from a $1\text{-}cm^{-1}$ to an $8\text{-}cm^{-1}$ resolution spectrum (see Figure 3.17A and B for comparison). When the transformation is made to the auto-reference spectrum, most of the propane spectrum is removed (Figure 3.17C), leaving only a couple of peaks that were narrower than $8\,cm^{-1}$ resolution. This figure clearly shows that only those spectral features that are narrower than $8\,cm^{-1}$ resolution will survive conversion to the auto-reference method.

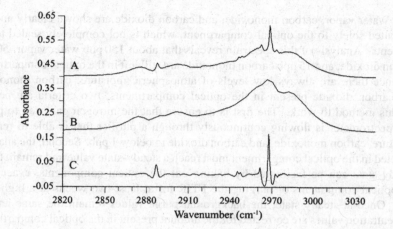

Figure 3.17   Auto-reference spectrum of a broad-line absorbing molecule: (A) propane at $1 cm^{-1}$ resolution, (B) propane at $8 cm^{-1}$ resolution, and (C) propane at the final result of $1 cm^{-1}$ ratioed to $8 cm^{-1}$ resolution.

One consequence of the auto-reference method is that unlike a standard FTIR spectrum, where the ratio of the sample against the background will remove the spectral features of water, carbon monoxide, and carbon dioxide present in the optical compartment, the spectral peaks of these components will still be present in all subsequent spectral samples and therefore must be compensated for in some manner. Figure 3.18 shows the result of the auto-reference method when flowing purified nitrogen purge gas into the optical compartment, as well as into the gas

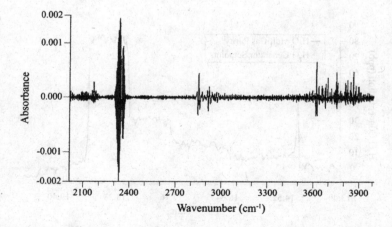

Figure 3.18   Auto-reference spectrum result when flowing nitrogen purge gas through the gas cell as well as the optical compartment. All narrow absorbing components in the optical compartment are still present when the final spectrum is displayed.

cell. Water vapor, carbon monoxide, and carbon dioxide are shown clearly and are attributed solely to the optical compartment, which is not completely sealed to the elements. Analysis of this spectrum reveals that about 150 ppb water vapor, 50 ppb carbon dioxide, and 15 ppb carbon monoxide are still left in the optical compartment.

Since there are always low levels of atmospheric moisture, carbon monoxide, and carbon dioxide present in the optical compartments, two criteria are needed for this method to work. The first is to ensure that the nitrogen gas used to purge the spectrometer is flowing continuously through a purifier that is able to remove moisture, carbon monoxide, and carbon dioxide to below 1 ppb. Second, the analytes detected in the optical compartment must reach a steady-state value and remain there. Steady state can be facilitated by baking out the optical components, evacuating the optical components or purging the FTIR for at least two weeks at a high flow rate. Once at steady state, the background purge values remain the same and all concentration values are corrected for the amount present in the optical compartment. Care must be taken to ensure that the optical components are tightly sealed and that the ambient air or temperature does not greatly affect the gases present in the optical compartment when attempting to quantify single-digit ppb water vapor levels in bulk gas analyses.

The auto-reference method has been used to obtain single-digit-ppb level detection of water vapor using a 5.11-m gas cell at 1 atm pressure. For low-level ppb detection of water vapor in 100 % ammonia, a background level of 150 ppb with a standard deviation of 8 ppb or better has been used successfully [11]. A series of water vapor dilution cut points in 100 % ammonia was performed using a 5.11-m gas cell and a high-purity FTIR system showing single-digit-ppb detection limits for water vapor (Figure 3.19). Part-per-trillion-level detection can be obtained by using longer-path gas cells and/or higher gas cell pressures.

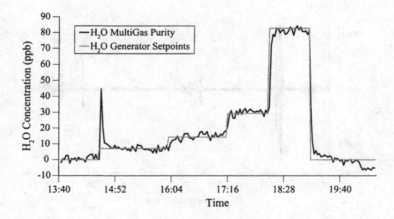

Figure 3.19   Dilution time plot showing ppb water vapor detection in 100 % ammonia using FTIR. Data used with permission from Air Products and Chemicals, Inc. and MKS Instruments.

There is a way of accounting automatically for the background purge components. If the high-purity nitrogen purge is held at steady state, a snapshot of a spectrum of purified 100 % ammonia gas containing less than 1 ppb water vapor, carbon monoxide, and carbon dioxide will contain the spectral peaks of the water, carbon monoxide, and carbon dioxide in the optical compartment (Figure 3.20A for a layout of this process). This information is then exploited in the subsequent analysis by CLS. The spectral-matching technique of CLS will remove this background content automatically during the analysis process, leaving behind the water, carbon monoxide, and carbon dioxide peaks, which are due only to the ammonia gas stream.

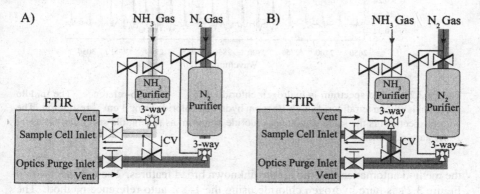

Figure 3.20 (A) Process by which a clean 100 % ammonia spectrum is collected using an ammonia purifier to remove water vapor, carbon monoxide, and carbon dioxide to below 1 ppb; (B) process by which the contents of the optical compartment can be determined using a nitrogen purifier to remove water vapor, carbon monoxide, and carbon dioxide to below 1 ppb.

If a clean 100 % bulk gas spectrum cannot be obtained or if the matrix gas does not have any spectral features in the IR region, the amount of water vapor, carbon monoxide, and carbon dioxide in the spectrometer will need to be determined. This is done by measuring the amount of water vapor in the spectrometer while purging the gas cell with a slipstream from the purified nitrogen purge gas (Figure 3.20B). From this, steady-state concentrations for water, carbon monoxide, and carbon dioxide are obtained and can be removed mathematically from the final results predicted.

Silicon tetrafluoride is a perfect example of the application of this technique without a purifier that can remove the contaminants from the bulk gas itself. Silicon tetrafluoride is an important precursor to the high-purity-grade silane used in both the semiconductor and solar industries. It is very difficult to obtain a pure sample of silicon tetrafluoride that does not contain hydrogen chloride or hydrogen fluoride. To determine the actual concentration of silicon tetrafluoride, the hydrogen chloride content must be quantified, and in some cases, it can be as high as 3 %. To complicate matters further, broad features in the silicon tetrafluoride spectrum occur where the hydrogen chloride peaks appear, and they must be accounted for or they will interfere with the analysis. In this case, use of the auto-reference method is key to providing low detection limits as well as accurate analysis of the hydrogen chloride content since

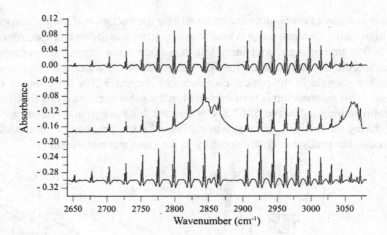

Figure 3.21   Top spectrum is hydrogen chloride using 1–8× auto-reference. The middle spectrum is silicon tetrafluoride with 100 ppm hydrogen chloride using 1 cm⁻¹ resolution. The bottom spectrum is the same silicon tetrafluoride spectrum as the middle spectrum but using 1–8× auto-reference.

the method automatically removes the unknown broad features. The top spectrum in Figure 3.21 is pure hydrogen chloride using the 1–8× auto-reference method. The middle spectrum is silicon tetrafluoride with about 100 ppm of hydrogen chloride, displayed in the standard FTIR mode using 1 cm⁻¹ resolution. The broad features are due to the silicon tetrafluoride and are not consistent from one batch to the other. The bottom spectrum is the same silicon tetrafluoride spectrum, but in this case the spectrum was rendered to the 1–8× auto-reference method. It can be seen clearly that the broad features that are present in the 1 cm⁻¹ resolution spectrum are no longer present, revealing a more complete hydrogen chloride spectrum. This results in a more robust analysis of the hydrogen chloride content within the silicon tetrafluoride gas sample.

Gas Cell Bypass Method   The second alternative background removal scheme also has no requirement for a blank or nitrogen gas stream to be passed through the gas cell to compensate for the carbon monoxide, carbon dioxide, and water in the optical compartment. Whereas the auto-reference method used software as a means to remove the background signal the gas cell bypass method uses hardware to physically bypass the gas cell and collect the light from the rest of the FTIR spectrometer.

The background is collected by automatically moving the mirrors that transfer the light from the modulator into the gas cell path so that the light now bypasses the gas cell (Figure 3.22, Background Stage). Once this background signal is collected, the mirrors automatically slide back so that the light again passes through the gas cell which contains the sample gas (Figure 3.22, Sampling Stage). Only those components present in the nitrogen purge gas are seen during the background collection.

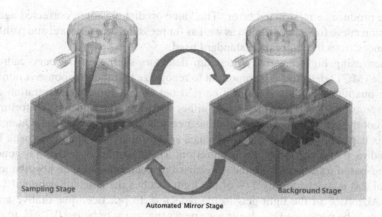

Figure 3.22   Mirror position when sampling the gas stream (left) and when sampling the background only (right).

Although this method allows for bypassing the gas cell, it does have a drawback in implementation. The background spectrum does not contain any of the spectral absorbance information, due to the gas cell optics or mirror coatings. Because this information is missing in the background spectrum when it is ratioed to the sample spectrum, the resulting baseline will not be flat around zero. This baseline issue can, however, easily be compensated for in the quantification method by adding the baseline collected as a component of the calibration to be measured. In addition, by tracking the fit of the original baseline to the sample provides an additional point that can be used to determine the health of the system, indicating when the mirrors are coated or degraded.

Furthermore, this process also has the benefit that by collecting the background constantly, the absorbance baseline is much more stable, and the same baseline can be used for longer periods of time than the standard 24 hours and can last weeks or months before it needs to be replaced.

3.3.4   Automatic Region Selection for CLS Methods

One of the largest sources of error is the uncertainty of the component concentration measured, in particular if any of the components in the method span a large concentration range. The ability to span a large concentration range automatically would greatly reduce the error in the concentrations predicted; however, it is difficult to do this since the nonlinearity of the response increases as the spectral peak intensity increases for many narrow absorbing components.

As mentioned earlier, for most gas-phase processes, CLS has been the method preferred for determining unknown concentrations using spectral peaks of known components and matching them to the unknown. In general, the method uses a spectrum of known concentration and scales it until the spectral peak-matching routine produces a minimized error. The value predicted is then corrected against a correction curve for concentration as well as for pressure, temperature, and pathlength differences from the calibration standard used.

When using high-sensitivity quantum detectors such as a mercury–cadmium–telluride (MCT) detector, it is important to recognize that their response is nonlinear. Before quantitative analysis can be carried out over a broad concentration range, this response must be corrected using either software or hardware. Unfortunately, this is not the only place where a nonlinear response can be generated. Quantitative analysis is performed using the absorbance spectrum since a change in peak height here is directly proportional to concentration. The absorbance spectrum is generated by a log-based transform from the transmission spectrum. Once an absorbance unit (AU) of 2.0 is reached, only 1 % of the light reaches the detector, while at a value of 1.0 AU, 10 % of the light gets to the detector. In practice, quantitative analysis should only be performed on absorbance peaks that are at or below 0.7 AU, to reduce the problem of a nonlinear response at such low energy throughput.

When quantifying a gas concentration that spans a very broad concentration range and unless the gas is a very broad absorber that has a linear calibration curve, one analysis region will not be sufficient to quantify it. At low concentrations, one generally relies on the strongest-absorbing peaks, but as the concentration increases, those peaks begin to rise above 0.7 AU and eventually go off scale. Narrow-absorbing-component responses are very nonlinear with respect to increases in concentration, so changes in the region of analyses are required if a broad concentration range is to be used in the analysis method.

Since it may not be possible for an analyst to be present at all times when the concentration analysis region is no longer valid for the concentrations present in the gas sample, an automatic region selection method can be deployed. In particular, an extended weighted CLS method has been used based on assigning higher weight to the more intense spectral bands and at the same time assigning less weight to those bands that result in a higher amount of spectral residuals [12]. Therefore, those bands that have the higher intensity and a better fit to the actual sample spectrum will provide the most information for the final analyte concentration prediction. Those spectral regions that have higher residuals can be minimized to avoid using regions that may contain unknown components or are at highly absorbing regions.

The automated method performs the quantification based on selecting portions of the spectrum where the bands have similar intensities. In this way the effects of any differences in nonlinear behavior will be minimized. Figure 3.23 shows an example of overlaid spectra of methane spanning a concentration range from 1 ppm up to 1 %. Superimposed on this are three regions: b_1, containing the largest peak in methane; and b_2 and b_3, which contain similar peaks in height but at a much lower intensity than that of b_1. When examining the results of the prediction analysis for all three bands spanning the full concentration profile, band b_1 has greater difficulty quantifying the higher concentration range. This is due to the intensity of the methane peak in region b_1, for a number of the spectra in the concentration profile are greater than 1.0 AU, and in traditional CLS this region would not be used at all. Regions b_2 and b_3 would be more appropriate for the higher-concentration analyses.

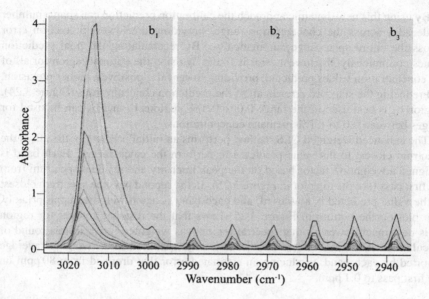

Figure 3.23   Spectral plots of methane absorbance at concentrations from 1 ppm to 1 % versus three region selections.

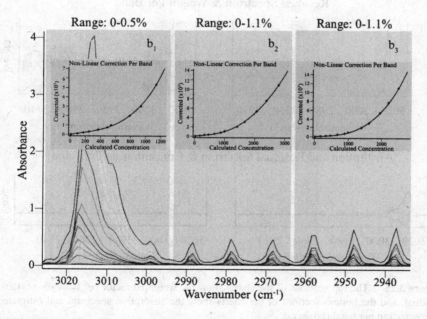

Figure 3.24   Enhanced CLS showing the dynamic range of predictions from ppm to percent methane for all three regions.

By using this new dynamic approach the calibration prediction can span a number of decades across the concentration range, providing the lowest prediction error across the entire span range automatically. By recalculating the final prediction values continuously, the lowest error is found by using the optimal region for all of the concentration values predicted, providing lower false positives, higher precision, and reducing the standard error in all of the prediction concentrations (Figure 3.24). Region b_1 is best used in the range 0.0 to 0.5 %; regions b_2 and b_3 can be used for ranges between 0.0 to 1.1 % methane concentrations.

The enhanced weighted CLS method performs an initial prediction, then uses the spectrum closest to the value predicted to perform the calculations. Each band is assigned a weighting factor based on the peak intensity and residual spectrum from the first pass (see the top plot in Figure 3.25). In the second pass the spectrum closest to the value predicted is now used, and each band is again weighted appropriately. The plot on the bottom in Figure 3.25 shows that the residual amount for region b_1 is now much lower, as that spectral region was weighted lower in this round of calculations. The final concentration of the analyte is based on the weighted average reported by each band, producing an estimated error that dropped from 80 ppm on the first pass to 0.1 ppm.

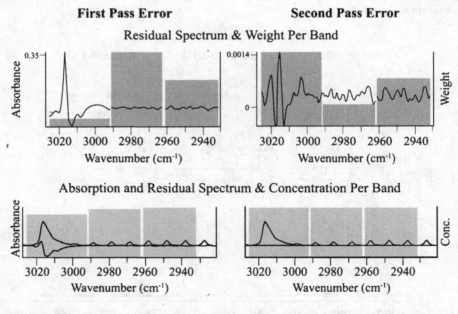

Figure 3.25   The top sections of the charts show the weighted factor per band (boxes) and residual, and the bottom sections of the charts show the absorption spectrum and estimated concentration per band (boxes).

3.4 Future Applications

The future of FTIR spectroscopy for applications in specialty gas analysis lies in its ability to provide detection of impurities at lower concentrations (sub-ppb level) and, particularly, in corrosive environments. The use of more powerful infrared sources such as quantum cascade lasers can greatly improve the ability to get more signal to the detector compared to standard silicon carbide sources. However, improvements in the materials used for the gas-wetted components are also needed so that they can withstand corrosive gases for long periods of time as well as not retard the optical light path. For instance, higher reflective gas cells that provide much longer effective pathlengths and also have corrosion-resistant mirrors would be needed to reach the ppt and lower concentration range. In addition, more software and hardware development is needed that allows for automatic spectral background corrections to be made without interrupting the sample stream. Clearly, the future lies in combining rugged and clever hardware with new software and algorithm developments to move FTIR to an even lower level of detection for the specialty gas industry.

REFERENCES

1. Haaland, D. M., & Easterling, R. G. (1982). Application of new least-square methods for the quantitative infrared analysis of multicomponent samples. *Applied Spectroscopy, 36,* 6, 665–673.

2. Wold, S., Sjostrom, M., & Eriksson, L. (2001). PLS-regression: a basic tool of chemometrics. *Chemometrics and Intelligent Laboratory Systems: An International Journal Sponsored by the Chemometrics Society, 58,* 2, 109.

3. Haaland, D. M., & Thomas, E. V. (1988). Partial least-squares methods for spectral analyses: 2. Application to simulated and glass spectral data. *Analytical Chemistry, 60,* 11, 1202–1208.

4. Buldyreva, J., Lavrentieva, N., & Starikov, V. (2011). *Collisional Line Broadening and Shifting of Atmospheric Gases: A Practical Guide for Line Shape Modelling by Current Semi-classical Approaches.* London: Imperial College Press.

5. Rosenmann, L., Hartmann, J. M., Perrin, M. Y., & Taine, J. (1988). Accurate calculated tabulations of IR and Raman CO_2 line broadening by CO_2, H_2O, N_2, O_2 in the 300 to 2400 °K temperature range. *Applied Optics, 27,* 18, 3902–3906.

6. Sinclair, P., Duggan, P., Berman, R., Drummond, J., & May, A. (1998). Line broadening in the fundamental band of CO in CO–He and CO–Ar mixtures. *Journal of Molecular Spectroscopy, 191,* 2, 258–264.

7. Haaland, D. M., & Melgaard, D. K. (2002). New augmented classical least squares methods for improved quantitative spectral analyses. *Vibrational Spectroscopy, 29,* 1/2, 171–175.

8. Marshik, B., Phillips, M., Eckman, C., Thorn, W., III, & Gameson, L. (2012). FTIR line broadening effects on CO, CO_2 and CH_4 in N_2 versus other matrix gases. *Pittsburgh Conference, Specialty Gas Session, Orlando, FL.*

9. Vahey, G., Honeywell, P., & Perez, J. E. (2004). Overcoming hydrogen-bonding in on-line spectroscopy of hydrofluoric acid. *International Forum on Process Analytical Chemistry, Washington, DC.*

10. Espinoza, L. H., Niemczyk, T. M., & Stallard, B. R. (1998). Generation of synthetic background spectra by filtering the sample interferogram in FT–IR. *Applied Spectroscopy, 52,* 3, 375–379.

11. MKS Instruments, Inc. (2003). *Ammonia Contaminant Detection Using MultiGas™ Purity Analyzer.* MKS Application Note 13/02–9/10. Andover, MA: MKS Instruments, Inc.

12. Bonano, E. (2009). Self-optimizing specialty gas calibration error-reduction software. *Pittsburgh Conference, Specialty Gas Session, Chicago.*

CHAPTER 4

EMERGING INFRARED LASER ABSORPTION SPECTROSCOPIC TECHNIQUES FOR GAS ANALYSIS

Frank K. Tittel,[1] Rafal Lewicki,[1] Robert Lascola,[2] and Scott McWhorter[2]

[1]Rice University, Electrical and Computer Engineering Department, Houston, Texas
[2]Savannah River National Laboratory, Analytical Development Division, Aiken, South Carolina

4.1 Introduction

Laser-based spectroscopic techniques are useful for the quantitative detection and monitoring of molecular trace gas species in the mid-infrared spectral region because many molecules have their fundamental absorption band in this region. The spectroscopic instrumentation generates a measurable signal that depends on absorption of the target medium. The choice of an optimum detection scheme depends on the requirements of the specific application and the characteristic features of the infrared laser source. Well-established detection methods include several types of multipass gas absorption cells, with the option to apply wavelength, frequency, and amplitude modulation to the laser source. Internal and external cavity-enhanced spectroscopies are two methods of increasing the magnitude of the molecular absorption signal. Photoacoustic and photothermal open-path monitoring (with and without a retro-reflector), such as light detection and ranging (LIDAR), differential optical absorption spectroscopy (DOAS), laser-induced fluorescence (LIF), laser breakdown

Trace Analysis of Specialty and Electronic Gases.
By William M. Geiger and Mark W. Raynor. Copyright © 2013 John Wiley & Sons, Inc.

spectroscopy (LIBS), and fiber optic or waveguide evanescent wave spectroscopy, are other useful mid-infrared detection schemes. A key optical component for laser absorption spectroscopy (LAS) has been the introduction and commercial availability of high-performance semiconductor lasers, in particular quantum cascade lasers (QCLs) since 1994 [1] and interband cascade lasers (ICLs) since 1995 [2,3]. The development of both QCLs and ICLs continues worldwide [4–9]. Quantum cascade lasers (QCLs) are useful compact mid-infrared sources for ultrasensitive and highly selective trace gas monitoring as the result of recent advances in their design and technology. They have been demonstrated to operate over a wide range of mid-infrared wavelengths from approximately 3 to 24 µm. Pulsed and continuous-wave (CW) QCL devices capable of thermoelectrically cooled room-temperature operation are available commercially in the spectral region 4 to 12 µm [5,10]. These devices have several important practical features, including single-mode emission with mode-hop free frequency tuning, high power (tens to hundreds of milliwatts), and intrinsic narrow emission line-widths. These spectral characteristics permit the development of robust and fieldable trace gas sensors [11–14]. For example, the Rice Laser Science Group has explored the use of several methods for carrying out infrared laser absorption spectroscopy (LAS) with mid-infrared QCL, ICL, and laser diode (LD) sources, which include LAS based on a multipass gas cell [15], cavity ring-down spectroscopy (CRDS) [16], integrated cavity output spectroscopy (ICOS) [17,18], photoacoustic spectroscopy (PAS), and quartz-enhanced photoacoustic spectroscopy (QEPAS) [19–23]. These spectroscopic techniques permit the detection and quantification of molecular trace gases with demonstrated detection sensitivities ranging from parts per million by volume (ppmv) to parts per trillion by volume (pptv) levels, depending on the specific gas species and the detection method employed.

4.2 Laser Absorption Spectroscopic Techniques

In conventional absorption spectroscopy (CAS), using broadband incoherent radiation sources, such as thermal emitters, the wavelength resolution is determined by the resolving power of the spectral analyzer or spectrometer. LAS, on the other hand, uses coherent light sources, whose linewidths can be ultranarrow and whose spectral densities can be made many orders of magnitude larger (ca. 10^9 W/cm^2 · MHz) than those of incoherent light sources. The key advantages of mid-infrared LAS include:

1. An absorption spectrum that can be acquired directly by scanning the laser source across a desired rotational–vibrational resolved feature of the target analyte

2. High detection sensitivity with maximum accuracy and precision

3. Improved spectral selectivity compared to CAS, due to the narrow linewidths of CW QCLs (i.e., ca. 0.1 to 3 MHz with a high-quality power supply or less than 10 kHz with frequency stabilization and for pulsed QCLs of ca. 300 MHz) [24]

4. Fast response time (<1 second in some applications)

5. Good spatial resolution (e.g., millimeter-scale imaging of plasmas) [25]

6. Detector noise that becomes negligible for sufficiently large laser intensities

7. Nonintrusive methods to suppress laser intensity fluctuations, such as balanced detection or zero air subtraction, that can be applied readily to increase the signal-to-noise ratio (SNR) and hence improve detection sensitivity

8. Spatially coherent laser light that can be collimated, which allows the use of long-pathlength gas absorption and CRDS cells

9. The ability to lock the laser frequency to the center of a resolved molecular absorption line in order to determine the concentration levels of the target analyte with ultrahigh precision and accuracy

10. Size, weight, electrical power, thermal management, gas and wavelength calibration, protection from a harsh environment, autonomous operation, and remote access for long periods of time

11. Ease of instrument operation and data acquisition and analysis

During the past 25 years, mid-infrared LAS techniques have become widely available commercially. The primary uses to date have been in the spectral region of approximately 3 to 12 μm, which covers a substantial portion of the spectrum of fundamental molecular vibrations. Wider spectral coverage from 2 to 24 μm has been demonstrated and reported in the literature using antimony-based LDs, ICLs, and QCLs [5,10,26,27]. Expansion into the vibrational overtone region (3 to 5 μm) is also possible with optical parametric oscillators (OPOs) and sources based on difference frequency generation (DFG). Mid-infrared absorption spectra of several small molecules of potential interest for trace gas monitoring are shown in Figure 4.1. In this figure, which highlights environmental applications, certain spectral regions are shown to be unavailable due to absorption by water and carbon dioxide. For other applications, such as specialty and electronic gas purity measurements, it will be other bulk gases that block different parts of the spectrum. Regardless, there is usually a transmission window where analyte absorption lines can be cleanly accessed. In some cases, these may be less intense overtone or combination lines rather than intense fundamental absorption lines. High-sensitivity (ppmv to pptv) measurements have been demonstrated using all types of lines in both laboratory and field applications.

The primary requirements for trace gas sensing are sensitivity, selectivity, and response time. For small molecules with resolved rotational structure, the selectivity is obtained by choosing an absorption line that is free of interference from other species that might be present in the analyzed sample and by implementing a laser source that possesses a sufficiently narrow linewidth. For small molecules, selectivity is enhanced further by reducing the sample pressure to sharpen the absorption line without reducing the peak absorption. This condition continues until the linewidth begins to approach the Doppler width [28].

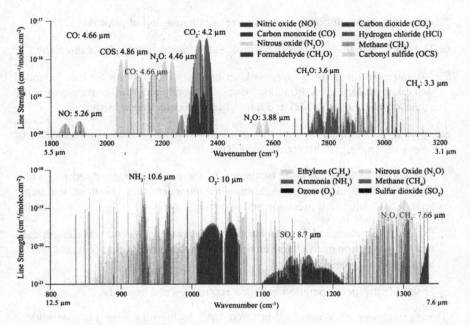

Figure 4.1   High-resolution transmission molecular absorption database simulation of absorption spectra in two mid-infrared atmospheric transmission windows.   With kind permission from Springer Science+Business Media: [21], fig. 1.

The fundamental behavior of laser absorption spectroscopy at a certain frequency ν (cm^{-1}) can be expressed by Beer–Lambert's law:

$$I(\nu) = I_0(\nu)e^{-\alpha(\nu)L} \tag{4.1}$$

where $I(\nu)$ and $I_0(\nu)$ are the intensities of the transmitted and incident laser light, respectively, $\alpha(\nu)$ is the absorption coefficient (cm^{-1}), and L is the effective optical pathlength (cm). Therefore, to obtain the optimum absorption sensitivity it is necessary to choose a strong molecular absorption line, use a platform with a long effective optical pathlength, and have a distinguishable absorption from baseline variations and laser power fluctuations. The first requirement is best met by choosing a target line associated with fundamental absorption bands, as these are stronger than overtone or combination bands. A sufficiently long pathlength can be obtained by using multipass gas cells or cavity-enhancement techniques. For sharp absorption lines, noise associated with laser power fluctuations can be reduced by averaging rapid scans over the time or by employing a modulation spectroscopy technique in the kHz regime [29,30]. In most applications, one detects the modulated absorption at twice the modulation frequency using a lock-in amplifier set to $2f$. The second harmonic signal is maximized at the absorption line center. The final requirement to distinguish absorption from baseline variations is the most challenging. Every long-pass arrangement exhibits accidental étalons, which typically have widths comparable to that of an absorption line. In principle, these can be removed by evacuating the cell, replacing the sample with "zero air" gas, which contains no trace gas species of interest, and then dividing the sample trace by this background trace. However, this approach assumes, often incorrectly, that these accidental étalons do not shift their pattern during the process of sample replacement [10]. Numerous research groups [31–35] have investigated and reported on the merits of rapid background subtraction, in particular wavelength-modulation spectroscopy (WMS) and frequency-modulation spectroscopy (FMS), to remove optical noise effectively.

For more complex multiatomic molecules, which do not have resolved rotational structure, the spectroscopic detection process is more demanding. A convenient method to distinguish sample and background absorptions is by pumping the sample out and replacing it with zero air, due to the absence of a nearby baseline for comparison. For weak absorption features, this imposes severe limits on the long-term power stability of the laser source, the absence of low-frequency laser noise, and baseline stability. Furthermore, in the mid-infrared fingerprint region, where many gases absorb, there may be other gases contributing to a broad absorption that will significantly decrease the selectivity of concentration measurements.

For open-path systems, difficulties for selective measurements will be encountered because (1) there is no way to replace the sample with zero air for providing a baseline trace, and (2) there is no way to reduce the linewidths (typically, $0.1\,\text{cm}^{-1}$ at atmospheric pressure), which in some cases might be too large to perform wavelength-modulation spectroscopy at an optimum modulation depth. LAS based on a multipass cell suffers from neither of these problems. The only issue with this approach is that long-path multipass cells are intrinsically bulky. Long optical pathlengths are obtained by employing multipass absorption cells where the optical beam is reflected back and forth between highly reflective spherical or cylindrical mirrors. Numerous implementations have been reported in the literature, but four fundamental designs (White, Herriott, astigmatic Herriott, and Chernin) have been used to achieve optical pathlengths of approximately 100 m for approximately a 0.5-m distance between the mirrors [36]. Furthermore, cavity-enhanced spectroscopy can increase the detection sensitivity even more. CRDS has been demonstrated to offer excellent trace gas detection sensitivity [16,37,38], using high-quality low-loss mirrors. Long pathlengths in small volumes can be achieved by off-axis ICOS, which has demonstrated considerable promise for trace gas sensing [17,18,39]. Other ultrasensitive and highly selective spectroscopic techniques for trace gas detection that have been reported by various research groups are: balanced detection [40], laser-induced breakdown spectroscopy (LIBS) [41], noise-immune cavity-enhanced optical heterodyne molecular spectroscopy (NICE–OHMS) [42,43], Faraday rotation spectroscopy [44–46], conventional photoacoustic spectroscopy (CPAS) [47,48], QEPAS [49,50], and direct frequency-comb spectroscopy (DFCS) [51–55]. The latter three techniques, in addition to CRDS and ICOS, are described later in the chapter.

4.2.1 Quantum and Interband Cascade Lasers

The expansion of LAS into a routinely deployable field-portable analytical instrument is attributed primarily to the development of QCL and ICL technology. Both QCLs and ICLs are smaller, easier to operate, and more rugged than other mid-infrared laser technologies, such as nonlinear optical techniques (OPO- and DFG-based instrumentation) and lead-salt diodes, which require cryogenic cooling. As cascade lasers are becoming the dominant technology for IR LAS, their properties and use are described briefly. More detailed descriptions may be found in recent review articles [5,10] and in vendor literature.

The unique property of QCLs compared to conventional diode lasers is the nature of the energy levels in which a population inversion is created. In diode lasers, the active region is a semiconductor with a bandgap that matches the desired output wavelengths. Electron–hole pairs are injected into the active region and emit light upon electron relaxation from the conduction to valence bands.

Energy levels in a QCL arise from a quantum mechanical effect, as illustrated in a simplified manner in Figure 4.2. The laser material consists of a repeating pattern of stacks of submicrometer layers of semiconductors. The pattern includes an active region and an injector region. Electrons in the active region will experience a few discrete energy levels. Light emission occurs when the electrons relax from upper to lower energy levels (state 3 to state 2) within a conductive band. To maintain the population inversion, the population in state 2 must be drained more rapidly than the lifetime associated with the 3–2 transition. This is accomplished by relaxation through resonant scattering of a phonon to state 1. Migration of the electron in state 1 to the next injector region is accomplished by providing a high density ("miniband") of states in the injector region and applying a potential across the stack that allows those states to be in resonance with state 1. The electron relaxes to the ground state of the miniband, which is in resonance with the excited energy level of the next active region. The potential also defines the energy separation of the miniband ground state and state 2 in the active region, suppressing thermal re-population of state 2. In this manner, a single electron can be propagated through multiple active regions, emitting a photon in each one, leading to the "cascade" effect. Practical cascade laser designs have additional energy levels in the active region to help maintain population inversion and to create a broader gain spectrum, but follow this general scheme.

Emitted light is captured in a waveguide that is coupled directly to the semiconductor stack. The waveguide is narrow, increasing the coupling efficiency, but resulting in a highly divergent beam that requires high numerical aperture optics for collection and collimation. Mid-infrared semiconductor lasers can be chosen to operate in either pulsed or continuous-wave (CW) mode. CW mode provides a narrower spectral linewidth (e.g., as low as 0.001 cm^{-1}) and reduced susceptibility to nonlinear optical effects, which can bias quantitative measurements if not considered carefully [10]. Laser powers of up to several hundred milliwatts are available commercially. Pulsed laser devices have fewer problems with heat management and can be operated at room temperature, whereas CW lasers may require additional chilled water and/or air cooling. Pulses are typically of nanosecond duration, and such QCLs can be

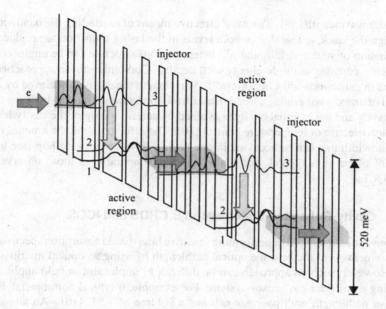

Figure 4.2   Schematic diagram indicating the quantum well–based energy levels that form the basis of QCLs. An electron injected into state 3 in the active region emits a photon when relaxing to state 2. After relaxing to state 1 (through emission of a phonon), the electron is coupled into a miniband of states in the injector, and eventually into the upper level of the next active region. The period of this structure is approximately 50 to 60 nm. Reprinted from [10] with permission from Elsevier.

operated at Hz to kHz repetition rates. QCLs, ICLs and LDs are available in the spectral range from 3 to 12 μm. The spectral range of an individual mid-infrared laser is determined by its mode of operation.

The optimum operating mode for a QCL, ICL, or LD for LAS is one that generates single-frequency radiation. This operating characteristic can be obtained by means of a distributed feedback (DFB) structure in which the lasing wavelength is selected by a grating that is etched directly onto the laser cavity. Spectral tuning is achieved through current or laser temperature changes, which alters the refractive index of the laser waveguide material and changes the Bragg scattering condition for the grating. The scanning range for DFB–QCLs is narrow, typically 5 to 10 cm^{-1}. Thus, detection, quantification, and monitoring of multiple analytes with a single laser is dependent on a fortuitous overlap of absorption features. One recent approach to expand the spectral range is to fabricate multiple QCLs on a single chip, exciting each one individually and adjusting chip temperature as required for accessing the desired wavelength/wavenumber. A spectral tuning range of approximately 85 cm^{-1} has been demonstrated [57].

Alternatively, a wider spectral coverage can be achieved by broadening the bandwidth of the active region and utilizing an external cavity (EC) to promote lasing at a particular wavelength [58]. The most effective means of expanding the bandwidth is to design the stack so that one or both levels in the lasing transition are replaced by a continuum of states. Additionally, a heterogeneous structure can be engineered so that a chip contains multiple energy well depths. Combining these approaches has resulted in gain bandwidths of $500 \, cm^{-1}$. Wavelength tunability is achieved by angle control of an external grating placed in a Littrow configuration. Mode hopping of the QCL cavity and the external cavity is avoided by careful tuning of the cavity lengths with piezoelectric or temperature control [10]. These factors limit the amount of the full bandwidth that can be accessed in one device; however, a mode hop-free tuning range of $60 \, cm^{-1}$ is typical, and linewidths are comparable to those observed for DFB–QCLs.

4.2.2   Cavity-Enhanced Spectroscopy: CRDS and ICOS

The conventional method of performing sensitive laser-based absorption spectroscopy measurements is to increase the optical pathlength by using an optical multipass gas cell. However, such an approach can be difficult to implement in field applications requiring a compact gas sensor system. For example, a typical commercial 100-m effective pathlength multipass gas cell has a volume of 3.5 L [10]. An alternative spectroscopic technique to obtain a long optical path is to make the light bounce along the same path between two ultralow-loss dielectric mirrors forming a high-finesse optical resonator that behaves as a ring-down cavity (RDC). Cavity ring-down spectroscopy is based on the principle of measuring the rate of decay of light intensity inside the RDC. The wave transmitted from an injected pulsed or CW laser into the RDC decays exponentially in time. The decay rate is proportional to the losses inside the RDC. For typical RDC mirrors having a reflectivity of 99.995 %, and spaced approximately 20 cm, an effective optical pathlength of 8 km is obtained. This exceeds the best performance of multipass cell spectroscopy and uses a smaller cell volume approximately ($25 \, cm^3$, depending on window/cell diameter and length). The light leaking out of the RDC can be used to characterize the absorption of the intracavity medium. The optical loss is the difference between total cavity losses and empty cavity losses. Once the absorption spectrum of the sample has been measured, the concentration of the target analyte can be determined using the absorption cross section and line shape parameters. A well-designed CRDS system can achieve a minimum detectable absorption limit of approximately 4×10^{-10}. Detailed mathematical treatment of CRDS is described in the literature [59,60].

Several platforms exist to perform CRDS [61–63] or ICOS and its variant, off-axis ICOS (OA–ICOS), a technique whereby one observes time-integrated ring-down events [18,39,64,65]. In these techniques the coupling efficiency of the laser radiation into the resonant cavity is critical and determines the amount of light that can be collected by a photodetector placed after the absorption cell. The CRDS technique is intrinsically background-free and requires only high-quality cavity mirrors, a fast detector, and appropriate data acquisition electronics. A conventional CRDS-based sensor platform is shown schematically in Figure 4.3. During a typical implementation, the CW laser is scanned slowly by dithering the CRD cavity length using a

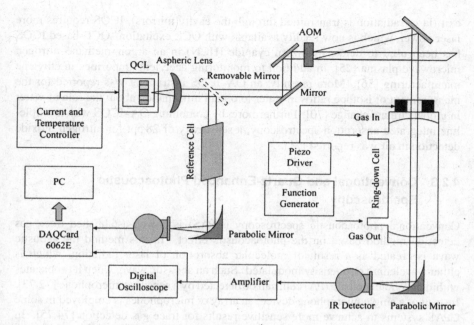

Figure 4.3 Typical CRDS-based sensor platform using an acousto-optic modulator as a high-speed beam chopper.

piezo microaccuator [16,37,66,67]. Commercial systems are available from several suppliers, such as Los Gatos Research, Inc., Tiger Optics, and Neoplas. For OA–ICOS, in which the optical sensor system is aligned in such a way that a maximum number of longitudinal and transverse modes is excited within the cavity, the typical optical throughput of the cavity is on the order of $\leq T/2$ (where T is transmission of the cavity mirrors) [68]. This method is related to absorption spectroscopy using a multipass cell; however, in ICOS, the beams are allowed to overlap on the mirrors after many cavity passes. The principal effect limiting sensitivity is output fluctuations caused by transmission noise due to the resonant mode structure. One approach for minimizing this noise is to arrange the laser spots on the mirrors in a circular pattern as in a Herriott multipass cell. If many passes of the cavity occur before any of these spots overlap, interference effects are minimal because the pathlength before overlap exceeds the coherence length of the laser. By introducing a small amount of astigmatism, the entire surface of the mirrors can be used [68]. Another approach for removing mode noise in OA–ICOS is to vibrate the mirrors [17,69]. This approach causes many mode hops to take place within the time required for light to travel its optical path in the cavity, effectively averaging out transmission noise. Thus, the trade-off between multipass absorption and ICOS is that in multipass absorption, this mode noise is not present because the spots never overlap, but for similar mirror size and separation, the total pathlength can be much greater in ICOS. Since there are no mirror holes in the ICOS cavity configuration to admit and allow the beam to exit (laser radiation is transmitted through the cavity mirrors), ICOS requires more laser power, which is now readily available with QCL excitation. QCL-based ICOS has been used to detect hydrogen cyanide (HCN) in an argon–methane–nitrogen microwave plasma [25], in addition to monitoring acetylene impurities in ethylene manufacturing [56]. More recently, an OA–ICOS instrument was reported for the measurement of isotope ratios in water to obtain information about the role of water in global climate change [70]. Furthermore, by combining OA–ICOS with multiple-line integrated absorption spectroscopy, a sensitivity of 28 ppt for nitrogen dioxide detection in air was reported [71].

4.2.3 Conventional and Quartz-Enhanced Photoacoustic Spectroscopy

Conventional photoacoustic spectroscopy (CPAS) is a well-established trace gas detection method based on the photoacoustic effect. In this method the acoustic wave is created as a result of molecular absorption of laser radiation, which is either wavelength or intensity modulated. Such an acoustic wave, when it propagates within the photoacoustic (PA) cell, can be detected by a sensitive microphone [72,73]. Instead of a single microphone device, an array of microphones is employed in some CPAS systems to achieve more sensitive results for trace gas detection [74,75]. In contrast to other infrared absorption techniques, CPAS is an indirect technique in which the effect on the absorbing medium, rather than the direct light attenuation, is analyzed. Therefore, no photodetector is required for the CPAS technique. A low-cost infrared detector located after the photoacoustic cell is usually employed for the purpose of monitoring laser power and performing a line-locking procedure of the laser frequency to the wavelength of the peak of the absorption selected.

To obtain an optimal acoustic signal for amplitude-modulated CPAS measurements, the laser modulation frequency is selected to match the first longitudinal acoustic resonance of the PA cell, given by the equation

$$f = \frac{\nu}{2L} \tag{4.2}$$

where ν is the speed of sound and L is the photoacoustic cell length. The resonance frequencies of PA cells are usually designed to be greater than 1 kHz, in order to make the CPAS technique immune to intrinsic $1/f$-type noise of the microphone and its preamplifier as well as to low-frequency external acoustic noise [48,76,77].

The photoacoustic signal (S_{PA}) detected is described by the following equation:

$$S_{PA} = CP\alpha cM \tag{4.3}$$

where C is the photoacoustic cell constant [Pa/(W · cm^{-1})], P is the optical power of the excitation laser (W), α is the absorption coefficient of the targeted gas [cm^{-1}/(mol · cm^{-3})], c is the analyte concentration (mol cm^{-3}), and M is the microphone responsivity (V/Pa).

Ideally, CPAS is a background-free technique because only the absorption of modulated laser radiation generates an acoustic signal. However, background signals can

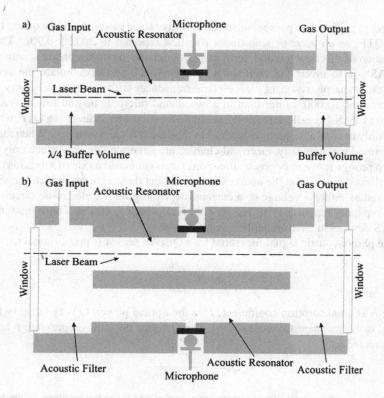

Figure 4.4   (a) Two-buffer-tube PA cell design; (b) ring differential resonance PA cell design.

originate from nonselective absorption of the gas cell windows (coherent noise) and external acoustic (incoherent) noise. Therefore, proper isolation of the photoacoustic cell from any mechanical vibrations will result in an improvement in the SNR measured. Further CPAS signal enhancement can be achieved by employing a different PAS cell design, such as a resonance photoacoustic cell with two buffer tubes [78] or a ring differential resonance photoacoustic cell [79,80]. Schematic diagrams of both designs are depicted in Figure 4.4. In the two-buffer-tube photoacoustic cell design (Figure 4.4a), the $\lambda/4$ buffer volume in line with the longitudinal acoustic resonator acts as an acoustic notch filter at the frequency of the resonator, to effectively suppress system flow noise [81]. In a differential photoacoustic cell (Figure 4.4b), two identical cylindrical channels are equipped with a microphone, which is placed in the middle of each channel, where the maximum pressure oscillations are observed. Because CPAS signal is proportional to the absorption coefficient and the laser power, it is possible for both designs to achieve minimum detectable concentrations at the sub-ppbv level by selecting the strongest absorption lines of the target gas and by using high-power laser sources, such as fiber amplifiers, CW DFB–QCLs, or EC–QCLs [82–89].

A novel approach to photoacoustic detection of trace gases utilizing a quartz tuning fork (QTF) as an acoustic transducer was first reported in 2002 [50,90]. The key innovation of this new method, termed quartz-enhanced photoacoustic spectroscopy (QEPAS), is to invert the common CPAS approach and accumulate the acoustic energy in a sharply resonant piezoelectric transducer with a very high quality factor (Q-factor) of 10,000, rather than in a broadband microphone and low Q (ca. 200) resonant CPAS gas cell. A suitable candidate for such a transducer is a QTF, which is commonly used as a frequency standard in digital clocks and watches. When the QTF is deformed mechanically, electrical charges are generated on its surface only when the two prongs move in opposite directions (antisymmetric mode of vibration). Thin silver films deposited on the quartz surfaces collect these charges, which can then be measured as either a voltage or a current, depending on the electronic circuit used. QTFs typically resonate at 32,768 Hz (2^{15} Hz), which results in a high immunity of QEPAS devices to environmental acoustic noise.

The photoacoustic signal measured by a QEPAS sensor is proportional to

$$S_0 \sim \frac{\alpha P Q}{f_0} \tag{4.4}$$

where α is an absorption coefficient, P is the optical power, Q is the quality factor, and f_0 is the resonant frequency [20]. The Q-factor depends on pressure p and can be expressed as

$$Q = \frac{Q_{\text{vac}}}{1 + Q_{\text{vac}} a p^b} \tag{4.5}$$

where Q_{vac} is the quality factor in vacuum and a and b are parameters dependent on a specific QTF design [50]. The pressure corresponding to the optimum sensitivity depends on the vibrational to translational (V–T relaxation) energy transfer cross section of the gas of interest. In addition, if the V–T relaxation rate is lower than the optical excitation modulation frequency, the amplitude of the optically induced acoustic signal is reduced. This effect is more significant for QEPAS, due to the high modulation frequency that is used. It is more likely to occur with small two- or three-atom molecules such as nitric oxide, carbon monoxide, or carbon dioxide, which do not have a dense ladder of energy levels to facilitate fast V–T relaxation. The optimum pressure for fast-relaxing molecules with resolved optical transitions is less than 100 torr, which also ensures Doppler-limited spectral resolution [50]. For slow-relaxing gases, this optimum pressure is higher and will lead to a broader linewidth than is desirable for the best detection selectivity.

In a typical QEPAS scheme the laser beam is focused between the QTF prongs as shown in Figure 4.5a. In this case, the probed optical path is only as long as the thickness of the QTF, or approximately 0.3 mm. Therefore, QEPAS is mostly sensitive to a sound source positioned in a 0.3-mm gap between the prongs. Sound waves from distant acoustic sources tend to move the QTF prongs in the same direction, which results in a zero net piezo-current and makes this element insensitive to such external excitation. The configuration depicted in Figure 4.5a can be useful when the excitation radiation cannot be shaped into a near-Gaussian beam, as for

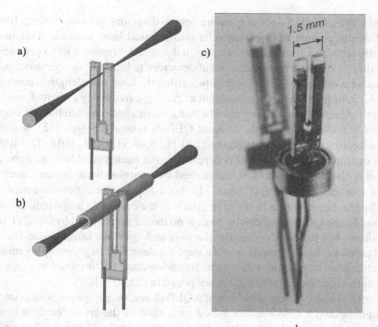

Figure 4.5 QTF-based spectrophones: (a) simplest spectrophone configuration, and (b) improved spectrophone configuration with an acoustic resonator formed by two tubes, and (c) typical quartz tuning fork geometry used in QEPAS trace gas measurements. Figure 4.5a and b with kind permission from Springer Science+Business Media: [22], fig. 1.

example, with spatially multimode lasers. The simplest QEPAS configuration with bare QTF was used in preliminary evaluation tests [91] and analyzed theoretically [92]. However, the trace gas detection sensitivity can be improved significantly using the configuration shown in Figure 4.5b, where a metal tube is added on each side of QTF to confine the optically generated acoustic vibrations in the gas and increase the effective interaction length between the radiation-induced sound and the QTF [93]. The two tubes act as an acoustic microresonator with respect to the QTF resonator. Recent experimental studies have shown that the optimum length for each microresonator tube is between $\lambda s/4$ and $\lambda s/2$ of the propagating sound wavelength (λs) [19]. Thus, for an optimal microresonator (mR) configuration, where each mR tube is 4.4 mm in length and 0.6 mm in inner diameter, an improvement of up to 30 times in the SNR is obtained in comparison to the bare QTF configuration [19]. Most QEPAS-based sensors utilize the configuration illustrated in Figure 4.5b. A typical quartz tuning fork, used in most QEPAS measurements to date, is illustrated in Figure 4.5c. Other QEPAS spectrophone configurations, such as off-beam QEPAS, are also possible [94,95]. Furthermore, two novel modifications of the QEPAS sensor architecture based on interferometric photoacoustic spectroscopy [96] and resonant optothermoacoustic detection [97] were reported recently.

As in CPAS, QEPAS does not require optical detectors and also benefits from high optical output powers from commercial mid-infrared laser sources. This feature is especially attractive for trace gas sensing in the spectral region 3 to 13 µm, where the availability of high-performance optical detectors is limited. In spectroscopic measurements based on the QEPAS technique, either the laser wavelength is modulated at $f_m = f_0/2$, or its intensity is modulated at $f_m = f_0$ frequency (where f_0 is the QTF resonant frequency), depending on whether a wavelength-modulation or amplitude-modulation technique is used. In most QEPAS sensor designs, a $2f$ wavelength-modulation spectroscopy has been used [20,21,28,49,91,93,98–101]. This technique provides complete suppression of coherent acoustic background that might be created when stray modulated radiation is absorbed by nonselective absorbers, such as the gas cell elements and the QTF itself. In this case, the noise floor is usually determined by the thermal noise of the QTF [102]. In the case of an amplitude-modulation (AM) mode, the QEPAS sensitivity limit is no longer determined by the QTF thermal noise alone, but by laser power fluctuations and spurious interference features as well. Therefore, the AM mode is often used for detecting large, complex molecules, when individual rotational–vibrational transitions are not resolved and applying a wavelength-modulation technique is not possible [23,103–106].

A direct, side-by-side comparison of a QEPAS sensor using a microresonator-QTF spectrophone and a CPAS sensor based on a state-of-the-art differential resonance PA cell was demonstrated by Dong et al. in 2012 [19]. Using a $2f$ wavelength-modulation technique, the detection sensitivity values obtained for both QEPAS and CPAS were within the same detection range for both fast (10 ppmv acetylene in nitrogen)- and slow (pure carbon dioxide)-relaxing molecules. A small practical advantage of the CPAS technique over the QEPAS technique for the analysis of pure carbon dioxide results from a lower modulation frequency and therefore a longer response time ($\tau = Q/\pi f$) of the CPAS spectrophone. However, in most cases the QEPAS detection sensitivity for slow relaxing molecules can be improved by adding a molecular species such as water [21,93,101] or sulfur hexafluoride [20], which eliminates the V–T relaxation bottleneck.

4.2.4 Cavity-Enhanced Direct Frequency-Comb Spectroscopy

A promising approach that addresses the issues of spectral coverage and analysis of multiple species while retaining the high spectral resolution and light intensity of lasers is DFCS [107]. The technique utilizes mode-locked lasers, which emit a train of ultrashort [<100 femtoseconds (fs)] optical pulses that, in the frequency domain, corresponds to a discrete comb of equally spaced and highly resolved lines. This relationship is illustrated in Figure 4.6. The specific frequencies of the comb, ν_m, are given by the equation

$$\nu_m = m f_{\text{rep}} + f_0 \tag{4.6}$$

where the comb repetition rate f_{rep} is related to the pulse envelope spacing τ by

$$f_{\text{rep}} = \frac{1}{\tau} \tag{4.7}$$

and the comb offset frequency f_0 is determined by the phase shift $\Delta\Theta_{ce}$ of the underlying electromagnetic wave between successive pulses,

$$f_0 = \frac{\Delta\Theta_{ce} f_{rep}}{2\pi} \tag{4.8}$$

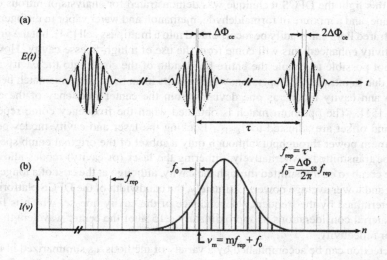

Figure 4.6 Time (a) and frequency (b)-domain pictures of the generation of a frequency comb from a train of femtosecond pulses from a mode-locked laser. Reprinted from [51] with permission from Annual Reviews, Inc.

The individual combs sample within the absorption profile of individual analyte transitions, and the bandwidth that can be covered is limited only by the output of the femtosecond laser, which can be as broad as hundreds of nanometers. As f_{rep} and f_0 are radio frequencies, they can be measured directly and thus provide a means for extremely accurate optical frequency measurements. Applications of the technique to atomic and molecular spectroscopy in the near-infrared began in 2004 [108]. More recent developments that expand the capability of the technique for trace gas analysis are (1) expansion of the spectral coverage from the near-infrared into the mid-infrared and (2) integration of high-finesse cavity enhancement [51,55].

Initial reports of DFCS for molecular spectroscopy utilized a Ti:sapphire laser, with a direct frequency output of 0.7 to 0.9 μm. Thus, spectroscopy involved analysis of vibrational overtones of small molecules: for example, acetylene, oxygen, water, and ammonia [109]. Coverage farther into the near-infrared (1.5 to 1.7 μm) was later obtained with erbium-doped fiber lasers [55], with analysis of overtones of methane, ethane, carbon dioxide, carbon monoxide, and ammonia. Extension of this technique to the mid-infrared (2.5 to 12 μm) will take advantage of the roughly 100-fold enhancement of the spectral line intensity of fundamental rotational–vibrational transitions. Although femtosecond lasers cannot access this region directly, nonlinear optical conversion techniques are relatively efficient, due to the short time scale and large peak power of the pulses. Both difference frequency generation (DFG) and optical parametric oscillators (OPOs) have been applied to DFCS in the mid-infrared spectral region. OPOs have the advantage of higher output powers, which is critical given that the power is distributed over many thousands of comb lines. For example, a ytterbium fiber-based OPO was used to obtain powers of 1 W from 2100 to 3600 cm^{-1}. With this light the DFCS technique was demonstrated for analysis of nitrous oxide, methane, and a mixture of formaldehyde, methanol, and water vapor in nitrogen [54].

Infrared light can readily be incorporated into a multipass cell [54], but the greatest sensitivity enhancements will come from the use of a high-finesse cavity. However, it is not possible to couple the entire bandwidth of the comb into the cavity at one time, due to intracavity dispersion, which results in an increasing mismatch between comb and cavity modes as one deviates from the center frequency of the cavity, f_{FSR} [51]. The optimum match is obtained when the frequency comb repetition rate and offset are adjusted to f_{FSR}. Locking the laser and cavity modes permits maximum power throughput, although only a subset of the original comb spectrum will be transmitted. Alternatively, dithering the laser (or cavity) modes allows the entire comb to be transmitted through the cavity, although at the cost of a longer duty cycle and lower average power. Ultimately, the bandwidth of the DFCS platform will be determined by the frequency dependence of the cavity finesse, which is limited by material considerations to approximately $\pm 15\%$ of the center wavelength of the mirror reflectivity.

Detection can be accomplished by a variety of methods, as summarized in Figure 4.7. Time-based detection methods (Figure 4.7a and b) have been largely supplanted by frequency-based detection (Figure 4.7c and d), which takes advantage of multiplex detection schemes. The specific frequency-based method chosen is dependent on the configuration of the rest of the DFCS platform. For example, if the laser modes are dithered to obtain transmission of every comb mode through the cavity, the detection mode depicted in Figure 4.7c is used. This mode combines high vertical dispersion of the light transmitted by an angled étalon [virtually imaged phased array (VIPA)] with horizontal dispersion of overlapping modes with a conventional grating and a two-dimensional array detection to permit a snapshot of the cavity transmittance. A detailed description of the technique, including conversion of the fringed two-dimensional image into a one-dimensional spectrum, is described by Thorpe et al. [55]. The spectrum at 1.5 to 1.7 μm is imaged as 8×25 nm sections, with spectral resolution of 0.025 to 0.03 cm^{-1}. Absorption is determined from the difference between images of reference and sample gases. Note that while the optical arrangement is potentially compact, the high dispersion of the VIPA means that calibration and reproducibility are very sensitive to small changes in alignment. Also, the method is restricted to the visible and near-infrared regions where two-dimensional array materials exhibit a response.

The configuration shown in Figure 4.7d, in which the light is analyzed with a Fourier transform spectrometer (FTIR), is suitable for when the laser and cavity modes are actively locked, or if mid-infrared light is generated. Compared to conventional FTIR spectroscopy with a blackbody emitter, the relative brightness of the a high-powered OPO allows spectra to be acquired approximately 10 times faster [51]. Since

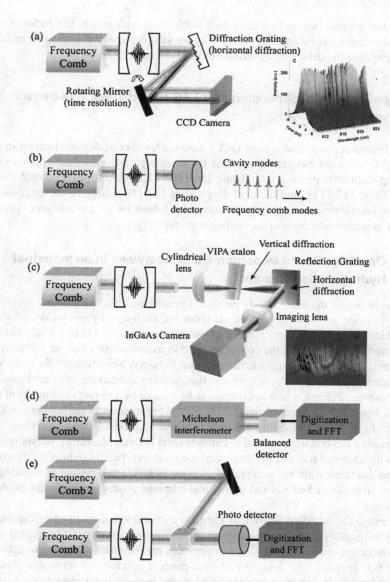

Figure 4.7   Detection schemes for cavity-enhanced DFCS. (a) Ring-down measurement, with horizontal frequency dispersion and vertical time dispersion provided by a rotating mirror. (b) Comb vernier spectroscopy, with time-decay monitoring of a single selected comb line. (c) Two-dimensional spectral resolution of multiple comb lines using a high-dispersion VIPA and a grating for horizontal resolution of multiple modes. (d) FT spectroscopy of cavity-locked comb lines. (e) Heterodyne detection of two comb sources. Techniques (c) and (d) are discussed in the text. Reprinted from [54] with permission from the Royal Society of Chemistry.

a phase-locked mid-infrared OPO emits a stable frequency comb, the addition of a high-finese femtosecond enhancement cavity for sample detection is feasible and can further improve the detection sensitivity of the system [51,55].

4.3 Applications of Semiconductor LAS-Based Trace Gas Sensor Systems

The performance, cost, and size of QCLs has generated considerable interest in developing LAS-based gas sensor systems for a variety of commercial applications, including industrial process monitoring [56,110,111,124], pollution control monitoring [32,33,112,113], and medical diagnostics [39,114–120]. In this section we provide a brief overview of the application of LAS-based trace gas analyzer systems relevant to semiconductor and specialty gas monitoring.

4.3.1 OA–ICOS Online Measurement of Acetylene in an Industrial Hydrogenation Reactor

Ethylene is one of the most widely produced petrochemicals and is a key precursor used in a variety of important industrial products, such as ethylene oxide, ethylene dichloride, polyethylene, and other high-volume derivatives [121]. Thus, careful control and optimization of the processes used to manufacture ethylene can provide significant economic advantages to the chemical industry. More than 97 % of ethylene is produced by pyrolysis of hydrocarbons, that is, thermal cracking of petrochemicals in the presence of steam, where acetylene can be a major by-product or impurity of this process, depending on the feedstock and quenching temperatures during the cracking process. In addition to causing a significant loss of product yield, acetylene acts as a poison to the end user's reaction catalyst used for manufacturing polyethylene, and downstream end users of the final ethylene product typically require ≤ 2 ppmv of acetylene contamination for acceptance. Consequently, rapid online monitoring of acetylene in both cracked gas and in the final ethylene product is important in olefin processing.

Currently, the standard analytical chemistry instrumentation used in the olefin industry to quantify acetylene in ethylene is gas chromatography (GC) coupled with flame ionization detection (FID), which provides measurement times up to approximately 600 seconds. To achieve real-time process control and minimize process down-time, an analyzer must provide much faster measurement rates (e.g., 1 to 30 seconds) while maintaining detection sensitivity and selectivity, which makes conventional GCs inadequate for this application. Micro-GCs with MEMS-based separation columns and detectors have demonstrated acetylene detection in pure ethylene with measurement times of 20 to 40 seconds [122,123]. No commercial deployment of such instruments has been reported. An alternative approach using a QCL-based OA–ICOS system (Los Gatos Research, Inc., Mt. View, CA), has been demonstrated successfully for acetylene detection in a multicomponent matrix from an olefin hydrogenation reactor [56,124]. The system provides long effective optical pathlengths (e.g., typically 500 to 10,000 m) and maintains the stable off-axis alignment that is essential for industrial applications.

A schematic of the Los Gatos Research OA–ICOS system implemented at the outlet of an industrial light hydrocarbon cracker is shown in Figure 4.8. The sensor employed a telecommunications-grade fiber-coupled DFB diode laser producing 20 to 30 mW centered near 1531.6 nm. Laser control was accomplished using a custom laser driver card where the temperature and current could be controlled to within ±0.01 °C and ±0.1 mA, sweeping approximately 50 GHz in 0.01 second. The beam was launched into the cavity in an off-axis approach to probe the P(11) transition of the $\nu_1 + \nu_3$ band of $^{12}C_2H_2$. A custom-amplified InGaAs detector with a gain of 3×10^6 V/A and a bandwidth of 30 kHz collected the light after traversing the cavity. One hundred transmission spectra were averaged (e.g., 1 second of data) to minimize measurement noise. A chemometric data analysis routine capable of fitting several absorbing species simultaneously was implemented and described in more detail by Le et al. [56].

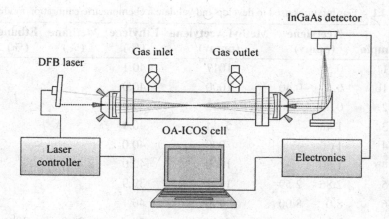

Figure 4.8 Schematic of an off-axis ICOS analyzer system.

In high-resolution diode laser applications, the intensity and shape of the absorbing features depends directly on the sample temperature and pressure and should be controlled precisely to match conditions under which calibration sets are obtained in order to maximize signal while maintaining adequate resolution between the absorption spectra of the matrix. In this case, Le et al. surrounded the cavity with a heated insulated blanket maintaining a sample temperature at 38 ± 1 °C. The sample pressure was controlled by a downstream pressure controller capable of maintaining the sample pressure range of 1.5 to 2.5 psi to within ±0.02 psi. Optimum temperature and pressure were determined by a series of comprehensive tests.

The sensor system was tested initially off-line at Dow Chemical Company's R&D laboratory (Freeport, TX) to determine the system's performance specifications empirically. The instrument's precision, detection limit, and stability were evaluated by continuously monitoring a flowing acetylene standard. The analyzer demonstrated a

1σ measurement precision of better than ± 0.025 ppmv for a 40-hour period. Based on these measurements, a detection limit of 0.050 ppmv in nitrogen was calculated on the basis of a signal-to-noise ratio of 3. The accuracy of this instrument was evaluated further for seven mixtures prepared by the British Oxygen Company (BOC) that mimic typical plant operating conditions during the steady-state, alarm level, and plant upset periods listed in Table 4.1. The mixtures were measured initially by BOC using GC with a relative uncertainty of $\pm 5\%$ for components below 5 ppmv and $\pm 2\%$ for components above 5 ppmv. Acetylene concentrations of three of the mixtures were verified by an optimized GC system at Dow Chemical Company and showed a slight deviation for samples 1b and 5. Additionally, the results of a successful comparison of the OA–ICOS system to typical online optimized GC monitoring at the Dow Chemical Company are summarized in Table 4.2. The slow response time of the GCs does not allow real-time monitoring of methyl acetylene under process conditions, and therefore comparison data are not available.

Table 4.1   Gas mixtures used to develop and validate a chemometric calibration model[a]

Sample	Acetylene (ppmv)		Methyl Acetylene (ppmv)	Ethylene (%)	Methane (%)	Ethane (%)
1	0.50		1017	40.1		
1b	0.50	0.40	1000	40.0		
2	0.96		3022	39.9		
3	1.48		5144	40.0		
4	1.50		990	40.0		
4b	1.50		1000	40.0		
5	2.86	2.59	1021	39.9		
6	8.0	8.00	999	40.1		
7	0.99		5024	40.0	20.0	30.0

Source: Reprinted from [56] with permission from the Society for Applied Spectroscopy.
[a] The mixtures span a wide range of anticipated process gas stream compositions. The acetylene concentrations of mixtures 1b, 5, and 6 were verified independently by the Dow Chemical Company.

Following a successful laboratory demonstration, the OA–ICOS system was installed in a Dow hydrocarbon plant (Oyster Creek, TX), where it was operated in parallel with two preexisting GC systems for five and a half months. Typical GC measurement times for acetylene and methyl acetylene were approximately 5 and 20 minutes, respectively. The OA–ICOS system routinely provided faster response times with comparable accuracy and, in some instances, tracked events that were too rapid for GC quantification, as demonstrated in Figure 4.9. Subsequently, Dow Chemical installed two OA–ICOS systems in a light hydrocarbon facility in Freeport, TX, and ran them in conjunction with newly installed GC systems capable of a 90-second measurement time. Results from the commercial implementation (Figure 4.10) showed that regardless of the GC response time, the OA–ICOS system outper-

Table 4.2   Measured values of several gas mixtures from both the OA–ICOS analyzer (6 s averaging) and a gas chromatograph[a]

Acetylene					
BOC Reported	**Off-Axis ICOS**	**3σ Precision ($\pm$ppmv)**		**Accuracy/Bias ($\pm$ppmv)**	
(ppmv)	**(ppmv)**	**ICOS**	**GC**	**ICOS**	**GC**
0.50	0.49	0.03	0.06	0.01	0.08
0.99	1.02	0.04	0.06	0.03	0.08
1.48	1.51	0.07	0.14	0.03	0.20
2.86	2.58	0.06	0.14	0.28	0.20
8.00	7.74	0.14	0.14	0.26	0.20

Methyl Acetylene			
BOC Reported (ppmv)	**OA–ICOS (ppmv)**	**3σ Precision ($\pm$ppmv)**	**Accuracy (%)**
1017	1020.9	5.8	0.4
1021	1021.1	13.9	0.0
3022	3082.3	63.0	2.0
5349	5144.0	130.7	3.8

Source: Reprinted from [56] with permission from the Society for Applied Spectroscopy.
[a] The gas chromatography data were obtained for a short-term study consisting of 30 consecutive analyses.

formed the optimized GC system and correctly reflected actual variations in reactor outlet composition.

4.3.2   Multicomponent Impurity Analysis in Hydrogen Process Gas Using a Compact QEPAS Sensor

High-purity hydrogen is used for various functions in semiconductor processing, such as a cover gas for furnace processing, a carrier gas for transport of active doping agents, or even reactive gas for material preparation. One such application is the microwave plasma-assisted low-pressure chemical vapor deposition preparation of single-crystal diamond as a wide-bandgap material via reaction of atomic hydrogen and a methyl radical [125]. To manufacture the diamond, high-purity hydrogen, methane, and argon are flowed through a chamber containing the deposition substrate. A 2.45-GHz microwave plasma is used to generate atomic hydrogen, which reacts with methane to yield a methyl radical and hydrogen gas. The carbon-containing radical then deposits onto the substrate as crystalline diamond. Improvements in the electronic properties of a single-crystal diamond can be attributed directly to

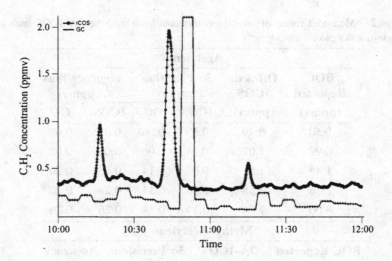

Figure 4.9   Measured online acetylene comparison data for an OA–ICOS analyzer and GC. During fast acetylene fluctuations, the GC either misses such an event entirely or misreports it after completion. In contrast, the ICOS analyzer reports the acetylene data with good accuracy. Reprinted from [56] with permission from the Society for Applied Spectroscopy.

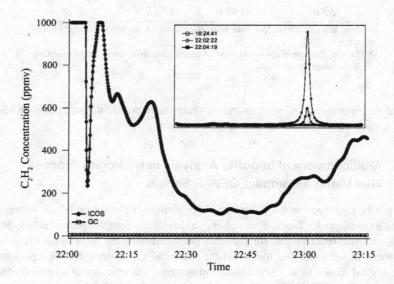

Figure 4.10   Demonstration of an OA–ICOS analyzer to measure high concentration levels (ca. 1000 ppmv), where the GC is limited to 10 ppmv. The inset depicts how the absorption spectra can be viewed for verification. The spectra indicate that the acetylene dip at 22:04 hours is a real process event. Reprinted from [56] with permission from the Society for Applied Spectroscopy.

enhanced crystalline quality and reduced defect concentration, which requires tight controls on the process gas mixtures and impurities formed in the processing, such as ammonia. Therefore, online monitoring of methane in the presence of hydrogen with simultaneous detection of impurity formation is a critical need for this process. The utility of a LAS-based sensor system employing a QCL for multicomponent process monitoring was demonstrated as a QEPAS sensor system designed for detection of ammonia and methane in a hydrogen carrier gas [126].

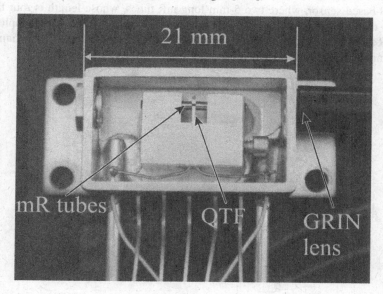

Figure 4.11 Fiber-coupled QEPAS ADM. With kind permission from Springer Science+Business Media: [126], fig. 1.

The QEPAS technique presents a unique new methodology for impurity analysis within hydrogen gas streams. An image of the QEPAS detection module is shown in Figure 4.11. As discussed in Section 4.2.3, the microresonator (mR) plays a crucial role in QEPAS sensors and acts similarly to the acoustic resonator in conventional PAS [72]. The reported speed of sound in hydrogen is 1330 m/s at room temperature [127], which is approximately four times faster than in air, since the density of hydrogen is only 1/14 of the density of air. The parameters and the performance of the mR are strongly dependent on the properties of the carrier gas, in particular the gas density and speed of sound within the gas. Thus, for this application the optimum length of the mR was calculated to be approximately 20 or 40 mm when accounting for the length of both mRs. This increased mR length presents a challenge to focus the excitation diode laser beam passing through the approximately 40-mm-long mR and the 300-μm gap between the prongs of the QTF without optical contact. In fact, any optical contact between the diode laser excitation radiation and the mR or QTF results in an undesirable, nonzero background, which is several to a few tens of times larger than the thermal noise level of QEPAS [101]. As a result, the optimized mR tube length of approximately 20 mm (which must be matched to the acoustic wavelength so that the acoustic energy can be accumulated efficiently in the mR tube) was not found suitable for QEPAS-based trace gas detection in hydrogen. However, in previous QEPAS-based sensor studies, it was observed that a nonmatched short-length mR can still increase the QEPAS sensitivity by a factor of 10 times or more [98]. In this case, the mR tubes act to confine the sound wave but do not exhibit well-defined resonant behavior. Therefore, a nonmatched mR configuration was adopted for the QEPAS-based sensor, where two 5-mm-long mR tubes, whose length is four times smaller than the approximately 20-mm optimum length evaluated, were employed. The mR tubes featured an inner tube diameter of 0.58 mm and an outer tube diameter of 0.9 mm, similar to the design shown in Figure 4.5b.

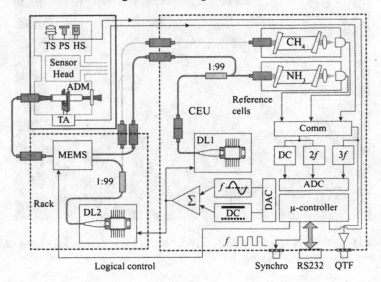

Figure 4.12 Compact two-gas QEPAS sensor. TS, PS, HS: temperature, pressure, and humidity sensors, ADM: acoustic detection module, TA: transimpedance amplifier, DL1, DL2: diode lasers, CEU: control electronics unit. With kind permission from Springer Science+Business Media: [126], fig. 2.

The schematic diagram of the multicomponent QEPAS sensor system shown in Figure 4.12 consists of three parts: (1) a control electronics unit (CEU), (2) an acoustic detection module (ADM), and (3) a switching module. The diode laser for ammonia detection (JDS Uniphase CQF 935.908-19570), along with two reference cells (Wavelength References, Inc.), were mounted inside the CEU, while the diode laser for methane detection (NEL, NLK1U5FAAA) and a 4×4 MEMS optical switch were mounted in the switching module. The CEU was responsible for measuring the QTF parameters, modulating the two diode lasers at the half resonant frequency of the QTF frequency, and locking the laser wavelength to an absorption line selected for the target analyte. The MEMS switch was controlled to direct either of the two diode lasers to the ADM via a parallel 4-bit binary code provided by the CEU. The sensor head consisted of the ADM, and miniature temperature, pressure, and humidity sensors mounted inside a compact enclosure. A notebook PC communicated with the CEU via a RS232 serial port for collection of the $2f$ harmonic data and gas temperature, pressure, and humidity parameters. Sensitivity optimization of the sensor system was carried out to determine the effect of pressure on the molecular line-broadening coefficient which determines the optimum laser wavelength-modulation amplitude. For methane detection, a $2\nu_3$ band near $6057.1 \, cm^{-1}$ was employed, while for ammonia detection a water and carbon dioxide interference-free transition at $6528.76 \, cm^{-1}$ was selected. A maximum signal for methane and ammonia was observed at 200 and 500 torr, respectively; however, when the two gases were measured simultaneously, a pressure of 100 torr was used, causing a decrease in SNR of approximately 8 % for each analyte. Sensor linearity and sensitivity were investigated for both molecules at their optimum pressure conditions and plotted in Figure 4.13. Based on these experiments, noise equivalent detection limits for methane and ammonia were calculated to be 3.2 and 1.27 ppmv, respectively, with a 1-second averaging time. The addition of water vapor did not significantly promote vibrational relaxation of methane, although a slight correction to the measured methane concentration was necessary by monitoring the moisture content. In addition, most industrial processes allow for longer measurement times. Therefore, the long-term stability of the instrument was characterized by measuring the Allan variance of the ammonia channel, shown in Figure 4.14. The data indicated that the sensor system followed a $1/\sqrt{t}$ dependence [34]. However, the Allan deviation started to drift when averaging exceeded 600 seconds. Hence, by signal averaging between 200 and 600 seconds, a detection sensitivity of 100 to 150 ppbv can be achieved. Further improvements in the detection sensitivity can be achieved by using higher-power diode laser sources or fiber-amplified diode lasers.

4.3.3 Analysis of Trace Impurities in Arsine by CE–DFCS at 1.75 to 1.95 μm

Low concentrations of impurities in source materials can have detrimental effects on the physical properties of semiconductors. For example, III–V semiconductors made from arsine and phosphine will be degraded if water or oxygen concentrations of greater than $10^{-8} \, mol/mol$ are present [53,111]. At these concentration levels, oxygen acts as a dopant, which creates additional energy levels in the semiconductor bandgap. Similar content levels must be managed for other impurities, such as hydrocarbons, carbon monoxide, carbon dioxide, and silicon- and sulfur-containing molecules [110]. Accurate detection of impurities at these levels is challenging and requires the development of new analytical techniques, especially for online monitoring during the growth phase.

Cavity-enhanced direct frequency-comb spectroscopy is an attractive candidate for this type of application, due to both its potential for sensitive measurements and its ability to access a broad spectral region with a single laser. These features are shown in a recent demonstration of the detection of trace levels of water in arsine [53]. Arsine is a difficult matrix for water detection, due to its congested and extensive

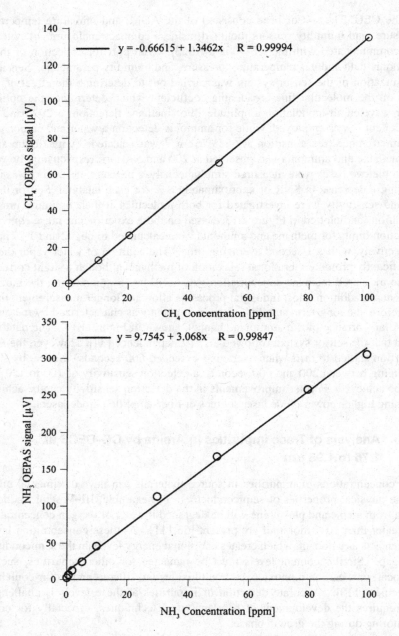

Figure 4.13   Plot of QEPAS response to various concentrations of certified (a) methane and (b) ammonia gas cylinders at the optimum pressure for each gas (i.e., 200 torr for methane and 50 torr for ammonia). With kind permission from Springer Science+Business Media: [126], figs. 7b and 11b.

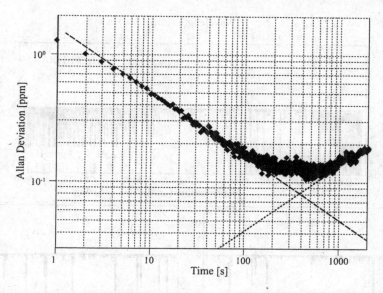

Figure 4.14 Allan deviation plotted as a function of the data-averaging period. Solid squares trace: laser is locked to the ammonia absorption line, data acquisition time 1 second. Dashed line: $1/\sqrt{t}$ slope. Dotted line: $\sqrt{t}$ slope. With kind permission from Springer Science+Business Media: [126], fig. 12.

vibrational spectrum. However, there is a window in the arsine spectrum between 5400 and 5500 cm^{-1} (1.82 to 1.85 μm in which water lines may be observed and quantified (see Figure 4.15).

A schematic of the experimental setup is shown in Figure 4.16. The output of a mode-locked Er-doped fiber ring laser (130 mW at 250 MHz) is amplified in a single-mode highly nonlinear fiber (HNF) amplifier, yielding 81-fs pulses with 400 mW of average power. The output is then broadened in a highly nonlinear silica fiber to generate spectral comb output in the region 4700 to 8300 cm^{-1} (1.2 to 2.1 μm). The light was coupled into a high-finesse (peak $F = 30{,}000$) Fabry–Perot cavity with one 2-m-radius concave mirror and one flat mirror. The cavity length of approximately 60 cm was adjusted to match the FSR to the comb repetition rate. The center wavelength of the cavity bandpass was approximately 1.85 μm, with a bandwidth of approximately 200 nm. The comb modes were dithered to overcome mode transmission issues associated with cavity dispersion. The cavity output is detected using a two-dimensionally dispersive VIPA spectrometer, with a 320 × 256 pixel indium antimonide (InSb) plane array detector. The spectrometer has a resolution of 0.031 cm^{-1} and can cover 50 cm^{-1} in a single image. Absorption spectra were obtained by alternating acquisitions of 20 sample and 20 reference gas transmission images (for a 150-ms integration time per image), with additional averaging as required. Total scanning time for a 700-cm^{-1} spectrum was approximately 2.5 hours,

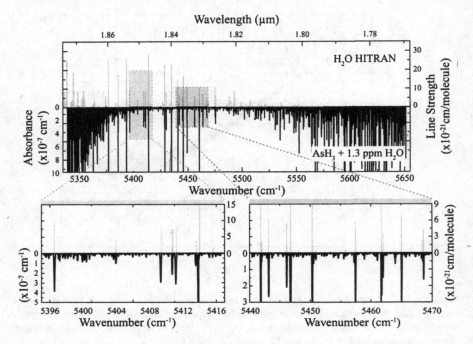

Figure 4.15   CE–DFCS spectra of arsine and 1.3 ppm water. Gray lines (extending upward) are water transitions from the HITRAN database. Black lines (extending downward) are the spectra observed. Insets show the coincidence of the observed and reference lines in the arsine spectral transparency window. With kind permission from Springer Science+Business Media: [53], fig. 2.

with additional time required to exchange sample and reference gases. The typical total gas pressure in the cavity was 200 torr.

With this system, Cossel et al. demonstrated trace detection of carbon dioxide, methane, hydrogen sulfide (manufacturer's sample, 10 ppmv ±10 % each), and water (2.5 ppmv) in nitrogen, and trace water in arsine [53]. For such a sample, concentrations of the manufacturer-supplied gases were determined within the specified uncertainty, and water was observed at 2.50 ± 0.12 ppm. Minimum detectable concentrations, based on the strongest lines in this spectral region, were: carbon dioxide, 325 ppmv; methane, 700 ppbv; hydrogen sulfide, 370 ppmv; water, 7 ppbv. For water, the range for linear measurement response extends over four orders of magnitude (7 ppbv to 100 ppmv).

As seen in Figure 4.15, the arsine background is strong below 5400 cm^{-1} and above 5500 cm^{-1}. In this spectral region, the methane transitions occur between 5500 to 5700 cm^{-1}, for hydrogen sulfide between 5050 and 5250 cm^{-1}, and for carbon dioxide between 5050 and 5130 cm^{-1}. Transitions for these molecules are obscured by arsine, and another spectral region is required for this application. The window at 5400 to 5500 cm^{-1} coincides with strong water lines. Figure 4.15 shows

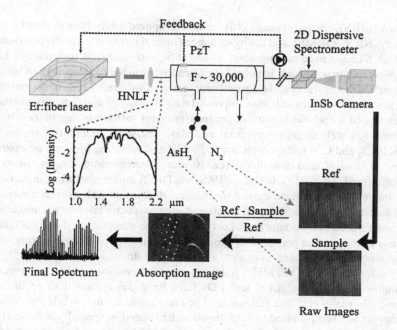

Figure 4.16   CE–DFCS experimental platform. With kind permission from Springer Science+Business Media: [122], fig. 2.

the coincidence of observed water lines (black/down) with lines from the HITRAN database (gray/up), and there are several lines that are distinguishable from the arsine background. Using these lines, the concentration in the sample was measured to be 1.26 ± 0.08 ppmv with an integration time of 600 seconds. From the spectral quality, a detection limit of 31 ppmv was calculated for this matrix.

To make the system more suitable for field (industrial) applications, it was noted that the following improvements could be implemented. Acquisition time can be decreased by setting up a double-beam configuration for simultaneous measurement of the sample and reference images. This would eliminate issues associated with gas delivery to and exchange in the cavity. Extension into other areas of the spectrum (1.2 to 1.75 μm and 1.95 to 2.1 μm) should be possible with the supercontinuum output of the Er:fiber laser, which would allow the use of stronger lines and spectral regions where background interference might not exist.

4.4   Conclusions and Future Trends

Spectroscopic techniques that include semiconductor-based laser absorption spectroscopy (LAS), cavity ring-down spectroscopy (CRDS), integrated cavity output spectroscopy (ICOS), conventional and quartz-enhanced photoacoustic spectroscopy (CPAS and QEPAS), evanescent wave spectroscopy, laser-induced breakdown spectroscopy (LIBS), noise-immune cavity-enhanced optical heterodyne molecular spectroscopy (NICE–OHMS), and cavity-enhanced optical frequency-comb spectroscopy (CE–OFCS) have become important as scientific and industrial techniques. Laser absorption spectroscopy has benefited significantly from the development of infrared laser sources that can access the desired near- and mid-infrared wavelengths and novel measurement techniques that have improved data acquisition and reduction methods. The choice of a specific spectroscopic measurement method is determined by the application as well as the commercial availability of semiconductor lasers such as QCLs, ICLs, and LDs with powers above 20 mW and operating lifetimes comparable to those of near-infrared laser diodes (ca. 10 years). Furthermore, improvements and innovations in LAS, CRDS, ICOS, QEPAS, and DFCS sensor platforms, in particular more stable, mass-produced optical/mechanical designs as well as data acquisition and reduction techniques, will lead to infrared semiconductor laser-based instruments that can be operated by nontechnical personnel and be manufactured at costs leading to sensor networks that permit both temporal and spatial trace gas monitoring.

Autonomously operated compact, reliable, real-time, sensitive ($<10^{-4}$), and highly selective (<3 to 500 MHz) trace gas sensors based on various spectroscopic techniques using QCLs, ICLs, and LDs have been demonstrated to be effective in numerous real-world and fundamental science applications. While the focus of this chapter has been limited to LAS-based sensors used in specialty and electronic gases such as chemical analysis and/or process control for manufacturing processes (e.g., petrochemical processing and exploration, alternative energy technologies and production, semiconductor wafer manufacture, pharmaceutical, metal processing, nuclear safeguards, and food and beverage industries), QCL-, ICL-, and LD-based sensors have been applied to a wide range of politically and economically important areas. These include such diverse fields as atmospheric chemistry and environmental monitoring (e.g., carbon monoxide, carbon dioxide, methane, and formaldehyde are important carbon gases in global warming), ozone depletion studies, acid rain, and photo smog formation), industrial emission measurements (e.g., quantification of smokestack emissions, fence line perimeter monitoring by the petrochemical industry, combustion incinerators, down-hole gas monitoring, and gas pipeline and industrial plant safety), urban (e.g., automobile, truck, aircraft, marine, and electrical power generation), and rural emissions (e.g., horticultural greenhouses, fruit storage, and rice agro-ecosystems). Furthermore, these types of sensors will also have a large impact on human health, with applications in the biomedical and life sciences, such as noninvasive medical diagnostics that involve the detection and monitoring of numerous exhaled-breath biomarkers (e.g., nitrogen oxide, carbon monoxide, carbon dioxide, ammonia, ethane, and acetone). With the development of efficient mid-infrared lasers [7,10,128–130], we envision a significantly improved performance coupled with a reduction in size and cost of thermoelectrically cooled semiconductor laser-based trace gas monitors that will lead to rapid increases in the implementation of sensor networks in many applications [112,113].

REFERENCES

1. Faist, J., Capasso, F., Sivco, D. L., Hutchinson, A. L., & Cho, A. Y. (1994). Quantum cascade laser. *Science, 264,* 553–556.

2. Meyer, J. R., Vurgaftman, I., Yang, R. Q., & Ram-Mohan, L. R. (1996). Type-II and type-I interband cascade lasers. *Electronics Letters, 32,* 45–46.

3. Yang, R. Q. (1995). Infrared laser based on intersubband transitions in quantum wells. *Superlattices and Microstructures, 17,* 77–83.

4. Bismuto, A., Beck, M., & Faist, J. (2011). High power Sb-free quantum cascade laser emitting at 3.3 μm above 350 °K. *Applied Physics Letters, 98,* 19, 191104.

5. Capasso, F. (2010). High-performance mid-infrared quantum cascade lasers. *Optical Engineering, 49,* 11, 111102.

6. Gupta, J. A., Ventrudo, B. F., Waldron, P., & Barrios, P. J. (2010). External cavity tunable type-I diode laser with continuous-wave singlemode operation at 3.24 μm. *Electronics Letters, 46,* 1218–1220.

7. Razeghi, M., Bai, Y., Slivken, S., & Darvish, S. R. (2010). High-performance InP-based midinfrared quantum cascade lasers at Northwestern University. *Optical Engineering, 49,* 111103.

8. Vurgaftman, I., Kim, M., Kim, C. S., Bewley, W. W., Canedy, C. L., Lindle, J. R., et al. (2010). Challenges for mid-IR interband cascade lasers. *Proceedings of the SPIE, 7616,* 761619.

9. Zeller, W., Naehle, L., Fuchs, P., Gerschuetz, F., Hildebrandt, L., & Koeth, J. (2010). DFB lasers between 760 nm and 16 μm for sensing applications. *Sensors, 10,* 2492–2510.

10. Curl, R. F., Capasso, F., Gmachl, C., Kosterev, A. A., McManus, B., Lewicki, R., Pushkarsky, M., et al. (2010). Quantum cascade lasers in chemical physics. *Chemical Physics Letters, 487,* 13, 1–18.

11. Kosterev, A. A., & Tittel, F. K. (2002). Chemical sensors based on quantum cascade lasers. *IEEE Journal of Quantum Electronics 38,* 582–591.

12. McManus, J. B., Zahniser, M. S., Nelson, J. D. D., Shorter, J. H., Herndon, S., Wood, E., & Wehr, R. (2010). Application of quantum cascade lasers to high-precision atmospheric trace gas measurements. *Optical Engineering, 49,* 111124–11.

13. Lee, B., Wood, E., Zahniser, M., McManus, J., Nelson, D., Herndon, S., et al. (2011). Simultaneous measurements of atmospheric HONO and NO_2 via absorption spectroscopy using tunable mid-infrared continuous-wave quantum cascade lasers. *Applied Physics B: Lasers and Optics, 102,* 417–423.

14. Lewicki, R., Kosterev, A. A., Thomazy, D. M., Risby, T. H., Solga, S., Schwartz, T. B., & Tittel, F. K. (2011). Real time ammonia detection in exhaled human breath using a distributed feedback quantum cascade laser based sensor. *Proceedings of SPIE, 7945,* 50K–2.

15. Weidmann, D., Wysocki, G., Oppenheimer, C., & Tittel, F. K. (2005). Development of a compact quantum cascade laser spectrometer for field measurements of CO_2 isotopes. *Applied Physics B: Lasers and Optics, 80,* 255–260.

16. Kosterev, A. A., Malinovsky, A. L., Tittel, F. K., Gmachl, C., Capasso, F., Sivco, D. L., et al. (2001). Cavity ringdown spectroscopic detection of nitric oxide with a continuous-wave quantum-cascade laser. *Applied Optics, 40,* 5522–5529.

17. Bakhirkin, Y. A., Kosterev, A. A., Curl, R. F., Tittel, F. K., Yarekha, D. A., Hvozdara, L., Giovannini, M., & Faist, J. (2006). Sub-ppbv nitric oxide concentration measurements using CW thermoelectrically cooled quantum cascade laser-based integrated cavity output spectroscopy. *Applied Physics B: Lasers & Optics, 82,* 149–154.

18. McCurdy, M. R., Bakhirkin, Y., Wysocki, G., & Tittel, F. K. (2007). Performance of an exhaled nitric oxide and carbon dioxide sensor using quantum cascade laser-based integrated cavity output spectroscopy. *Journal of Biomedical Optics, 12.*

19. Dong, L., Lewicki, R., Liu, K., Tittel, F. K., Buerki, P. R., & Weida, M. J. (2012). Ultra-sensitive carbon monoxide detection by using EC–QCL based quartz-enhanced photoacoustic spectroscopy. *Applied Physics B: Lasers and Optics, 107,* 2, 275–283.

20. Kosterev, A. A., Bakhirkin, Y. A., & Tittel, F. K. (2005). Ultrasensitive gas detection by quartz-enhanced photoacoustic spectroscopy in the fundamental molecular absorption bands region. *Applied Physics B: Lasers and Optics, 80,* 133–138.

21. Kosterev, A. A., Wysocki, G., Bakhirkin, Y. A., So, S., Lewicki, R., Fraser, M., Tittel, F., & Curl, R. F. (2008). Application of quantum cascade lasers to trace gas analysis. *Applied Physics B: Lasers and Optics, 90,* 165–176.

22. Kosterev, A. A., Dong, L., Thomazy, D., Tittel, F., & Overby, S. (2010). QEPAS for chemical analysis of multi-component gas mixtures. *Applied Physics B: Lasers and Optics, 101,* 649–659.

23. Lewicki, R., Wysocki, G., Kosterev, A. A., & Tittel, F. K. (2007). QEPAS based detection of broadband absorbing molecules using a widely tunable, CW quantum cascade laser at 8.4 μm. *Optics Express, 15,* 7357–7366.

24. Zaugg, C. A., Lewicki, R., Day, T., Curl, R. F., & Tittel, F. K. (2011). Faraday rotation spectroscopy of nitrogen dioxide based on a widely tunable external cavity quantum cascade laser. *Proceedings of SPIE, 7945,* 500–501.

25. Welzel, S., Hempel, F., Hübner, M., Lang, N., Davies, P. B., & Röpcke, J. (2010). Quantum cascade laser absorption spectroscopy as a plasma diagnostic tool: an overview, *Sensors, 10,* 6861–6900.

26. Belenky, G., Shterengas, L., Kipshidze, G., & Hosoda, T. (2011). Type-I diode lasers for spectral region above 3 μm. *IEEE Journal on Selected Topics in Quantum Electronics, 17,* 5, 1426–1434.

27. Christensen, L. E., Mansour, K., & Yang, R. Q. (2010). Thermoelectrically cooled interband cascade laser for field measurements. *Optical Engineering, 49,* 11, 111119.

28. Kosterev, A. A., Bakhirkin, Y. A., Tittel, F. A., McWhorter, S., & Ashcraft, D. (2008). QEPAS methane sensor performance for humidified gases. *Applied Physics B: Lasers and Optics, 92,* 103–109.

29. Schilt, S., Thévenaz, L., & Robert, P. (2003). Wavelength modulation spectroscopy: Combined frequency and intensity laser modulation. *Applied Optics, 42,* 6728–6738.

30. Schilt, S., & Thévenaz, L. (2006). Wavelength modulation photoacoustic spectroscopy: Theoretical description and experimental results. *Infrared Physics and Technology, 48,* 154–162.

31. Fried, A., Sewell, S., Henry, B., Wert, B. P., Gilpin, T., & Drummond, J. R. (1997). Tunable diode laser absorption spectrometer for ground-based measurements of formaldehyde. *Journal of Geophysics Research, 102*, D5, 6253–6266.

32. Fried, A., Henry, B., Wert, B., Sewell, S., & Drummond, J. R. (1998). Laboratory, ground-based, and airborne tunable diode laser systems: performance characteristics and applications in atmospheric studies. *Applied Physics B: Lasers and Optics, 67*, 317–330.

33. Fried, A., & Richter, D. (2007). Infrared absorption spectroscopy in analytical techniques for atmospheric measurement. In D. Heard (Ed.), *Analytical Techniques for Atmospheric Measurement*. Malden, MA: Blackwell Publishing, pp. 72–146.

34. Werle, P. (2011). Accuracy and precision of laser spectrometers for trace gas sensing in the presence of optical fringes and atmospheric turbulence. *Applied Physics B: Lasers and Optics, 102*, 313–329.

35. Zahniser, M. S., Nelson, D. D., McManus, J. B., Kebabian, P. L., & Lloyd, D. (1995). Measurement of trace gas fluxes using tunable diode laser spectroscopy. *Philosophical Transactions: Physical Sciences and Engineering, 351*, 371–382.

36. McManus, J. B., Kebabian, P. L., & Zahniser, W. S. (1995). Astigmatic mirror multipass absorption cells for long-path-length spectroscopy. *Applied Optics, 34*, 3336–3348.

37. Rao, G. N., & Karpf, A. (2010). High sensitivity detection of NO_2 employing cavity ringdown spectroscopy and an external cavity continuously tunable quantum cascade laser. *Applied Optics, 49*, 4906–4914.

38. Tittel, F. K., Kosterev, A. A., Bakhirkin, Y. A., Roller, C. B., Weidmann, D., & Curl, R. F. (2002). Chemical sensors based on quantum cascade lasers. *IEEE Journal of Quantum Electronics, 38*, 582–591.

39. Bakhirkin, Y. A., Kosterev, A. A., Roller, C., Curl, R. F., & Tittel, F. K. (2004). Mid-infrared quantum cascade laser based off-axis integrated cavity output spectroscopy for biogenic nitric oxide detection. *Applied Optics, 43*, 11, 2257–2266.

40. Sonnenfroh, D. M., Rawlins, W. T., Allen, M. G., Gmachl, C., Capasso, F., Hutchinson, A. L., et al. (2001). Application of balanced detection to absorption measurements of trace gases with room-temperature, quasi-CW quantum-cascade lasers. *Applied Optics, 40*, 812–820.

41. Gottfried, J., De Lucia, F., Munson, C., & Miziolek, A. (2009). Laser-induced breakdown spectroscopy for detection of explosives residues: a review of recent advances, challenges, and future prospects. *Analytical and Bioanalytical Chemistry, 395*, 283–300.

42. Foltynowicz, A., Schmidt, F. M., Ma, W., & Axner, O. (2008). Noise-immune cavity-enhanced optical heterodyne molecular spectroscopy: current status and future potential. *Applied Physics B: Lasers and Optics, 92*, 313–326.

43. Ye, J., Ma, L.-S., & Hall, J. L. (1998). Ultrasensitive detections in atomic and molecular physics: demonstration in molecular overtone spectroscopy. *Journal of the Optical Society of America B, 15*, 6–15.

44. Ganser, H., Urban, W., & Brown, A. M. (2003). The sensitive detection of NO by Faraday modulation spectroscopy with a quantum cascade laser. *Molecular Physics, 101*, 545–550.

45. Lewicki, R., Doty, J. H., Curl, R. F., Tittel, F. K., & Wysocki, G. (2009). Ultrasensitive detection of nitric oxide at 5.33 μm by using external cavity quantum cascade laser-based Faraday rotation spectroscopy. *Proceedings of the National Academy of Sciences USA, 106*, 12587–12592.

46. Litfin, G., Pollock, C. R., Curl, R. F., & Tittel, F. K. (1980). Sensitivity enhancement of laser absorption spectroscopy by magnetic rotation effect. *Journal of Chemical Physics, 72*, 6602–6605.

47. Elia, A., Lugará, P. M., Di Franco, C., & Spagnolo, V. (2009). Photoacoustic techniques for trace gas sensing based on semiconductor laser sources. *Sensors, 9*, 616–628.

48. Lima, J. P., Vargas, H., Miklós, A., Angelmahr, M., & Hess, P. (2006). Photoacoustic detection of NO_2 and N_2O using quantum cascade lasers. *Applied Physics B: Lasers and Optics, 85*, 279–284.

49. Kosterev, A. A., & Tittel, F. K. (2004). Ammonia detection by use of quartz-enhanced photoacoustic spectroscopy with a near-IR telecommunication diode laser. *Applied Optics, 43*, 6213–6217.

50. Kosterev, A. A., Tittel, F. K., Serebryakov, D. V., Malinovsky, A. L., & Morozov, I. V. (2005). Applications of quartz tuning forks in spectroscopic gas sensing. *Review of Scientific Instruments, 76*, 43105.

51. Adler, F., Thorpe, M. J., Cossel, K. C., & Ye, J. (2010). Cavity-enhanced direct frequency comb spectroscopy: technology and applications. *Annual Review of Analytical Chemistry, 3*, 175–205.

52. Adler, F., Maslowski, P., Foltynowicz, A., Cossel, K. C., Briles, T. C., Hartl, I., & Ye J. (2010). Mid-infrared Fourier transform spectroscopy with a broadband frequency comb. *Optics Express, 18*, 21861.

53. Cossel, K. C., Adler, F., Bertness, K. A., Thorpe, M. J., Feng, J., Raynor, M. W., & Ye, J. (2010). Analysis of trace impurities in semiconductor gas via cavity-enhanced direct frequency comb spectroscopy. *Applied Physics B: Lasers and Optics, 100*, 4, 917–924.

54. Foltynowicz, A., Masłowski, P., Ban, T., Adler, F., Cossel, K. C., Briles, T. C., & Ye, J. (2011). Optical frequency comb spectroscopy. *Faraday Discussions, 150*, 23–31.

55. Thorpe, M. J., Balslev-Clausen, D., Kirchner, M. S., & Ye, J. (2008). Cavity-enhanced optical frequency comb spectroscopy: application to human breath analysis. *Optics Express, 16*, 4, 2387–2397.

56. Le, L. D., Tate, J. D., Seasholtz, M. B., Gupta, M., Owano, T., Baer, D., et al. (2008). Development of a rapid on-line acetylene sensor for industrial hydrogenation reactor optimization using off-axis integrated cavity output spectroscopy. *Applied Spectroscopy, 62*, 59–65.

57. Lee, B. G., Belkin, M. A., Audet, R., MacArthur, J., Diehl, L., Pflugl, C., & Capasso, F. (2007). Widely tunable single-mode quantum cascade laser source for mid-infrared spectroscopy. *Applied Physics Letters, 91*, 231101.

58. Caffey, D., Radunsky, M. B., Cook, V., Weida, M., Buerki, P. R., Crivello, S., & Day, T. (2011). Recent results from broadly tunable external cavity quantum cascade lasers. *Proceedings of SPIE Photonics West*, 7953–7954.

59. Busch, W. K., & Busch, A. M. (1999). Introduction to cavity-ringdown spectroscopy. In K. W. Busch & M. A. Busch (Eds.), *Cavity-Ringdown Spectroscopy*. New York: Oxford University Press, pp. 7–19.

60. Paldus, B. A., & Kachanov, A. A. (2005). An historical overview of cavity-enhanced methods. *Canadian Journal of Physics, 83*, 975–999.

61. O'Keefe, A., & Deacon, D. A. G. (1988). Cavity ring-down optical spectrometer for absorption measurements using pulsed laser sources. *Reviews of Scientific Instruments, 59*, 2544–2551.

62. Ramponi, A. J., Milanovich, F. P., Kan, T., & Deacon, D. (1988). High sensitivity atmospheric transmission measurements using a cavity ringdown technique. *Applied Optics, 27*, 4606–4608.

63. Scherer, J. J., Paul, J. B., O'Keefe, A., & Saykally, R. J. (1997). Cavity ringdown laser absorption spectroscopy: History, development, and application to pulsed molecular beams. *Chemical Review, 97*, 25–51.

64. O'Keefe, A. (1998). Integrated cavity output analysis of ultra-weak absorption. *Chemical Physics Letters, 293*, 331–336.

65. O'Keefe, A., Scherer, J. J., & Paul, J. B. (1999). CW integrated cavity output spectroscopy. *Chemical Physics Letters, 307*, 343–349.

66. Paldus, B. A., Harb, C. C., Spence, T. G., Zare, R. N., Gmachl, C., Capasso, F., et al. (2000). Cavity ringdown spectroscopy using mid-infrared quantum-cascade lasers. *Optics Letters, 25*, 666–668.

67. Sukhorukov, O., Lytkine, A., Manne, J., Tulip, J., & Jager, W. (2006). Cavity ring-down spectroscopy with a pulsed distributed feedback quantum cascade laser (R. Manijeh and J. B. Gail, Trans.). *SPIE*, 61270A.

68. Paul, J. B., Lapson, L., & Anderson, J. G. (2001). Ultrasensitive absorption spectroscopy with a high-finesse optical cavity and off-axis alignment. *Applied Optics, 40*, 4904–4910.

69. Engel, G. S., Drisdell, W. S., Keutsch, F. N., Moyer, E. J., & Anderson, J. G. (2006). Ultrasensitive near-infrared integrated cavity output spectroscopy technique for detection of CO at 1.57 µm: new sensitivity limits for absorption measurements in passive optical cavities. *Applied Optics, 45*, 9221–9229.

70. Sayres, D. S., Moyer, E. J., Hanisco, T. F., Clair, J. M., Keutsch, F. N., O'Brien, A., et al. (2009). A new cavity based absorption instrument for detection of water isotopologues in the upper troposphere and lower stratosphere. *Review of Scientific Instruments, 80*, 44102–44102.

71. Rao, G. N., & Karpf, A. (2011). Extremely sensitive detection of NO_2 employing off-axis integrated cavity output spectroscopy coupled with multiple line integrated absorption spectroscopy. *Applied Optics, 50*, 1915–1924.

72. Miklós, A., Hess, P., & Bozoki, Z. (2001). Application of acoustic resonators in photoacoustic trace gas analysis and metrology. *Review of Scientific Instruments, 72*, 1937–1955.

73. Rossi, A., Buffa, R., Scotoni, M., Bassi, D., Iannotta, S., & Boschetti, A. (2005). Optical enhancement of diode laser–photoacoustic trace gas detection by means of external Fabry–Perot cavity. *Applied Physics Letters, 87*, 041110.

74. Elia, A., Lugará, P. M., & Giancaspro, C. (2005). Photoacoustic detection of nitric oxide by use of a quantum-cascade laser. *Optics Letters, 30*, 988–990.

75. Hofstetter, D., Beck, M., Faist, J., Nägele, M., & Sigrist, M. W. (2001). Photoacoustic spectroscopy with quantum cascade distributed-feedback lasers. *Optics Letters, 26*, 887–889.

76. da Silva, M. G., Vargas, H., Miklós, A., & Hess, P. (2004). Photoacoustic detection of ozone using a quantum cascade laser. *Applied Physics B: Lasers and Optics, 78*, 6, 677–680.

77. Pushkarsky, M. B., Tsekoun, A., Dunayevskiy, I. G., Go, R., & Patel, C. K. N. (2006). Sub-parts-per-billion level detection of NO_2 using room-temperature quantum cascade lasers. *Proceedings of the National Academy of Sciences USA, 103*, 10846–10849.

78. Rey, J. M., & Sigrist, M. W. (2008). New differential mode excitation photoacoustic scheme for near-infrared water vapour sensing. *Sensors and Actuators B: Chemical, 135*, 161–165.

79. Lee, C.-M., Bychkov, K. V., Kapitanov, V. A., Karapuzikov, A. I., Ponomarev, Y. N., Sherstov, I. V., & Vasiliev, V. A. (2007). High-sensitivity laser photoacoustic leak detector. *Optical Engineering, 46*, 064302.

80. Miklós, A., Hess, P., Mohacsi, A., Sneider, J., Kamm, S., & Schafer, S. (1999). Improved photoacoustic detector for monitoring polar molecules such as ammonia with a 1.53 µm DFB diode laser. *AIP Conference Proceedings, 463*, 126–128.

81. Bijnen, F. G. C., Reuss, J., & Harren, F. J. M. (1996). Geometrical optimization of a longitudinal resonant photoacoustic cell for sensitive and fast trace gas detection. *Review of Scientific Instruments, 67*, 8, 2914–2923.

82. Bernegger, S., & Sigrist, M. W. (1990). CO-laser photoacoustic spectroscopy of gases and vapours for trace gas analysis. *Infrared Physics, 30*, 5, 375–429.

83. Costopoulos, D., Miklós, A., & Hess, P. (2002). Detection of N_2O by photoacoustic spectroscopy with a compact, pulsed optical parametric oscillator. *Applied Physics B: Lasers and Optics, 75*, 2, 385–389.

84. Mukherjee, A., Prasanna, M., Lane, M., Go, R., Dunayevskiy, I., Tsekoun, A., & Patel, C. K. N. (2008). Optically multiplexed multi-gas detection using quantum cascade laser photoacoustic spectroscopy. *Applied Optics, 47*, 4884–4887.

85. Ng, J., Kung, A. H., Miklós, A., & Hess, P. (2004). Sensitive wavelength-modulated photoacoustic spectroscopy with a pulsed optical parametric oscillator. *Optical Letters, 29*, 1206–1208.

86. Pushkarsky, M. B., Webber, M. E., & Patel, C. K. N. (2003). Ultra-sensitive ambient ammonia detection using CO_2-laser-based photoacoustic spectroscopy. *Applied Physics B: Lasers and Optics, 77*, 381–385.

87. Pushkarsky, M. B., Dunayevskiy, I. G., Prasanna, M., Tsekoun, A. G., Go, R., & Patel, C. K. N. (2006). High-sensitivity detection of TNT. *Proceedings of the National Academy of Sciences USA, 103*, 19630–19634.

88. Sigrist, M. W., & Thoeny, A. (1993). Atmospheric trace gas monitoring by CO_2 laser photoacoustic spectroscopy (I. S. Harold and P. Ulrich, Trans.). *SPIE*, 174–184.

89. Webber, M. E., Pushkarsky, M., & Patel, C. K. N. (2003). Fiber-amplifier-enhanced photoacoustic spectroscopy with near-infrared tunable diode lasers. *Applied Optics, 42*, 2119–2126.

90. Kosterev, A. A., Bakhirkin, Y. A., Curl, R. F., & Tittel, F. K. (2002). Quartz-enhanced photoacoustic spectroscopy. *Optics Letters, 27*, 1902-1904.

91. Wysocki, G., Kosterev, A. A., & Tittel, F. K. (2006). Influence of molecular relaxation dynamics on quartz-enhanced photoacoustic detection of CO_2 at $\lambda = 2$ µm. *Applied Physics B: Lasers and Optics, 85*, 301–306.

92. Petra, N., Zweck, J., Kosterev, A. A., Minkoff, S. E., & Thomazy, D. (2009). Theoretical analysis of a quartz-enhanced photoacoustic spectroscopy sensor. *Applied Physics B: Lasers and Optics, 94*, 673–680.

93. Lewicki, R., Wysocki, G., Kosterev, A. A., & Tittel, F. K. (2007). Carbon dioxide and ammonia detection using 2 μm diode laser based quartz-enhanced photoacoustic spectroscopy. *Applied Physics B: Lasers and Optics, 87*, 157–162.

94. Liu, K., Guo, X. Y., Yi, H. M., Chen, W. D., Zhang, W. J., & Gao, X. M. (2009). Off-beam quartz-enhanced photoacoustic spectroscopy. *Optics Letters, 34*, 1594–1596.

95. Liu, K., Yi, H., Kosterev, A. A., Chen, W., Dong, L., Wang, L., et al. (2010). Trace gas detection based on off-beam quartz enhanced photoacoustic spectroscopy: optimization and performance evaluation. *Review of Scientific Instruments, 81*, 103103.

96. Köhring, M., Pohlkötter, A., Willer, U., Angelmahr, M., & Schade, W. (2011). Tuning fork enhanced interferometric photoacoustic spectroscopy: a new method for trace gas analysis. *Applied Physics B: Lasers and Optics, 102*, 133–139.

97. Kosterev, A. A., & Doty, J. H., III. (2010). Resonant optothermoacoustic detection: Technique for measuring weak optical absorption by gases and micro-objects. *Optics Letters, 35*, 3571–3573.

98. Horstjann, M., Bakhirkin, Y. A., Kosterev, A. A., Curl, R. F., Tittel, F. K., Wong, C. M., Hill, C. J., & Yang, R. Q. (2004). Formaldehyde sensor using interband cascade laser based quartz-enhanced photoacoustic spectroscopy. *Applied Physics B: Lasers and Optics, 79*, 799–803.

99. Kosterev, A. A., Bakhirkin, Y. A., Tittel, F. K., Blaser, S., Bonetti, Y., & Hvozdara, L. (2004). Photoacoustic phase shift as a chemically selective spectroscopic parameter. *Applied Physics B: Lasers and Optics, 78*, 673–676.

100. Kosterev, A. A., Mosely, T. S., & Tittel, F. K. (2006). Impact of humidity on quartz-enhanced photoacoustic spectroscopy based detection of HCN. *Applied Physics B: Lasers and Optics, 85*, 295–300.

101. Spagnolo, V., Kosterev, A., Dong, L., Lewicki, R., & Tittel, F. (2010). NO trace gas sensor based on quartz-enhanced photoacoustic spectroscopy and external cavity quantum cascade laser. *Applied Physics B: Lasers and Optics, 100*, 125–130.

102. Grober, R. D., Acimovic, J., Schuck, J., Hessman, D., Kindlemann, P. J., Hespanha, J., et al. (2000). Fundamental limits to force detection using quartz tuning forks. *Review of Scientific Instruments, 71*, 2776–2780.

103. Bauer, C., Willer, U., Lewicki, R., Pohlkötter, A., Kosterev, A., Kosynkin, D., et al. (2009). A mid-infrared QEPAS sensor device for TATP detection. *Journal of Physics: Conference Series, 157*, 012002.

104. Bauer, C., Willer, U., & Schade, W. (2010). Use of quantum cascade lasers for detection of explosives: progress and challenges. *Optical Engineering, 49*, 11, 111126.

105. Kosterev, A. A., Buerki, P., Dong, L., Reed, M., Day, T., & Tittel, F. (2010). QEPAS detector for rapid spectral measurements. *Applied Physics B: Lasers and Optics, 100*, 173–180.

106. Wojcik, M. D., Phillips, M. C., Cannon, B. D., & Taubman, M. S. (2006). Gas-phase photoacoustic sensor at 8.41 μm using quartz tuning forks and amplitude-modulated quantum cascade lasers. *Applied Physics B: Lasers and Optics, 85*, 307–313.

107. Cundiff, S. T., & Ye, J. (2003). Colloquium: femtosecond optical frequency combs. *Reviews of Modern Physics, 75*, 1, 325–342.

108. Marian, A., Stowe, M. C., Lawall, J. R., Felinto, D., & Ye, J. (2004). United time-frequency spectroscopy for dynamics and global structure. *Science, 306*, 2063–2068.

109. Thorpe, M. J., Moll, K. D., Jones, R. J., Safdi, B., & Ye, J. (2006). Broadband cavity ringdown spectroscopy for sensitive and rapid molecular detection. *Science, 311*, 1595–1599.

110. Funke, H. H., Grissom, B. L., McGrew, C. E., & Raynor, M. W. (2003). Techniques for the measurement of trace moisture in high-purity electronic specialty gases. *Review of Scientific Instruments, 74*, 3909–3933.

111. Lehman, S. K., Bertness, K. A., & Hodges, J. T. (2003). Detection of trace water in phosphine with cavity ring-down spectroscopy. *Journal of Crystal Growth, 250*, 262–268.

112. So, S., Amiri Sani, A., Zhong, L., & Tittel, F. (2009). Laser spectroscopic trace-gas sensor networks for atmospheric monitoring applications. *ESSA Workshop 2009, San Francisco, CA*.

113. So, S., Jeng, E., Smith, C., Krueger, D., & Wysocki, G. (2010). Next generation infrared sensor instrumentation: remote sensing and sensor networks using the open PHOTONS repository (S. Marija and P. Gonzalo, Trans.). *SPIE*, 780818.

114. Cheng, W.-H., & Lee, W.-J. (1999). Technology development in breath microanalysis for clinical diagnosis. *Journal of Laboratory and Clinical Medicine, 133*, 3, 218–228.

115. Dweik, R. A., & Amann, A. (2008). Exhaled breath analysis: the new frontier in medical testing. *Journal of Breath Research, 2*.

116. McCurdy, M. R., Bakhirkin, Y., Wysocki, G., Lewicki, R., & Tittel, F. K. (2007). Recent advances of laser-spectroscopy-based techniques for applications in breath analysis. *Journal of Breath Research 1*, 014001.

117. Mürtz, P., Menzel, L., Bloch, W., Hess, A., Michel, O., & Urban, W. (1999). LMR spectroscopy: a new sensitive method for on-line recording of nitric oxide in breath. *Journal of Applied Physiology, 86*, 1075–1080.

118. Mürtz, M., & Hering, P. (2008). Online monitoring of exhaled breath using mid-infrared laser spectroscopy. In M. and I. T. Sorokina (Eds.), *Mid-infrared Coherent Sources and Applications*. Amsterdam,: Springer-Verlag, pp. 535–555.

119. Risby, T., & Tittel, F. K. (2010). Current status of mid-infrared quantum and interband cascade lasers for clinical breath analysis. *Optical Engineering, 49*, 111123.

120. Shorter, J. H., Nelson, D. D., McManus, J. B., Zahniser, M. S., & Milton, D. K. (2010). Multicomponent breath analysis with infrared absorption using room-temperature quantum cascade lasers. *IEEE Sensors Journal, 10*, 76–84.

121. Lee, S. H. (2004). *Partial catalytic hydrogenation of acetylene in ethylene production*. Johor, Malaysia: KLM Technology Group. Retrieved from http://www.klmtechgroup.com/PDF/Articles/acetylene_converter.pdf.

122. Manginell, R. P., Okandan, M., Kottentstette, R. J., Lewis, P. R., Adkins, D. R., Shul, R. J., Bauer, J. M., et al. (2003). Monolithically-integrated microchemlab for gas-phase chemical analysis. *Proceedings of the u-TAS 2003 Workshop, Squaw Valley, CA, September*, pp. 1247–1250.

123. Robinson, A. (2003). Sandia microsensors for industrial process monitoring. *DOE Sensors and Automation FY03 Annual Review Meeting, San Francisco, CA*.

124. Le, L. D., Tate, J. D., Seasholtz, M. B., Gupta, M., Baer, D., Knittel, T., et al. (2006). Rapid online analysis of acetylene for improved optimization of hydrogenation reactors. *Proceedings of the 2006 AIChE Ethylene Producers Conference*, p. 218E.

125. Isberg, J., Hammersberg, J., Johansson, E., Wikström, T., Twitchen, D. J., Whitehead, A. J., Coe, S. E., et al. (2002). High carrier mobility in single-crystal plasma-deposited diamond. *Science, 297*, 5587, 1670–1672.

126. Dong, L., Tittel, F. K., Wright, J., Peters, B., Ferguson, B. A., & McWhorter, S. (2012). Compact QEPAS sensor for trace methane and ammonia detection in impure hydrogen. *Applied Physics B: Lasers and Optics, 107*, 2, 459–467.

127. Weast, R. C., & Astle, M. J. (1981). *61st CRC Handbook of Chemistry and Physics*. New York: CRC Press.

128. Bewley, W., Canedy, C., Kim, C. S., Kim, M., Lindle, J. R., Abell, J., et al. (2010). Ridge-width dependence of mid-infrared interband cascade laser characteristics. *Optical Engineering, 49*, 11, 111116.

129. Lyakh, A., Maulini, R., Tsekoun, A. G., & Patel, C. K. N. (2010). Progress in high-performance quantum cascade lasers. *Optical Engineering, 49*, 111105.

130. Troccoli, M., Wang, X., & Fan, J. (2010). Quantum cascade lasers: high-power emission and single-mode operation in the long-wave infrared (λ >6 µm). *Optical Engineering, 49*, 111106.

CHAPTER 5

ATMOSPHERIC PRESSURE IONIZATION MASS SPECTROMETRY FOR BULK AND ELECTRONIC GAS ANALYSIS

Daniel R. Chase and Glenn M. Mitchell

Matheson, Advanced Technology Center, Longmont, Colorado

5.1 Introduction

Atmospheric pressure ionization coupled with mass spectrometry (APIMS) was developed in the early 1970s by Horning et al. [1–3] at the Baylor College of Medicine (Houston, TX) [4]. The term *atmospheric pressure chemical ionization mass spectrometry* (APCIMS) has also been used for this technique but has not been widely adopted by the specialty and electronics gas industry. APIMS was initially applied to the analysis of (1) liquid analytes introduced via a liquid chromatograph [5] and (2) gaseous impurities in air [6,7]. A review of atmospheric pressure ion sources by Covey et al. [5] is an excellent source of information regarding liquid delivery to an APCI source. With time, the semiconductor industry provided the impetus for APIMS to reach its full potential as a trace analysis technique for gases.

The authors dedicate this chapter in memoriam to Dr. Alan R. Bandy, a pioneer in trace species detection by APIMS.

Trace Analysis of Specialty and Electronic Gases.
By William M. Geiger and Mark W. Raynor. Copyright © 2013 John Wiley & Sons, Inc.

Impurities in bulk and specialty gases can be incorporated into microelectronic devices during manufacture and affect the final device performance negatively. Mitsui et al. report that atmospheric impurities in bulk gases can be responsible for breakdown voltage issues, junction leaks, interference of selective deposition, and decrease in conductance [8]. With electronic specialty gases, a number of issues, including uncontrolled doping, lattice defects, and generation of deep levels, have been attributed to the presence of trace impurities, resulting in reduced yields. Therefore, depending on the application, analysis and control of impurities at low concentrations can be very important in both process gas types [9]. Due to improvements in purifier technology and a better understanding of the importance of purifier placement within delivery systems, the purity of the bulk gas source is often less critical than the actual purity delivered to the process tool [10]. Many process tool specifications require that impurity concentrations in the gases delivered to the tool be less than 100 ppt for each impurity. Therefore, APIMS can be employed in this application to monitor impurity levels downstream of bulk purifiers to ensure that only gases of consistently high purity reach the process tool. In addition, APIMS has an important role to play in the development of new purification media, where it is used downstream of the purifier to characterize the efficiency of impurity removal and/or to determine the capacity of the purifier material for removing specific impurities.

5.2 APIMS Operating Principle

The principle behind APIMS involves primary ionization of a matrix gas and subsequent charge or proton transfer to an analyte at or near atmospheric pressure (secondary ionization). At this pressure the mean free path between molecules is very short and results in a large number of collision-induced charge transfers where nearly all analyte molecules are ionized if the reaction chemistry is favorable [5]. This serves to differentiate APIMS from electron impact (EI) ionization mass spectrometry, where ionization occurs at very low pressures, resulting in longer mean free path, lower ionization efficiencies, and decreased sensitivity. Once generated, ions and ion clusters are manipulated using ion optics lenses at pressures in the range of 10^{-4} torr and transferred into the quadrupole mass filter region, which is pumped down to pressures of 10^{-5} to 10^{-6} torr using turbo molecular pumps.

The most commonly used type of mass spectrometer with atmospheric pressure ionization (API) is the linear quadrupole. This type of mass spectrometer is used more often because it allows high scan speeds and has a high transmission rate for ions, which lead to high sensitivity. The linear quadrupole consists of four hyperbolically or cylindrically shaped rod electrodes. For spatial orientation, the rods can be thought of as occupying the four corners of a rectangular prism extending in the z-direction. Ions flow through the center of the quadrupole in the z-direction (along the length of the rods). Two opposing rods are oriented along the x-axis and the other two opposing electrodes are oriented along the y-axis. The voltage consisting of a direct-current (DC) and a radio-frequency (RF) component is applied to each rod set. Opposite rods are connected electrically, with one pair (x-axis) being positive and the other (*y*-axis) negative. The voltage of each set is increased while keeping the DC-to-RF voltage ratio constant. This has an attraction–repulsion effect on the ions, causing them to follow an oscillating path in the *z*-direction. There is always a small average force in the direction of the *z*-axis, and if the ion path is stable, the ion can make it through the quadrupole and to the detector. If the path is not stable, the ions collide with the electrodes and are neutralized. For a given set of DC voltage, RF voltage, and frequency, ions of a given mass-to-charge ratio (*m/z*) will pass through to the detector. If running in a selected ion monitoring mode, the user selects a specific mass (actually, a mass-to-charge ratio) to monitor, and the correct voltage and frequency are applied by the electronics. A typical quadrupole mass filter can resolve ions that differ by 1 atomic mass unit (amu).

The most common type of detector used for the linear quadrupole is the channel electron multiplier. As an ion enters the detector, it impinges on the surface and secondary electrons are emitted. A large potential is applied across the front and rear of the detector, causing electrons to be accelerated toward the end of the detector. As the electrons impinge on the walls, each one causes several more electrons to be emitted. This cascade of electrons continues to the end of the detector, and a current proportional to the number of impinging ions is sent to a preamplifier (for more gain) and then to an analog-to-digital converter (ADC).

The operation of the mass spectrometer and detector is fairly straightforward. More detailed information on the principle and operation of the mass spectrometer can be found in references such as that of Gross [11]. The primary difficulty in APIMS analysis relates to the API source. It is important to understand how impurities are ionized under specific conditions and in specific matrix gases. The focus of this chapter, therefore, is on API techniques applied to analysis of impurities in the electronic-grade bulk and specialty gases commonly used in the semiconductor industry. The two most common types of API sources used for trace analysis in these types of gases are point-to-plane corona discharge and ^{63}Ni β foil. The information presented here is limited to these two sources.

5.3   Point-to-Plane Corona Discharge Ionization

In a point-to-plane corona discharge source (Figure 5.1), a potential of 3 to 5 kV is typically applied to a small-diameter needle positioned several millimeters from an aperture. Potentials necessary for optimum ionization are dependent on the ionization potential (IP) of the matrix gas and are, therefore, specific to each matrix gas. The current of the discharge is typically on the order of a few microamperes and is controlled by the discharge voltage [12]. Sample is introduced into the corona discharge region at a rate of 1 to 5 L/min using a mechanical pump. Flow into and out of the source may be controlled with mass flow controllers (MFCs), and the inlet pressure may be controlled with a back-pressure regulator, as shown in the sample delivery manifold in Figure 5.2. As total inlet flow can change due to upstream dilutions, an electronically controlled back-pressure regulator is recommended. Because inlet pressure affects the pressure in the declustering region, any changes in the inlet pres-

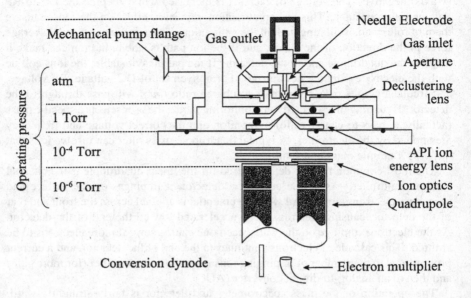

Figure 5.1 Point-to-plane corona discharge APIMS source from an Extrel CMS Attospec-1 including a simplified probe. Copyright ©2012 Extrel CMS, LLC.

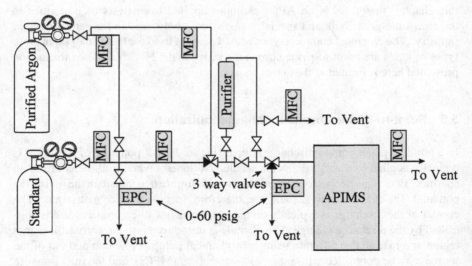

Figure 5.2 APIMS manifold with back-pressure dilution capability, inline purification or purifier testing capability, upstream back-pressure regulation via electronic pressure controllers (EPCs), and mass flow controlers (MFCs) on outlet.

sure will affect the ionization (see Section 5.4.1 on sensitivity). Primary matrix ions quickly react with neutrals in the discharge region to produce secondary ions. These ions can be positive or negative and are formed by charge transfer from primary ions, protonation, or deprotonation [5]. For charge transfer to occur, the primary ion must have a higher IP than that of the impurity of interest. IPs for select matrix gases and impurities are given in Table 5.1. For proton transfer to occur, the impurities must have a greater proton affinity than that of the primary reagent ion (often protonated water cluster ions).

Table 5.1 IPs and reference sources for select matrices and impurities

Gas		IP (eV)	Refs.
Ammonia	NH_3	10.16	
Argon	Ar	16.68	[13]
Arsine	AsH_3	9.89	
Carbon dioxide	CO_2	14.4, 13.79	[13,14]
Carbon monoxide	CO	14.1, 14.013	[13–15]
Helium	He	24.58	[14]
Hydrogen	H_2	15.426	[14,15]
Hydrogen chloride	HCl	12.7	
Methane	CH_4	14.5, 12.98	[13,14]
Nitrogen	N_2	15.51, 15.58	[13–15]
Oxygen	O_2	12.5, 12.075	[13–15]
Phosphine	PH_3	9.87	
Silane	SiH_4	11.65	
Water	H_2O	12.56, 12.59	[13,14]

The ions formed in the discharge region proceed through the orifice and into the declustering region, under the influence of a strong electric field operating at a pressure close to 1 torr [16]. As ions exit the last aperture of the declustering region, they are focused into the quadrupole by a series of optics lenses. A tungsten filament may be incorporated into the ion optics region and can be useful for diagnostic purposes.

5.4 Factors Affecting Sensitivity in Point-to-Plane Corona Discharge APIMS

5.4.1 Effects of Pressure

As mentioned in the introduction, ionization at atmospheric pressure reduces the free path between molecules, resulting in a large number of collision-induced charge transfers. This pressure effect is illustrated by an experiment conducted by Ishihara et al. [17]. The effect of increased ionization chamber (discharge region) pressure on APIMS response for oxygen, water vapor, and carbon dioxide in nitrogen is demonstrated in Figure 5.3. As would be expected, ion intensity increases with increasing pressure in the discharge region.

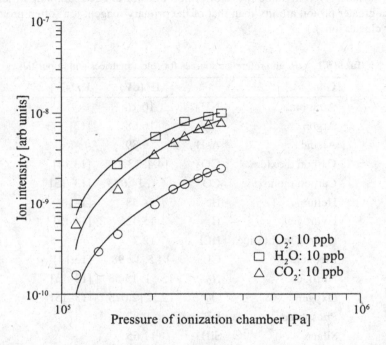

Figure 5.3   Ion intensity for 10 ppb oxygen, 10 ppb carbon dioxide, and 10 ppb water vapor in nitrogen as a function of ionization chamber pressure [17]. Copyright ©1997 The Japan Society of Applied Physics.

This pressure effect is also transferred into the declustering region (between the aperture and the declustering lens). Ions are transferred through the aperture and into this region by electric field effects and by the vacuum suction effect [5]. At this point a vacuum suction effect dominates and the ions are accelerated through the aperture at supersonic velocities. Behind the aperture, the gas and ions form a barrel-shaped expanding jet that terminates in a shock wave (or Mach disk). The declustering lens, when placed within the Mach disk, serves to focus and decluster the ions. Positioning the lens entrance so that it interfaces with the Mach disk optimizes the transmission of ions from the declustering region, through the API lens, and into the ion optics region [5]. Increasing or decreasing the APIMS inlet pressure will not only affect ionization in the discharge region but will affect ion transmission in the declustering region as well. To obtain the optimum ionization, it is necessary to adjust (and hold constant) the inlet pressure for each specific impurity and in each matrix gas.

As the inlet pressure increases, the pressure in the detector region also increases, leading to an increase in neutral molecules in the detector region. The result is an increase in detector noise. When increasing the pressure at the inlet it is necessary to find the pressure that will optimize the signal-to-noise ratio. Ishihara et al. found that the signal-to-noise ratio for oxygen, water, and carbon dioxide in nitrogen was optimized (maximum) at an inlet pressure of 1875 torr (250 kPa) [17].

5.4.2   Effects of Declustering Lens Voltage

In addition to pressure, the voltage potential between the aperture and the declustering lens (declustering voltage) is used to affect declustering. As the declustering voltage is increased, ions and clusters of ions can be further fragmented or dissociated (see Figure 5.9). If ion clusters are desired, declustering voltage should be kept to a minimum but high enough to provide an electric field effect. This parameter should be optimized for each impurity and in each matrix gas.

5.4.3   Effects of Coexisting Analytes

Because APIMS sensitivity relies on charge transfer or proton transfer, the presence of other impurities that can compete for charge or protons is a critical factor to be considered. Figure 5.4 demonstrates the sensitivity of an impurity (X) in the presence of a competing impurity (Y) at varying concentrations of Y and for differing plasma densities. As the presence of competing impurity Y increases, the sensitivity of impurity X decreases. In addition, as the plasma density increases, the sensitivity of impurity X is less affected by the increase in concentration of competing impurity Y.

Although these data represent secondary ionization resulting from proton transfer, the concept can be carried through to charge exchange ionization by using charge-exchange reaction rate constants rather than proton-transfer reaction rate constants [18]. Point-to-charge exchange-rate constants may be found in the *Atomic Data and Nuclear Data Tables* [19]. Using these constants, it should be possible to perform a similar rate equation analysis for various impurities ionized by charge exchange reactions. Experimental data for the effect of water on APIMS response to varying concentrations of carbon dioxide have been demonstrated by Extrel CMS (Figure 5.5).

Water has a lower IP than carbon dioxide and, as predicted, causes a decreased response for carbon dioxide as the water vapor concentration increases. An additional factor in this example is related to a rapid clustering reaction between water vapor and carbon dioxide. The combination of IP and cluster formation not only reduces the response to carbon dioxide but also decreases the sensitivity (decreased slope). The essence of the result is that competing impurities having lower IPs or higher proton affinities can, at high enough concentration, reduce the ion density for the impurity of interest, resulting in decreased response. In addition, cluster reactions can complicate the issue by either increasing or decreasing sensitivity for the molecular

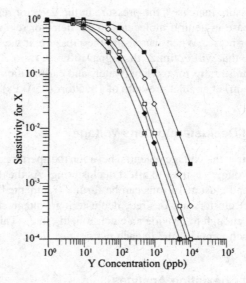

Figure 5.4 Sensitivity of X as a function of concentration of Y for various plasma densities: (⊡) $5 \times 10^8 \, \text{cm}^{-3}$; (◆) $8 \times 10^8 \, \text{cm}^{-3}$; (□) $1 \times 10^9 \, \text{cm}^{-3}$; (◇) $2 \times 10^9 \, \text{cm}^{-3}$; (■) $5 \times 10^9 \, \text{cm}^{-3}$ [18]. Copyright ©American Chemical Society.

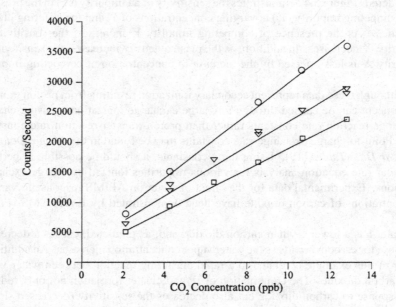

Figure 5.5 Effect of water vapor impurity on carbon dioxide concentration in positive-ion corona discharge APIMS. Upper curve: water vapor = 2.24 ppb. Middle curve: water vapor = 12.3 ppb. Lower curve: water vapor = 22.4 ppb [12]. Copyright ©2012 Extrel CMS, LLC.

ion (M^+ or MH^+) for impurity M. If possible, measures should be taken to reduce the concentration of the competing impurity via purification.

An alternative method to purification, described by Bandy [20], is the addition of an isotopically labeled internal standard. This method removes errors associated with competing ions and is an excellent approach to APIMS analysis. The drawback, however, is that it can be extremely difficult or impossible to obtain isotopically labeled standards for all of the impurities of interest and in the appropriate balance gases. Cost would certainly be an issue as well. This method is discussed in more detail in Section 5.4.4.

The sensitivities for various impurities in nitrogen, argon, and hydrogen are given in Table 5.2. API sources have changed little with respect to design over the years, and therefore sensitivities have not improved dramatically.

Table 5.2   Sensitivity of APIMS for various contaminants in nitrogen, argon, and hydrogen matrices (based on a signal-to-noise ratio of 2).

| | Matrix | | |
Impurity	**Nitrogen**	**Argon**	**Hydrogen**
Argon	2 ppm	—	NA[a]
Carbon dioxide	4 ppt	6 ppt	4 ppt
Carbon monoxide	3 ppb	10 ppt	4 ppt
Methane	30 ppt	40 ppt	1.4 ppt
Nitrogen	—	300 ppb	2.5 ppt
Oxygen	40 ppt	200 ppt	19 ppb
Water vapor	2 ppt	9 ppt	0.7 ppt

Source: [13]
[a] Not applicable.

5.4.4 Isotopic Dilution APIMS Measurements

The basic principle behind an isotopic dilution (ID) method is the addition of a stable isotopically labeled analyte of the trace compound of interest to the sample as an internal standard [20–23]. The top three advantages of this technique for APIMS measurement of trace impurities are:

1. Insensitivity to sampling variations

2. Insensitivity to changes in the APIMS instrumental sensitivity

3. Internal calibration in every sample

Perhaps the most important feature of the isotopic dilution technique is the inclusion of the isotopomer of the analyte in every measurement, allowing for an internal calibration on every sample. Changes in instrument performance affect both the standard and analyte compound equally, thereby having no effect on the accuracy of the measurement. Moreover, the APIMS sensitivity and lower limit of detection can be calculated accurately in realtime, allowing the performance of the instrument to be checked with every measurement.

Another key benefit to the isotopic dilution technique with an APIMS relates to surface conditioning of the manifold by an ever-present isotopomer of the analyte. This conditioning reduces analyte interaction with the manifold surface, resulting in a more stable concentration over time at low levels. Less fluctuation at trace levels results in a higher signal-to-noise ratio and therefore a lower limit of detection.

The concentration of the trace analyte is determined relative to the internal standard after correcting for the trace analyte isotopomers in the standard cylinder and isotopic standard isotopomers present in a sample through use of the following equation [20]:

$$C_a = C_s \frac{K_{ss}R - K_{as}}{K_{aa} - K_{sa}R} \tag{5.1}$$

where C_a is the trace analyte concentration, C_s is the manifold concentration of the isotopic standard, R is the ratio of trace analyte signal counts to isotopic standard signal counts (or I_a/I_s), and the K·values are the percentage contribution of the species (i.e., how much isotope standard is in the analyte and how much analyte isotope is in the standard). The percentage isotope contribution for the trace analyte is known based on natural isotope distributions and can be calculated. The percentage of isotopic material in the standard, as well as analyte isotopomer in the standard, is based on the purity of the standard gas samples used.

Recently, Bandy [24] reported use of this technique in positive and negative APIMS analysis of specialty and industrial gases. In his paper, Bandy reported that the use of ID-APIMS (positive ion) assisted in specialty gas purifier testing where one of the analytes, methane, was not effectively removed and was causing a reduction in ionization and sensitivity. Bandy went on to further illustrate the effectiveness of the ID technique by explaining how it could be used to determine the concentration of an analyte (carbon dioxide in his example) when the signal is in saturation (negative ion). Overall, as long as a cost-effective and stable isotopomer of an analyte of interest is available, use of the ID method could be quite powerful in determining very low concentrations of the analyte in complicated matrices by either corona discharge or ^{63}Ni APIMS.

5.5 Applications of Point-to-Plane Corona Discharge APIMS in Bulk and Electronic Gases

By the late 1990s APIMS was being used routinely for trace analysis of bulk gases used in the manufacture of semiconductor materials [25]. At that time, impurities had been quantified successfully at ppt levels in argon, nitrogen, helium, and even oxygen matrices. To date, other bulk and specialty gases have been added to that list, including hydrogen, methane, hydrogen chloride, silane, and germane.

5.5.1 Bulk Gas Analysis

Argon Matrix Gas When argon is used as the matrix gas, the primary ionization by point-to-plane corona discharge in positive-ion mode follows

$$Ar + e^- \rightarrow Ar^+ + 2e^-$$
$$Ar^+ + Ar + Ar \rightarrow Ar_2^+ + Ar$$

Secondary ionization of impurities occurs via the charge transfer reactions

$$Ar^+ + M \rightarrow Ar + M^+$$
$$Ar_2^+ + M \rightarrow 2Ar + M^+$$

Oxygen, Carbon Monoxide, and Carbon Dioxide Impurity Measurements in Argon Independent oxygen, carbon monoxide and carbon dioxide impurity measurements in an argon matrix are fairly straightforward, keeping in mind the effect on ionization from competing impurities discussed previously. Because water vapor has a lower IP than carbon monoxide and carbon dioxide, it will be ionized preferentially, or steal charge from CO^+ and CO_2^+, and may cause the response for carbon monoxide and carbon dioxide to be reduced at high enough water concentrations. An example of APIMS response to oxygen in purified argon (water vapor concentration below 50 ppt) using point-to-plane corona discharge APIMS in positive-ion mode is shown in Figure 5.6.

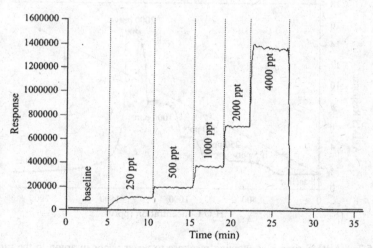

Figure 5.6 APIMS time plot showing response to oxygen in argon from 250 to 4000 ppt using an Extrel CMS Attospec-1 with a point-to-plane corona discharge source, single-quadrupole MS, and positive-ion detection ($m/z = 32$).

Water Vapor Impurity Measurement in Argon For the measurement of water in any matrix, it is critical to begin with a very dry system. The matrix gas should be purified using a heated gas purifier (getter) to remove impurities to sub-ppb levels. Heating manifold lines to at least 60 °C will decrease dry-down time dramatically and help to desorb water off internal surfaces. Valve and MFC materials may prohibit heating to higher temperatures. Dead volumes in the manifold should be eliminated or kept to a minimum to prevent water vapor from migrating into the sample stream. Further, heating the source to at least 100 °C will prevent water vapor from condensing or adsorbing onto surfaces.

An example of a water vapor calibration in argon matrix in the range 125 to 2000 ppt is shown in Figure 5.7. The time plot shows the response of an Extrel CMS Attospec-1 corona discharge APIMS to increasing water vapor concentration in purified argon at $m/z = 18$. Water vapor standards were generated using a KIN–TEK SP61-H2O-L permeation tube moisture generator which delivered a known concentration of water to a dual-stage dilution manifold (Figure 5.2). The water vapor was then diluted into the range desired and delivered to the APIMS inlet at 2 slpm. It is important to note that for this application, the declustering voltage needs to be high enough to dissociate water vapor clusters in order to obtain a maximum response and therefore maximum sensitivity at $m/z = 18$. This voltage can be found experimentally by monitoring clusters in real time and observing cluster dissociation with increasing declustering voltage.

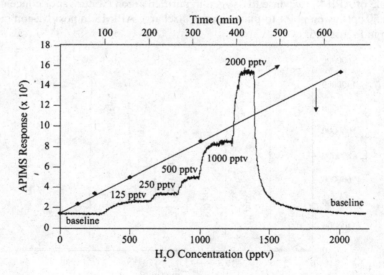

Figure 5.7 APIMS time plot showing response to water vapor in argon in the range 125 to 2000 pptv and resulting calibration plot generated from the data. Instrument: Extrel CMS Attospec-1 with a point-to-plane corona discharge source, single-quadrupole MS, and positive-ion detection ($m/z = 18$).

Nitrogen Impurity Measurement in Argon Until 1998, APIMS had not been used successfully to quantify ppt levels of nitrogen in argon, due to an unfavorable rate constant for the charge transfer reaction between Ar^+ and nitrogen [26]. At that time, Hunter et al. investigated the use of hydrogen to protonate nitrogen through reactions proposed in the following equations:

$$Ar + e^- \rightarrow Ar^+ + 2e^-$$
$$Ar^+ + Ar + Ar \rightarrow Ar_2^+ + Ar$$
$$Ar_2^+ + H_2 \rightarrow ArH^+ + H$$
$$ArH^+ + H_2 \rightarrow H_3^+ + Ar$$
$$H_3^+ + N_2 \rightarrow N_2H^+ + H_2$$
$$ArH^+ + N_2 \rightarrow N_2H^+ + Ar$$

By monitoring the N_2H^+ ion at $m/z = 29$, nitrogen can be detected down to 25 ppt in argon [26]. Through experimentation, it was determined that a concentration of 2.4 % hydrogen (see Figure 5.8).

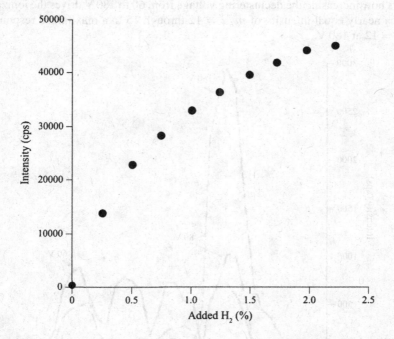

Figure 5.8   APIMS response at $m/z = 29$ for 1 ppb nitrogen spiked into argon as a function of varying hydrogen concentration. Reprinted with permission from [26]. Copyright ©American Vacuum Society.

When calibrating, it is necessary to maintain consistent (preferably low ppt) water levels for subsequent quantitative measurements. As discussed in Section 5.4.3 regarding effects of coexisting analytes, any water vapor present in the source will compete for protons from the primary ions H_3^+ and ArH^+:

$$H_3^+ + H_2O \rightarrow H_3O^+ + H_2$$
$$ArH^+ + H_2O \rightarrow H_3O^+ + Ar$$

As demonstrated in Figure 5.8, a reduction in available hydrogen can result in a dramatic reduction in response for the N_2H^+ ion ($m/z = 29$).

Methane Impurity Measurement in Argon Methane has a low IP relative to argon and is readily ionized in that matrix. Using point-to-plane corona discharge, methane can be ionized to CH_4^+ ($m/z = 16$), CH_3^+ ($m/z = 15$), CH_2^+ ($m/z = 14$), CH^+ ($m/z = 13$), and C^+ ($m/z = 12$) using collision-induced dissociation (CID) [27]. CID is the process of accelerating secondary ions through the aperture, where they collide with molecules of the matrix gas. The intensity of each of these ions is dependent on the declustering voltage. The effect of declustering voltage on the ionization of methane has been demonstrated by Ridgeway et al. [27]. Figure 5.9 shows how increasing the declustering voltage from 60 to 180 V drives the ionization from a nearly equal intensity of $m/z = 12$ through 15 to a maximized response at $m/z = 12$ at 180 V.

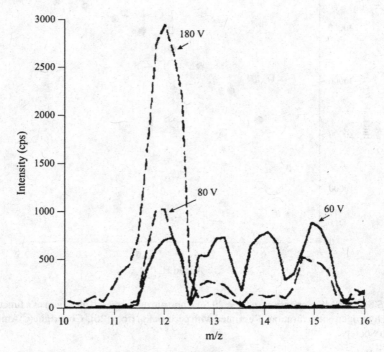

Figure 5.9 Effect of declustering voltage on an APIMS spectrum of methane [27]. Reproduced by permission of The Electrochemical Society.

These results demonstrate the importance of the declustering voltage on secondary ionization and how it can be manipulated to maximize ion intensity and to selectively ionize impurities to minimize interferences.

Nitrogen Matrix Gas   When nitrogen is used as the matrix gas, the primary ionization by point-to-plane corona discharge in the positive-ion mode proceeds via [8]

$$N_2 + e^- \rightarrow N_2^+ + 2e^-$$
$$N_2 + 2e^- \rightarrow 2N^+ + 4e^-$$
$$N^+ + N_2 + N_2 \rightarrow N_2 + N_3^+$$
$$N_2^+ + N_2 + N_2 \rightarrow N_2 + N_4^+$$

The intensity of each of these ionized species is related to the declustering voltage. At lower declustering voltages the N_3^+ and N_4^+ species dominate. These ions transfer charge to impurities having lower IPs to form secondary ions [8]:

$$N_3^+ + M \rightarrow N_2 + N + M^+ \tag{5.2}$$
$$N_4^+ + M \rightarrow 2N_2 + M^+ \tag{5.3}$$

Nitrogen has an IP similar to that of argon (Table 5.1) and its secondary ions will transfer charge to impurities in a similar manner. Reaction rates will differ and can be found in the *Atomic Data and Nuclear Data Tables* [19]. A list of experimental rate constants for the formation of secondary impurity ions, including O_2^+, H_2O^+, and CH_4^+, can be found in a publication by Ketkar et al. regarding ion–molecule reactions in APIMS [28]. Rate constants for coexisting impurity–impurity reactions are also given and provide insight into reactions that will dominate.

Hydrogen Impurity Measurement in Nitrogen   The quantitative analysis of hydrogen in nitrogen can be complicated by the presence of water [8,29]. Because water has a higher proton affinity than nitrogen, it is reasonable to assume that the ionization of N_2H^+ would be reduced in the presence of water. Experimental results contradict this assumption, however, and N_2H^+ ion intensity increases with an increase in water [8]. Therefore, the water vapor concentration must be consistent between calibration and subsequent quantitative analysis or should be removed to sub-ppb levels if at all possible. Techniques for the selective removal of water vapor from the matrix gas and sample gas to sub-ppb levels have been developed and should be investigated [13].

Carbon Monoxide Impurity Measurement in Nitrogen   A method for the trace analysis of carbon monoxide in nitrogen suggested by Kambara et al. uses krypton as a clustering agent. At a high enough concentration, krypton can cluster with carbon monoxide and quench the formation of the Kr^+N_2 ion [14]. The method is very dependent on krypton concentration, and formation of the Kr^+N_2 is difficult to monitor. Due to the ability to resolve masses at thousandths of an amu, time-of-flight

(TOF) APIMS and triple-quadrupole APIMS might be a better choice for trace carbon monoxide analysis in nitrogen.

Hydrocarbon Impurity Measurement in Nitrogen   The same CID technique applied by Ridgeway et al. for methane impurity analysis in argon can be applied to a selection of hydrocarbon impurities in nitrogen [27]. C_1 to C_3 hydrocarbons and acetone can only be monitored qualitatively at concentrations below 1 ppm for $m/z = 12$. Lower declustering voltages are necessary to quantify the impurities. Table 5.3 lists the impurities, limits of detection (LOD), ionic species monitored, fragment ion, and declustering voltages investigated.

Table 5.3   Limits of detection for various impurities in nitrogen at specified m/z values and declustering voltages

Impurity	LOD (ppt)	m/z	Fragment Ion	Declustering Voltage (V)
Acetone	125	43	CH_3CO^+	31
Ethane	50	26	$C_2H_2^+$	60
Isopropyl alcohol	200	45	CH_3CHOH^+	50
Methane	20	16	CH_4^+	31
Propane	110	27	C^+	40

Source: [27].

Oxygen Matrix Gas

Water Vapor Impurity Measurement in Oxygen   Due to its low IP, oxygen is a poor choice for charge transfer reactions. Oxygen ions will, however, readily cluster with water vapor to form fragile adducts. In 1990, Mitsui et al. [8] proposed the main ionization reactions that occur, and in 1998, Scott et al. reported the possibility of sub-ppb analysis of water in oxygen using a cluster reaction [25]. The primary ionization of oxygen with added water progresses as follows:

$$O_2 + e^- \rightarrow O_2^+ + 2e^-$$
$$O_2^+ + O_2 + O_2 \rightarrow O_4^+ + O_2$$
$$O_2^+ + H_2O + O_2 \rightarrow O_2^+ \cdot H_2O + O_2$$
$$O_4^+ + H_2O \rightarrow O_2^+ \cdot H_2O + O_2$$

A reduction in the declustering voltage results in a dissociation of the ion clusters, indicating a weak association. Based on the rate of dissociation, the $O_2^+ \cdot H_2O$ ion has the strongest association of the three. The calibration data in Figure 5.10 were generated by adding 1 to 5.1 ppb water vapor to dry oxygen and monitoring the $O_2^+ \cdot H_2O$ ion [25]. Care must be taken to control the declustering energy so as not to dissociate the fragile $O_2^+ \cdot H_2O$ ion.

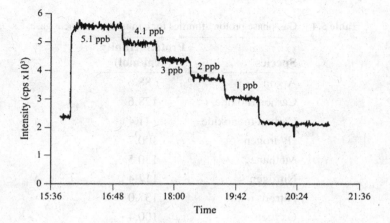

Figure 5.10   Change in the APIMS response at $m/z = 50$ for changing concentrations of added water vapor in oxygen. Reprinted with permission from [25]. Copyright ©American Chemical Society.

Hydrogen Matrix Gas   While hydrogen does have an IP close to that of nitrogen (15.43 and 15.58 eV, respectively), the primary ionization of hydrogen leads to the formation of H_3^+, and secondary ionization proceeds through proton transfer rather than charge transfer. The primary ionization and subsequent proton transfer reactions in the following equations are described by Mitsui et al. [8].

$$H_2 + e^- \rightarrow H_2^+$$
$$H_2^+ + H_2 \rightarrow H_3^+ + H\cdot$$
$$H_2^+ + N_2 \rightarrow N_2H^+ + H\cdot$$
$$H_2^+ + CO_2 \rightarrow CO_2H^+ + H\cdot$$
$$H_2^+ + H_2O \rightarrow H_3O^+ + H\cdot$$

The following proton transfer reactions are reported by Ridgeway et al. as observed in mass spectral data obtained from APIMS analysis of bulk hydrogen [30]:

$$H_3^+ + N_2 \rightarrow N_2H^+ + H_2$$
$$H_3^+ + H_2O \rightarrow H_3O^+ + H_2$$
$$H_3^+ + CH_4 \rightarrow CH_5^+ + H_2$$
$$H_3^+ + CO_2 \rightarrow CO_2H^+ + H_2$$

Because ion residence time in the discharge region is so short, the rate of proton transfer to the impurity can be a limiting factor for sensitivity. Proton affinities for some molecules of interest are listed in Table 5.4.

With the exception of oxygen and argon, hydrogen has the lowest proton affinity given in Table 5.4. Sensitivities for water, methane, carbon dioxide, carbon monoxide, and nitrogen in hydrogen are 0.7, 1.4, 4.0, 4.0 and 2.5 ppt, respectively [13]. A

Table 5.4   Gas-phase proton affinities for common permanent gases

Species	Proton Affinity (kcal/mol)
Argon	88.6
Carbon dioxide	128.6
Carbon monoxide	141.4
Hydrogen	100.7
Methane	130.5
Nitrogen	117.4
Nitrous oxide	137.0
Oxygen	100.4
Water vapor	173.0

Source: [30].

triple-quadrupole APIMS instrument with point-to-plane corona discharge was used to collect these data. The ions monitored for these results and specific instrument conditions such as lens voltages, were not reported.

The effects of coexisting impurities are very significant and, once ionized, impurities of interest can transfer a proton to competing impurities that have a higher proton affinity [27]. Water has the highest proton affinity in the list and will act as a "proton sponge" in the presence of other impurities [30]. Once again, it is very beneficial to remove or control the amount of competing impurities, especially water, during calibration and subsequent quantitative analysis.

Helium Matrix Gas   Although helium has the highest IP of the matrix gases discussed, there is limited information in the literature about its application as a matrix gas. Pusterla et al. report secondary ionization for hydrogen clustered with helium at $m/z = 5$ (HeH^+) [31]. The cluster He_2H^+ at $m/z = 9$ was also observed but only for very specific conditions and was, therefore, not used. Charge transfer reactions were demonstrated for nitrogen/carbon monoxide ($m/z = 28$), water ($m/z = 18$), methane/^4He ($m/z = 16$), oxygen ($m/z = 32$), and argon ($m/z = 40$). APIMS-TOF was used for the experiment. Instrument conditions and sensitivities were not reported. Heated purifiers (getters) were used to purify the matrix gas to sub-ppb levels of impurities.

5.5.2 Electronic Specialty Gas Analysis

Hydrogen Chloride Matrix Gas   Corrosive gases such as hydrogen chloride can react with metal surfaces, resulting in reduced yield in device manufacturing as well as "premature failure of sensitive gas system components" [32]. In an effort to characterize the effect of water vapor and hydrogen chloride on metal surfaces,

Ma et al. used APIMS to monitor water in a hydrogen chloride/nitrogen stream. Hydrogen chloride was run initially through 316L tubing and then blended with nitrogen at the APIMS source [32]. A calibration curve was generated by varying the water vapor concentration in the nitrogen stream while maintaining constant flow for hydrogen chloride and nitrogen. A nonlinear fit was applied to the data points. The configuration of the source and mass spectrometer was not reported, and no ion reaction mechanisms were speculated or reported either. This technique is included to demonstrate that it is possible to detect water in a corrosive gas, but further studies would need to be conducted to demonstrate sensitivity and the ability of the source to withstand the effects of corrosive materials.

Silane Matrix Gas   Trace analysis of impurities in silane is limited by its low IP (11.65 eV) [33]. As oxygen will cluster with water vapor to form ions that can be quantitated, so will silane.

A number of research groups have investigated impurity analysis in silane by APIMS. Kitano et al. used a novel APIMS source developed by Hitachi Tokyo Electronics Co. Ltd. that employs a two-chamber configuration (Figure 5.11). The first chamber is used to ionize the matrix gas, which in this case is argon. The second chamber is used to combine the ionized argon with silane where charge transfer can occur. Figure 5.12 shows the ion–molecule reactions that occur in the two-compartment source. This design prevents silane from being deprotonated in the discharge region, whereby silicon particles can coat the needle and plug the aperture. Figure 5.13 shows spectra for purified and unpurified silane. In purified silane, the ions and ion clusters are SiH_3^+ and silane clusters, with the exception of $SiH_3^+ \cdot H_2O$. In the unpurified silane, disiloxane ($SiH_3–O–SiH_2^+$) and clusters of disiloxane are present ($m/z = 77$ and $m/z = 109$). Kitano et al. conclude that disiloxane is the major impurity of concern in silane.

Siefering and Whitlock also investigated the use of a two-chamber APIMS source for trace impurity analysis in silane [33]. The $SiH_3^+ \cdot H_2O$ ion was also seen in this experiment, but the $m/z = 77$ peak observed by Kitano et al. was not seen. Instead, an ion with $m/z = 79$ was observed and attributed to a disilane moisture cluster $Si_2H_5^+ \cdot H_2O$. An ion at $m/z = 67$ was also observed and attributed to $SiH_3^+ \cdot (H_2O)_2$. Instrument conditions such as lens voltages were not reported, so a conclusion cannot be drawn with regard to the differences observed by the two groups.

Experiments conducted by Ohki and Ohmi during the same time period and with the same two-chamber source configuration confirmed the presence of the $m/z = 49$ $SiH_3^+ \cdot H_2O$ cluster ion [35]. Ions were also observed at $m/z = 67, 79, 81,$ and 97. The ions at $m/z = 67$ and 79 were consistent with the observation by Siefering and Whitlock and were attributed to silane/water vapor clusters.

Further studies conducted by Gotlinsky et al. working out of Tadahiro Ohmi's laboratory at Tohoku University supported reaction chemistries observed previously [36]. In this case, both $m/z = 77$ and $m/z = 79$ were observed, indicating that both the disiloxane ion and the silane/hydroxyl silane clusters could be formed under the same conditions. They were able to demonstrate the ability to remove siloxane and

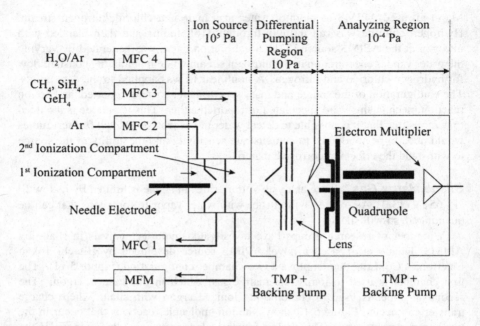

Figure 5.11 APIMS instrument with a two-compartment ion source [34]. Copyright ©The Japan Society of Applied Physics.

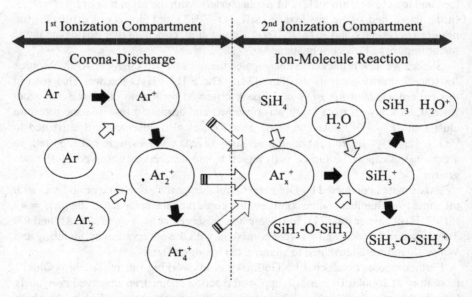

Figure 5.12 Ion–molecule reaction model in a two-compartment ion source [34]. Copyright ©The Japan Society of Applied Physics.

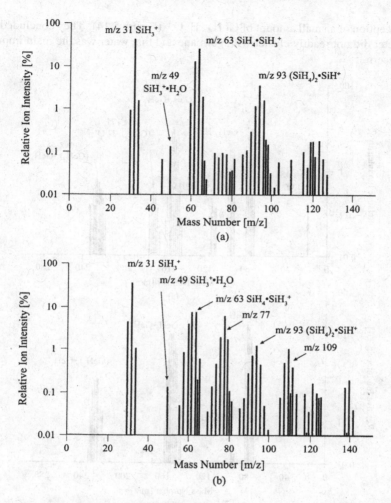

Figure 5.13 (a) Spectrum of silane passed through a Li–H purifier. Impurities reported to be less that 1 ppb. (b) Spectrum of unpurified silane from a cylinder with water present [34]. Copyright 2001 ©Japan Society of Applied Physics.

siloxane clusters using a new purification technology. The result of using purified silane on wafer production was a dramatic reduction in particles that can "adversely affect the quality of thin films" [36].

Germane Matrix Gas   Using the two-chamber APIMS source described earlier, Kitano et al. studied the secondary ionization reactions for germane in an attempt to identify impurities. When water was present at the high concentrations typically found in cylinders (ca. 1 to 2 ppm), water vapor clusters with germane were observed. When water was removed to 2 ppb, the water clusters were no longer observed, with the exception of a small amount of $GeH_4 \cdot H_2O^+$ (Figure 5.14). They concluded that the water did not readily cluster with germane and that water was the main impurity of concern.

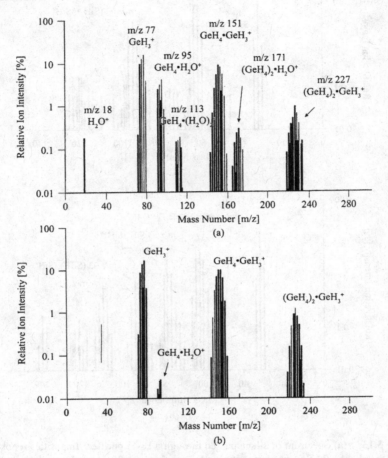

(a)

(b)

Figure 5.14 (a) Spectrum of germane with no purification. The concentration of water vapor is 1860 ppb. (b) Spectrum of purified germane with water at 2 ppb. Water clusters are absent except for a small amount of $GeH_4 \cdot H_2O^+$ [34]. Copyright ©2001 The Japan Society of Applied Physics.

5.6 Nickel-63 Beta Emitter APIMS

As seen in earlier in the chapter, corona discharge ionization has a predominant role in the measurement of trace impurities in a variety of inert matrix gases. However, direct measurement of impurities in electronics specialty gases via positive ionization is sometimes difficult, due to competing proton affinities of the matrix gas and the impurities to be measured. One such example is the measurement of water vapor impurity in high-purity ammonia.

Alternative ionization sources have recently been used [21,24,37] to overcome competing proton affinities through the use of negative-ion clusters. In this section we describe the development and use of ^{63}Ni radioactive beta emitters for the determination of impurities through the clustering of negative ions. Although methods for overcoming proton affinity barriers using emitter sources in the positive-ion mode are discussed, the main focus is on negative-ion clustering. Concluding this section is a discussion regarding isotopic dilution as a means to remove dependency from daily calibrations, instrumentation drift, and sampling variability when making trace impurity measurements using APIMS.

5.6.1   Nickel-63 Source Design

The ^{63}Ni atmospheric pressure ionization (API) source for an APIMS is similar to the design of an electron capture detector [38,39] used in gas chromatography, as demonstrated by Horning et al. [1,2,40,41] and Grimsrud et al. [42]. ^{63}Ni sources are usually nickel-alloy foil coated with low-level millicuries of ^{63}Ni, and are thus considered low-level radioactive β emitters. It should be noted that the use of ^{63}Ni in the U.S. is licensed and governed under the United States Nuclear Regulatory Commission, and the source requires secure storage, annual testing, and appropriate signage for the laboratory. An example of a ^{63}Ni API source is shown in Figure 5.15.

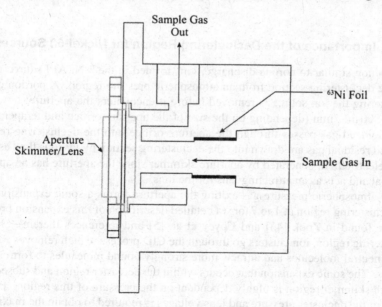

Figure 5.15 ^{63}Ni β emitter API source [21].

In this example, the ^{63}Ni foil is 1 cm × 3 cm, and was coiled to form a cylinder 0.24 cm^3 in volume. The source was designed to provide transit times of 1 to 10 ms to the aperture of the declustering region. Once past the aperture, the remaining source design is similar to the corona discharge source design shown in Figure 5.1.

5.6.2   Ion Formation from a Nickel-63 Source

Sample gas inside the source volume is in constant interaction with the radioactive foil, where it is ionized by the spontaneous β particles from the decay of the foil. Ions are produced from the ^{63}Ni foil by nuclear decay to ^{63}Cu, which subsequently emits a β particle:

$$^{63}\text{Ni} \rightarrow {}^{63}\text{Cu} + \beta$$

On average, the half-life for ^{63}Ni is on the order of 100 years.

The β particle can then interact with a matrix gas M to produce a secondary electron and a positive ion from the matrix gas:

$$M + \beta \rightarrow M^+ + e^-$$

Secondary electrons produced in this way can also interact with impurities or additives (Y) in the matrix gas to produce negative ions used for clustering:

$$Y + e^- \rightarrow Y^-$$

5.6.3   Importance of the Declustering Region for Nickel-63 Sources

In a fashion similar to corona discharge, ions formed in the ^{63}Ni API source travel into the declustering section from an atmospheric pressure region. A portion of the gas entering the ion source is removed before it encounters the aperture; however, 600 to 800 mL/min (depending on the size of the aperture orifice and temperature) of the ionized gas passes through the aperture orifice into the declustering region. Ions and residual gas are drawn into the declustering section by viscous flow as well as the electric field generated by aperture skimmer since the aperture has an applied potential and acts as an attracting lens for the ions.

The atmospheric pressure gas exiting the aperture forms a sonic expansion into the declustering region at 1 to 7 torr (a detailed description of this expansion process may be found in Zook [43] and Covey et al. [5] and references therein). In the declustering region, ion clusters go through the CID process, which removes weakly bound neutral molecules and attracts more strongly bound molecules to form stable clusters. The sonic expansion that occurs within the decluster region and subsequent aperture skimmer region is highly dependent on the pressure of this region. Proper tuning of the decluster pressure and lens voltages is required to obtain the maximum formation of ion clusters and, ultimately, the sensitivity of the measurement using ^{63}Ni ionization [21,43].

5.6.4   Overcoming Competing Positive-Ion Proton Affinities

The formation of positive ions of various trace gas species of interest occurs through proton transfer reactions with other protonated species having lower proton affinities. An example of this effect can be seen in the proton transfer from protonated water or water clusters. The positive-ion reaction chemistry to form protonated water and water clusters in an APIMS is well documented in the literature [1,7,28,38,39,44–46] and is similar for both the corona discharge and ^{63}Ni ionization sources [40]. The dominant gas-phase APIMS reactions forming protonated water clusters and protonated water are the following:

$$H^+(H_2O)_{n-1} + H_2O + N_2 \rightarrow H^+(H_2O)_n + N_2$$

and

$$H_3O^+(OH) + H_2O \rightarrow H_3O^+(H_2O) + OH$$

where the initial species are formed from equations (5.2) and (5.3). Species with a gas-phase proton affinity [47] greater than that of water (691 kJ/mol, 165 k/mol), can abstract a proton from the protonated water cluster by the following reactions [48,49]:

$$H_3O^+(H_2O)_n + X \rightarrow XH^+(H_2O)_m + (n + 1 - m)H_2O \qquad (5.4)$$

$$YH^+(H_2O)_n + X \rightarrow XH^+(H_2O)_m + (n + 1 - m)H_2O \qquad (5.5)$$

Here X is a species having a higher proton affinity than that of both water vapor and Y [18], where $n > m$. However, for some species, the gas-phase hydration energetics (i.e., stability) of the newly formed protonated species are small, and thus equation (5.4) will shift to the left. An overabundance of water vapor will also shift equation (5.5) to the left. Therefore, in this example, to create a long-lived protonated species, the amount of water must be reduced and the water clusters dehydrated, either through purification, drying, dilution, or by heating the gas [50], to achieve high yields for the protonated molecules. This effect is also true for other molecules having high proton affinity, such as ammonia. Similar methods would need to be devised to favor the creation of other protonated species.

As mentioned above and in Section 5.4.3, compounds with a gas-phase proton affinity higher than that of water can extract a proton from protonated water or water clusters to form a positive ion, and the same principle is also true for coexisting analytes [18] that have a larger proton affinity than the impurity of interest. Even species with a lower proton affinity, but with higher gas-phase hydration energies, can interfere significantly. These co-analytes and their hydrate clusters reduce the sensitivity for the detection of species of interest in positive-ion APIMS either as a persistent background signal or by reducing ionization through the removal of available transfer agents [21].

5.6.5   Negative-Ion Cluster Formation

Negative-ion corona discharge is less often used for negative-ion formation, due to the fact that negative-ion corona discharge can produce a high background due to the abundance of electrons being produced and allowed through the quadrupole. ^{63}Ni ionization has the ability to create negative ions without such background interference. Because of this, much lower detectable limits are achievable with ^{63}Ni negative ionization [24].

The initial production of negative ions in an APIMS with a ^{63}Ni source begins with the electron capture process, whereby oxygen captures an electron to create the O_2^- ion or another species with a high electron affinity [1,7]. Once the initial negative ion is created, the process of cluster formation occurs for those species with an affinity for negative ions. Cluster formation can also occur from secondary ion–molecule reactions [51–55]. An example of the ionization and cluster formation for trace-level sulfur dioxide in air is provided below to demonstrate the negative-ion cluster mechanisms [21]:

Direct clustering:

$$O_2^- + SO_2 \rightarrow SO_2^- + O_2$$
$$SO_2^- + O_2 + N_2 \rightarrow SO_4^- + N_2$$

Ion–molecule reactions:

$$O_2 + e + (O_2 \text{ or } N_2) \rightarrow O_2^- + (O_2 \text{ or } N_2)$$
$$O_2^- + O_3 \rightarrow O_3^- + O_2$$
$$O_3^- + CO_2 \rightarrow CO_3^- + O_2$$
$$CO_3^- + SO_2 \rightarrow SO_3^- + CO_2$$
$$SO_3^- + O_2 + N_2 \rightarrow SO_5^- + N_2$$

However, as will be demonstrated for the case of water measurements in high-purity ammonia, the competition for clustering and ionization in negative-ion formation is similar to that for positive-ion formation (discussed in Section 5.6.4).

5.7 Specialty Gas Analysis Application: Determination of Oxygenated Impurities in High-Purity Ammonia

There has been significant growth recently in the light-emitting diode (LED) market, specifically in the manufacture of gallium nitride (GaN) and aluminum gallium nitirde (AlGaN) devices for use in high-brightness blue, ultraviolet (UV), and white LED products. These nitride devices are critically sensitive to impurities present in the ammonia process gas, which can become incorporated in the nitride films. They are particularly susceptible to oxygen incorporation through volatile oxygenated species such as water, as well as silicon, germanium, or carbon incorporation from hydrides species such as silane, germane, and hydrocarbons. Incorporation of these impurities leads to reduced performance and shorter lifetimes of the GaN and AlGaN devices [56].

This need for high-purity has led to stringent product specifications for anhydrous ammonia that have, until recently, exceeded the lower detectable limits of analytical instrumentation. Table 5.5 demonstrates the oxygenated impurity specifications for high-purity ammonia versus the current industry specification.

Table 5.5   Trace oxygenated impurity specifications for high-purity anhydrous ammonia

Impurity	Current Specification (ppb)	High-Purity Specification (ppb)
Carbon dioxide	100	50
Carbon monoxide	50	10
Oxygen	50[a]	10
Water vapor	50	1[b]

Source: Adapted from [56,57].
[a] Oxygen and argon are a combined specification.
[b] 1 ppb water in the vapor phase, measured as 100 ppb in the liquid phase.

To provide improved quality assurance to the semiconductor industry, new analytical methodologies have been developed for the determination of trace impurities, with oxygenated species being of special interest [24,37,56–62]. Part of this development work was the determination of water and other oxygenated species through negative-ion clusters formed by ^{63}Ni APIMS [24,37,61,62].

5.7.1 Difficulty of Determining Water Vapor in Ammonia by Positive-Ion APIMS

Most often, trace gas analysis by APIMS is performed by positive ionization of the trace gas species by charge and proton transfer reactions. Electrons from a corona discharge or from β-particle decay in ^{63}Ni interact with a matrix gas M to create the initial charge for the reaction. An example of the multistep matrix gas ionization for nitrogen is given in Section 5.4.2. A simplified version is

$$M + e^- \rightarrow M^+ + 2e^-$$

Interaction of a trace gas impurity with the ionized matrix gas M^+ results in a charge transfer, as demonstrated for water vapor in equation (5.6). As long as no competing analytes are present in the sample to remove the charge, equation (5.6) is quickly followed by a proton transfer between the ionized trace gas and another molecule of the same chemical species, as seen in equation (5.7) [24].

$$M^+ + H_2O \rightarrow H_2O^+ + M \tag{5.6}$$

$$H_2O^+ + H_2O \rightarrow H_3O^+ + OH \tag{5.7}$$

If an analyte having greater proton affinity is present, an additional proton transfer reaction occurs. For example, if ammonia is present in the sample to be analyzed, protonated water will transfer a proton to the ammonia molecule as follows:

$$H_3O^+ + NH_3 \rightarrow NH_4^+ + H_2O$$

Depending on the decluster pressure and lens voltages, cluster reactions can be used to detect water (see Section 5.6.4). The ammonium ion can readily cluster with water to form a stable complex [equation (5.8)] [63]; however, the clustering of protonated ammonia with another ammonia molecule quickly removes the pathway for clustering of water [equation (5.9)].

$$NH_4^+ + H_2O \rightarrow NH_4^+ \cdot H_2O \tag{5.8}$$

$$NH_4^+ + NH_3 \rightarrow NH_4^+ \cdot NH_3 \tag{5.9}$$

As Bandy mentions in his report on water vapor in ammonia, determining trace gas species with lower proton affinities than ammonia in anhydrous ammonia by positive-ion APIMS with any ionization method is not possible, due to the removal of charging species by the matrix gas itself [24]. It is for this reason that an alternative ionization source and ionization pathways were investigated for trace water vapor in ammonia.

5.7.2 Low-Level Water Vapor Measurements in Ammonia by Negative-Ion Nickel-63 APIMS

As mentioned previously, negative-ion corona discharge produces a high background at all masses (due to electrons passing through the APIMS mass filter) and has limited sensitivity and high detection limits. Use of a ^{63}Ni source to produce negative ions and clusters with little or no background at all masses can produce much lower limits of detection. Moreover, negative-ion clustering via ^{63}Ni ionization offers a greater level of flexibility since the ionization and clustering can be tailored by using additives with different electron affinities and reactions rates with primary ions [24]. Although the example presented here is for ammonia, these techniques may be extended to impurities in other hydride gases.

The use of oxygen as an additive into a nitrogen matrix gas to assist in the creation of the negative-ion cluster of the trace gas impurities in ammonia was a technique co-developed by Air Products and Chemicals, Inc. and Drexel University [37,61,62,64]. The generation of negative ions occurs as electrons interact with oxygen to form negative oxygen molecular ions:

$$O_2 + e^- \rightarrow O_2^-$$

For the work cited, 100 ppb oxygen in nitrogen was used as the additive to purified nitrogen. In the presence of water, the oxygen molecular ion will react to form various hydrated adducts or clusters, as demonstrated in equation (5.10) and shown in Figure 5.16.

$$O_2^- + nH_2O \rightarrow O_2^- \cdot (H_2O)_n, \ n = 1, 2, 3 \tag{5.10}$$

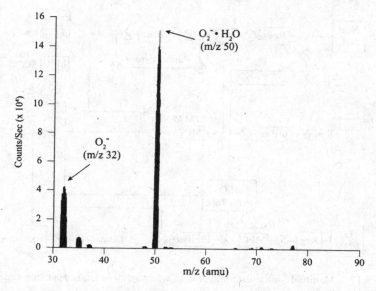

Figure 5.16   Negative-ion cluster formation of $O_2^- \cdot H_2O$ ($m/z = 50$) in nitrogen [64].

The $O_2^- \cdot (H_2O)_n$ negative-ion clusters can occur at $m/z = 50$ ($n = 1$), $m/z = 68$ ($n = 2$), and $m/z = 86$ ($n = 3$). Similar work was reported previously for positive-ion $O_2^+ \cdot (H_2O)_n$ clusters in bulk oxygen (see Section 5.5.1 and references noted).

In an ammonia matrix, the water clusters compete with ammonia based on the following reaction:

$$O_2^- + NH_3 \rightarrow O_2^- \cdot NH_3 \tag{5.11}$$

Creation of the $O_2^- \cdot NH_3$ reduces significantly the sensitivity for water vapor determination. To overcome this deleterious effect, the concentration of ammonia had to be determined to optimize the water cluster formation and, thus, sensitivity [37,61,62]. Concentrations of 5 % or less of ammonia in a purified nitrogen matrix will allow equation (5.12) to occur. Although ammonia will continue to cluster with the oxygen molecular ion, it has been shown experimentally to be less stable than the $O_2^- \cdot (H_2O)_n$ clusters at the concentration noted above, or less:

$$O_2^- \cdot NH_3 + H_2O \rightarrow O_2^- \cdot H_2O + NH_3 \tag{5.12}$$

The $O_2^- \cdot NH_3$ cluster is detected at $m/z = 49$ and is differentiated from the $O_2^- \cdot H_2O$ cluster ($m/z = 50$).

The manifold used to quantify trace water vapor in ammonia is shown in Figure 5.17. In this work, approximately 4 L/min of purified nitrogen (<1 ppb water vapor) was used to dilute high-purity ammonia down to 5 % or less. The oxygen additive was also added to the manifold at 100 mL/min. Quantitative results are obtained by adding known concentrations of water from a permeation tube to the manifold for calibration. Figure 5.18 demonstrates the APIMS response to a 50 ppb water

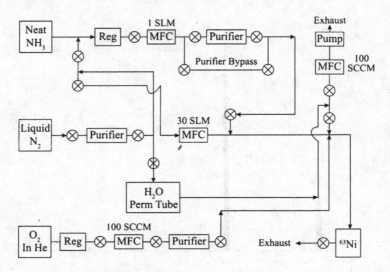

Figure 5.17 Manifold for the determination of water vapor in ammonia using negative-ion APIMS [64].

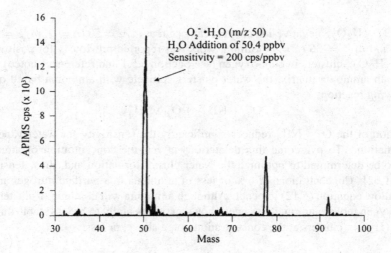

Figure 5.18 APIMS response to a 50 ppb water vapor challenge in purified ammonia using [63]Ni APIMS negative ions [62].

challenge in purified ammonia. The response at $m/z = 50$, $m/z = 68$, or the sum of both masses is linear with respect to water vapor concentration and can be used for the determination of total water vapor content in anhydrous ammonia. In the presence of an interfering analyte, such flexibility makes it possible to calibrate at a mass that is not interfered with. Figure 5.19 presents data for the $O_2^- \cdot H_2O$ cluster

($m/z = 50$) with a weighted least squares (WLS) method detection limit (MDL) of 8 ppb water vapor in ammonia.

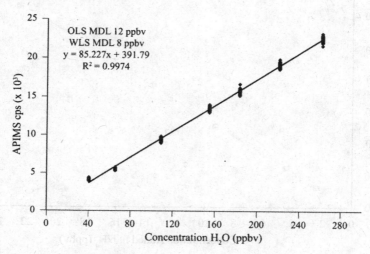

Figure 5.19 Calibration curve for water vapor in ammonia using an $O_2^- \cdot H_2O$ ($m/z = 50$) cluster. The weighted least squares method detection limit is 8 ppb water vapor in ammonia for the negative ion APIMS technique [62].

5.7.3 Measurement of Other Impurities in Ammonia by Negative-Ion Nickel-63 APIMS

The negative-ion adduct or cluster method can also be used for the determination of other impurities in ammonia. Two examples from Bandy [20] demonstrate this very effectively for the detection of extremely low levels of oxygen and sulfur dioxide, particularly when employing the isotopic dilution method described in Section 5.4.4.

For trace oxygen in ammonia, a manifold similar to that shown in Figure 5.17 was used. In the absence of other impurities, such as water vapor, the $O_2^- \cdot NH_3$ adduct from equation (5.11) is free to form and can be calibrated on $m/z = 49$. Figure 5.20 is the calibration of low ppb levels of oxygen in an ammonia–nitrogen matrix. By this method, the detection limit for oxygen in ammonia, particularly if using the isotopic dilution method making a ratio to $^{16}O/^{18}O$, is well below 0.1 ppb [20].

Similarly, trace levels of sulfur dioxide can be detected in ammonia using the $SO_2 \cdot O_2^-$ ion cluster, which is generated by the following reactions:

$$O_2 + e^- \rightarrow O_2^-$$
$$O_2^- + SO_2 \rightarrow SO_2 \cdot O_2^- \text{ (or } SO_4^- \text{)}$$

Figure 5.21 shows the nonlinear calibration plot for sulfur dioxide in ammonia for the $SO_2 \cdot O_2^-$ cluster ($m/z = 96$). The challenge concentrations did not exceed

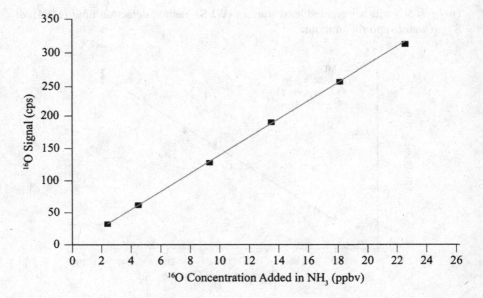

Figure 5.20 Calibration of ppb levels of oxygen in ammonia using an $O_2^- \cdot NH_3$ ($m/z = 49$) cluster. The limit of detection when using the isotopic dilution ratio of $^{16}O/^{18}O$ is below 0.1 ppb [20].

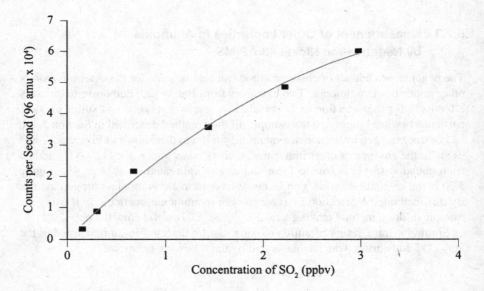

Figure 5.21 Calibration of ppb levels of sulfur dioxide in ammonia using an $SO_2 \cdot O_2^-$ ($m/z = 96$) cluster. The limit of detection when using the isotopic dilution ratio of $^{32}S/^{34}S$ is below 0.2 ppb [20].

3 ppb, and the lower limit of detection using the $^{32}S/^{34}S$ isotopic dilution ratio is below 0.2 ppb.

As demonstrated by Air Products and Chemicals, Inc. [62], carbon dioxide can also be detected as a negative-ion cluster by the following pathway:

$$O_2^- + CO_2 \rightarrow CO_4^-$$

$$CO_4^- + NH_3 \rightarrow NH_3 \cdot CO_4^-$$

The ion $NH_3 \cdot CO_4^-$ ($m/z = 93$) has been shown experimentally to be stable. Another form of this species, ammonium carbamate ($NH_4CO_2NH_2$), was studied previously as an impurity in semiconductor-grade ammonia [60].

Other oxygenated impurities, such as alcohols, amines, and oxyamines, could also complex with O_2^- to form stable adducts. This is best demonstrated by Figure 5.22, which is a negative-ion APIMS spectrum of unpurified ammonia. In this plot, negative-ion clusters for ammonia, water vapor, methanol, and carbon dioxide are easily seen. Other peaks observed have yet to be identified positively but could be any of the above-mentioned classes of compounds.

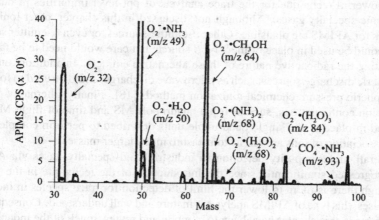

Figure 5.22 Negative-ion APIMS spectrum of unpurified ammonia showing O_2^- clusters of ammonia, water vapor, methanol, carbon dioxide, and other impurities [62].

5.7.4 Additional Electronegative Additives for Cluster Formation in Ammonia by Negative-Ion Nickel-63 APIMS

As mentioned previously, cluster reactions could be tuned by using additives having different electon affinities and reaction rates. Other oxygen-containing compounds, such as NO_x, could provide cluster ionization similar to that of oxygen. Additionally, halogen species could be used as additives to form negative-ion clusters. For example, if a halogen Z is any of the halogens with stable isotopes (i.e., chloride, bromide, or iodide), an adduct could form with water through the following negative-ion adduct:

$$^{a}Z^{-} + H_2O \rightarrow \,^{a}Z^{-} \cdot H_2O$$
$$^{b}Z^{-} + H_2O \rightarrow \,^{b}Z^{-} \cdot H_2O$$
$$^{i}Z^{-} + H_2O \rightarrow \,^{i}Z^{-} \cdot H_2O$$

where $a, b, \ldots, i$ are the stable isotopes of Z. The APIMS signal for each water cluster of ^{a}Z, ^{b}Z, $\ldots$, ^{i}Z, or the sum of all masses, could be used for the determination of total water vapor content in a specialty gas. In the case with halogens, use of the natural isotopes could be advantageous to calibration in terms of avoiding interference from other analytes with the same mass.

5.8 Conclusions

APIMS employing point-to-plane corona discharge and ^{63}Ni β ionization can be a very powerful technique for the trace analysis of ppt-level impurities in bulk and electronic specialty gases. Although not discussed in this chapter, other ionization sources for APIMS are plausible. Other β-emitter sources, or even α-emitter sources [65], could be used in place of the ^{63}Ni β source, but care would need to be taken in evaluating the radioactive nature of those alternative emitters. In addition, other atmospheric discharge sources, such as microwave discharge, could be used to focus on atmospheric pressure chemical ionization methods [18]. Finally, alternative instrumentation configurations, such as triple quadrupole MS and time-of-flight MS, can be used in place of the single-quadrupole units described to perform complexation and dissociation of ions as well as to measure much larger masses.

Overall, trace impurity detection in industrial and specialty gases via APIMS has progressed significantly since the introduction of the technique in the 1970s. Although there is room to lower the limit of detection for trace species in inert and bulk gases, this type of APIMS analysis is mature and well understood. Consequently, our efforts in this chapter have been to compile and review much of the independent research that has been sited globally with regard to important analytical aspects of the APIMS technique and gas applications.

In contrast, the area of trace detection in electronic specialty gases is just emerging. Therefore, although various applications (including important source hardware, operating conditions, and ionization mechanisms) have been discussed here, more work is warranted in terms of detection limits, clustering chemistries, and ionization within the matrix gas. Direct analysis of electronic specialty gases for the determination of trace species without dilution is still difficult to achieve, due to the detrimental effects those gases have on components in the APIMS instrumentation, as well as the inherent safety hazards of those gases. It is probably simply a matter of time until novel methods are developed to allow direct electronic specialty gas analysis by APIMS.

REFERENCES

1. Horning, E. C., Horning, M. G., Carroll, D. I., Dzidic, I., & Stillwell, R. N. (1973). New picogram detection system based on a mass spectrometer with an external ionization source at atmospheric pressure. *Analytical Chemistry, 45*, 936–943.

2. Carroll, D. I., Dzidic, I., Stillwell, R. N., Horning, M. G., & Horning, E. C. (1974). Subpicogram detection system for gas phase analysis based upon atmospheric pressure ionization (API) mass spectrometry. *Analytical Chemistry, 46*, 706–710.

3. Horning, E. C., Carroll, D. I., Dzidic, I., Haegele, K. D., Horning, M. G., & Stillwell, R. N. (1974). Atmospheric pressure ionization mass spectrometry: solvent-mediated ionization of samples introduced in solution and in a liquid chromatographic effluent stream. *Journal of Chromatographic Science, 12*, 725–729.

4. Bruins, A. P. (1991). Mass spectrometry with ion sources operating at atmospheric pressure. *Mass Spectrometry Reviews, 10*, 53–77.

5. Covey, T. R., Thomson, B. A., & Schneider, B. B. (2009). Atmospheric pressure ion sources. *Mass Spectrometry Reviews, 28*, 870–897.

6. French, J. B., Thomson, B. A., Davidson, W. R., Reid, N. M., & Buckley, J. A. (1984). Atmospheric pressure chemical ionization mass spectrometry. In F. W. Karasek, O. Hutzinger, and S. H. Safe (Eds.), *Mass Spectrometry in Environmental Sciences.* New York: Plenum Press, pp. 101–121.

7. Siegel, M. W., & Fite, W. L. (1976). Terminal ions in weak atmospheric pressure plasmas: applications of atmospheric pressure ionization to trace impurity analysis in gases. *Journal of Physical Chemistry, 80*, 26, 2871–2881.

8. Mitsui, Y., Irie, T., & Mizokami, K. (1990). Mass spectrometer for ppt-trace analysis. *Ultra Clean Technology, 1*, 1, 3–12.

9. Mitsui, Y., Irie, T., Iijima, S., & Mizokami, K. (1994). Development of a new APIMS for the detection of trace impurities in special gases. In *Proceedings of the 40th Annual Technical Meeting of the Institute of Environmental Sciences.* Mount Prospect, IL: Institute of Environmental Sciences, pp. 246–253.

10. Lee, M. Y. M. (2010). Process control through minimizing variations in gas purity: part 1. Contamination control begins at the point of supply. *Gases and Instrumentation, 4*, 6, 23–29.

11. Gross, J. H. (2011). *Mass Spectrometry: A Textbook*, 2nd Ed. New York: Springer-Verlag.

12. Extrel CMS (2012). *Atmospheric Pressure Ionization Mass Spectroscopy: Operation Principles for Attospec-1.* Pittsburgh, PA: Extrel CMS.

13. Siefering, K., Berger, H., & Whitlock, W. (1993). Quantitative analysis of contaminants in ultrapure gases at the parts-per-trillion level using atmospheric pressure ionization mass spectroscopy. *Journal of Vacuum Science Technology A, 11*, 4, 1593–1597.

14. Kambara, Z. H., Ogawa, Y., Mitsui, Y., & Kanomata, I. (1980). Carbon monoxide detection in nitrogen gas by atmospheric pressure ionization mass spectrometry. *Analytical Chemistry, 52*, 1500–1503.

15. Wilkinson, P. G. (1963). Diatomic molecules of astrophysical interest: ionization potentials and dissociation energies. *Astrophysical Journal, 138*, 778–800.

16. Ketkar, S. N., Dulak, J. G., Fite, W. L., Buchner, J. D., & Dheandhanoo, S. (1989). Atmospheric pressure ionization tandem mass spectrometric system for real-time detection of low-level pollutants in air. *Analytical Chemistry, 61*, 260–264.

17. Ishihara, Y., Umehara, H., Nishina, A., & Kimijima, T. (1997). The analysis of trace impurities in nitrogen gas by high pressure ionization mass spectrometry. *Japanese Journal of Applied Physics, 36*, 6999–7003.

18. Ketkar, S. N., Penn, S. M., & Fite, W. L. (1991). Influence of coexisting analytes in atmospheric pressure ionization mass spectrometry. *Analytical Chemistry, 63*, 924–925.

19. Albritton, D. L. (1978). Ion-neutral reaction-rate constants measured in flow reactors through 1977. *Atomic Data and Nuclear Data Tables, 22*, 1, 1–89.

20. Bandy, A. R. (2004). Determination of trace impurities in ammonia using isotope dilution atmospheric pressure ionization mass spectrometry. In *Semicon West 2004, SEMI Technology Symposium: Innovations in Semiconductor Manufacturing.*

21. Mitchell, G. M. (2001). *Determination of vertical fluxes of sulfur dioxide and dimethyl sulfide in the remote marine atmosphere by eddy correlation and an airborne isotopic dilution atmospheric pressure ionization mass spectrometer.* Retrieved from ProQuest Dissertations and Theses (3004038).

22. Bandy, A. R., Thornton, D. C., Ridgeway, R. G., Jr., & Blomquist, B. W. (1992). Key sulfur-containing compounds in the atmosphere and ocean: determination by gas chromatography–mass spectrometry and isotopically labeled internal standards. In J. A. Kaye (Ed.), *Isotope Effects in Gas-Phase Chemistry.* Washington, DC: American Chemical Society, pp. 409-422.

23. Bandy, A. R., Thornton, D. C. , & Driedger, A. R., III (1993). Airborne measurements of sulfur dioxide, dimethyl sulfide, carbon disulfide, and carbonyl sulfide by isotope dilution gas chromatography/mass spectrometry. *Journal of Geophysical Research, 98*, 23423–23442.

24. Bandy, A. R. (2001). Identification and determination of contamination in industrial gases using negative ion APIMS and isotope dilution. *Gases and Technology, 1*, 18–24.

25. Scott, A. D., Jr., Hunter, E. J., & Ketkar, S. N. (1998). Use of a clustering reaction to detect low levels of moisture in bulk oxygen using an atmospheric pressure ionization mass spectrometer. *Analytical Chemistry, 70*, 1802–1804.

26. Hunter, E. J., Homyak, A. R., & Ketkar, S. N. (1998). Detection of trace nitrogen in bulk argon using proton transfer reactions. *Journal of Vacuum Science Technology A, 16*, 5, 3127–3130.

27. Ridgeway, R. G., Ketkar, S. N., & Martinez de Pinillos, J. V. (1994). Use of atmospheric pressure ionization mass spectrometry to monitor hydrocarbon type impurities in bulk nitrogen. *Journal of the Electrochemical Society, 141*, 9, 2478–2482.

28. Ketkar, S. N., Ridgeway, R. G., & Martinez de Pinillos, J. V. (1992). Ioniimolecule reactions in the source of an atmospheric pressure ionization mass spectrometer. *Journal of the Electrochemical Society, 139*, 12, 3675–3678.

29. Kambara, H., Mitsui, Y., & Kanomata, I. (1979). Identification of clusters produced in an atmospheric pressure ionization process by a collisional dissociation method. *Analytical Chemistry, 51*, 9, 1447–1451.

30. Ridgeway, R. G., Ketkar, S. N., & Martinez de Pinillos, J. V. (1992). APIMS determination of impurities in bulk H_2. *Proceedings of the 38th Annual Meeting of the Institute of*

Environmental Sciences. Mount Prospect, IL: Institute of Environmental Sciences, pp. 68–73.

31. Pusterla, L., Solcia, C., & Succi, M. (1997). Study of critical analysis and possible interference problems by means of APIMS Technology. *Proceedings of the 43rd Annual Meeting of the Institute of Environmental Sciences, Los Angeles, CA, May 4–8.* Mount Prospect, IL: Institute of Environmental Sciences, pp. 336–346.

32. Ma, C., Athalye, A., Fruhberger, B., & Ezell, E. (1998). Moisture dry-down in high purity hydrogen chloride. *Proceedings of the 44th Annual Meeting of the Institute of Environmental Sciences.* Mount Prospect, IL: Institute of Environmental Sciences, p. 285.

33. Siefering, K., & Whitlock, W. (1995). Analysis of ppb-level impurities in silane gas using APIMS: Comparison of purified and unpurified silane products. *Proceedings of the 41st Annual Meeting of the Institute of Environmental Sciences.* Mount Prospect, IL: Institute of Environmental Sciences.

34. Kitano, M., Shirai, Y., Ohki, A., Babasaki, S., & Ohmi, T. (2001). Impurity measurement in specialty gases using atmospheric pressure ionization mass spectrometry with a two-compartment ion source. *Japanese Journal of Applied Physics, 40,* 2688–2693.

35. Ohki, A., & Ohmi, T. (1995). *Measurement and Behavior of Moisture and Siloxane in Silane Gas.* Technical Report SDM 95-106. Tokyo: Institute of Electronics, Information, and Communication Engineers.

36. Gotlinsky, B., O'Sullivan, J., Horikoshi, M., & Babasaki, S. (2000). Eliminating siloxane impurities from silane process gas using next generation purification. *MICRO,* July/August. Retrieved from http://micromagazine.fabtech.org/archive/00/07/gotlinsky.html.

37. Bandy, A. R., Ridgeway, R. G., & Mitchell, G. M. (2005). *Negative ion atmospheric pressure ionization mass spectrometry using a ^{63}Ni electron source.* U.S. Patent No. 6,956,206. Washington, DC: U.S. Patent and Trademark Office.

38. Hill H. H., Jr., & Baim, M. A. (1982). Ambient pressure ionization detectors for gas chromatography: II. Radioactive source ionization detectors. *Trends in Analytical Chemistry, 1,* 232–236.

39. Pellizzari, E. D. (1974). Electron capture detection in gas chromatography. *Journal of Chromatography, 98,* 323–361.

40. Dzidic, I., Carroll, D. I., Stillwell, R. N., & Horning, E. C. (1976). Comparison of positive ions formed in nickel-63 and corona discharge ion sources using nitrogen, argon, isobutane, ammonia and nitric oxide as reagents in atmospheric pressure ionization mass spectrometry. *Analytical Chemistry, 8,* 1763–1768.

41. Carroll, D. I., Dzidic, I., Stillwell, R. N., & Horning, E. C. (1975). Identification of positive reactant ions observed for nitrogen carrier gas in plasma chromatograph mobility studies. *Analytical Chemistry, 47,* 1956–1959.

42. Grimsrud, E. P., Kim, S. H., & Gobby, P. L. (1979). Measurement of ions within a pulsed electron capture detector by mass spectrometry. *Analytical Chemistry, 51,* 223–229.

43. Zook, R. D. (1990). *Mass spectrometric sampling from the atmospheric pressure ion source.* Retrieved from Proquest Dissertations and Theses (9109705).

44. Good, A., Durden, D. A., & Kebarle, P. (1970). Ion–molecule reactions in pure nitrogen and nitrogen containing traces of water at total pressures 0.5–4 torr: kinetics of clustering reactions forming $H^+ (H_2O)_n$. *Journal of Chemical Physics, 52*, 212–232.

45. Kambara, H., & Kanomata, I. (1977). Determination of impurities in gases by atmospheric pressure ionization mass spectrometry. *Analytical Chemistry, 49*, 2, 270–275.

46. Stimac, R. M., Cohen, M. J., & Wernlund, R. F. (1982). Water vapor measurements in small volumes using atmospheric pressure chemical ionization–mass spectrometry. *20th International Reliability Physics Symposium*. New York: Electron Devices and Reliability Societies of the Institute of Electrical and Electronics Engineers, pp. 260–263.

47. Hunter, E. P. L., & Lias, S. G. (1998). Evaluated gas phase basicities and proton affinities of molecules: An update. *Journal of Physical and Chemical Reference Data, 27*, 3, 413–656.

48. Sunner, J., Nicol, G., & Kebarle, P. (1988). Factors determining relative sensitivity of analytes in positive mode atmospheric pressure ionization mass spectrometry. *Analytical Chemistry, 60*, 13, 1300–1307.

49. Arnold, S. T., Thomas, J. M., & Viggiano, A. A. (1998). Reactions of $H_3O^+(H_2O)_n$ and $H^+(H_2O)_n(CH_3COCH_3)_m$ with CH_3SCH_3. *International Journal of Mass Spectrometry, 179–180*, 1/3, 243–251.

50. Sunner, J., Ikonomou, M. G., & Kebarle, P. (1988). Sensitivity enhancements obtained at high temperatures in atmospheric pressure ionization mass spectrometry. *Analytical Chemistry, 60*, 13, 1308–1313.

51. Albritton, D. L., Dotan, I., Streit, G. E., Fahey, D. W., Fehsenfeld, F. C., & Ferguson, E. E. (1983). Energy dependence of the O^- transfer reactions of O_3^- and CO_3^- with NO and SO_2. *Journal of Chemical Physics, 78*, 6614–6620.

52. Benoit, F. M. (1983). Detection of nitrogen and sulfur dioxides in the atmosphere by atmospheric pressure ionization mass spectrometry. *Analytical Chemistry, 55*, 13, 2097–2099.

53. Eisele, F. L., & Berresheim, H. (1992). High-pressure chemical ionization flow reactor for real-time mass spectrometric detection of sulfur gases and unsaturated hydrocarbons in air. *Analytical Chemistry, 64*, 3, 283–288.

54. Möhler, O., Reiner, T., & Arnold, F. (1992). The formation of SO_5 by gas phase ion-molecule reactions. *Journal of Chemical Physics, 97*, 11, 8233–8240.

55. Arnold, S. T., Morris, R. A., Viggiano, A. A., & Jayne, J. T. (1995). Ion chemistry relevant for chemical ionization detection of SO_3. *Journal of Geophysical Research, 100*, 14141–14146.

56. Mitchell, G. M., Vorsa, V., Milanowicz, J. A., Ragsdale, D. J., Marhefka, K. L., Wagner, M. D., Ketkar, S. N., Booth, T. M., & Conway, T. E. (2003). Trace impurity detection in ammonia for the compound semiconductor market. *Gases and Technology, 2*, 4, 8–13.

57. Mitchell, G. M., Vorsa, V., Ryals, G. L., Milanowicz, J. A., Ragsdale, D. J., Marhefka, K. L., & Ketkar, S. N. (2002). Trace impurity detection in ammonia for the compound semiconductor market 2002. *SEMI Technology Symposium: Innovations in Semiconductor Manufacturing*.

58. Funke, H., Raynor, M., Yeucelen, B., & Houlding, V. (2001). Impurities in hydride gases: 1. Investigation of trace moisture in the liquid and vapor phase of ultra-pure ammonia by FTIR spectroscopy. *Journal of Electronic Materials, 30*, 11, 1438–1447.

59. Funke, H. H., Grissom, B. L., McGrew, C. E., & Raynor, M. W. (2003). Techniques for the measurement of trace moisture in high-purity electronic specialty gases. *Review of Scientific Instruments, 74*, 9, 3909–3933.

60. Funke, H. H., Welchhans, J., Watanabe, T., Torres, R., Houlding, V. H., & Raynor, M. W. (2004). Impurities in hydride gases: 2. Investigation of trace CO_2 in the liquid and vapor phases of ultra-pure ammonia. *Journal of Electronic Materials, 33*, 8, 873–880.

61. Bandy, A. R., Tu, F. H., Mitchell, G. M., & Ridgeway, R. G. (2002). Determination of moisture in anhydrous ammonia at the ppbv level using negative ion APIMS. *SEMI Technology Symposium: Innovations in Semiconductor Manufacturing.*

62. Mitchell, G. M., Vorsa, V., Milanowicz, J. A., Ragsdale, D. J., Marhefka, K. L., Wagner, M. D., Ketkar, S. N., Booth, T. M., Conway, T. E. (2003). Trace impurity detection in ammonia for the compound semiconductor market. *2003 Eastern Analytical Symposium, Somerset, NJ.*

63. Nishina, A., Umehara, H., & Kimijima, T. (1999). *Method for analyzing impurities in aas and its analyzer.* U.S. Patent No. 6,000,275. Washington, DC: U.S. Patent and Trademark Office.

64. Bandy, A. R. (2004). Personal communication.

65. Ketkar, S. N., & Dheandhanoo, S. (2003). *Total impurity monitor for gases.* U.S. Patent No. 6,606,899 B1. Washington, DC: U.S. Patent and Trademark Office.

CHAPTER 6

GC/MS, GC/AED, AND GC–ICP–MS ANALYSIS OF ELECTRONIC SPECIALTY GASES

DANIEL COWLES,[1] MARK W. RAYNOR,[2] AND WILLIAM M. GEIGER[3]

[1]Air Liquide Balazs NanoAnalysis, Dallas, Texas
[2]Matheson, Advanced Technology Center, Longmont, Colorado
[3]Consolidated Sciences, Pasadena, Texas

6.1 Introduction

Gases used in the manufacture of semiconductor devices can be divided roughly into two groups: electronic specialty gases (ESGs) and bulk gases. *ESGs* (sometimes referred to as *specialty gases* or *spec. gases*) include halogens (e.g., fluorine, chlorine, bromine, chlorine trifluoride), hydrogen halides (e.g., hydrogen chloride, hydrogen bromide, hydrogen fluoride), halocarbons [e.g., perfluorocarbons (PFCs), chlorofluorocarbons (CFCs), and hydrochlorofluorocarbons (HCFCs)], hydrides (silane, diborane, phosphine, arsine, ammonia, germane, hydrogen selenide), and organometallic molecules. *Bulk* (or *inert*) *gases* include the noble elements (helium, argon, neon, xenon, krypton) as well as homonuclear diatomics (hydrogen, nitrogen, oxygen). The distinguishing characteristic of ESGs is their reactivity. Due to their weaker bond energies, ESGs can readily participate in chemical reactions, often at moderate temperatures and without assistance from an electrical discharge. This reactivity makes ESGs essential actors in many steps of chip making, from production of polysilicon

Trace Analysis of Specialty and Electronic Gases.
By William M. Geiger and Mark W. Raynor. Copyright © 2013 John Wiley & Sons, Inc.

to metal-layer and chemical vapor deposition, to etching, to chamber cleaning. By comparison, bulk gases are more often used for dilution, heat transfer, chemical transport, and inerting. Unsurprisingly, the reactivity of ESGs, which makes them useful as manufacturing reagents, also creates handling and analysis headaches. Many ESGs react immediately with water vapor and can corrode metal containment materials. Some ESGs are highly flammable and, in some cases, pyrophoric. Others are relatively inert at room temperature and ambient pressure but are highly reactive at elevated temperatures or in a plasma environment (e.g., nitrogen trifluoride). Finally, the majority of ESGs have boiling-point temperatures at or above room temperature (Figure 6.1) and are delivered as two-phase, liquefied gases in cylinders or tanks. The headspace of a liquefied gas container is fully wetted by the condensed gas. Liquefied-gas ESGs, such as ammonia, present additional handling challenges, such as unintentional re-liquefaction of the gas in the sampling manifold.

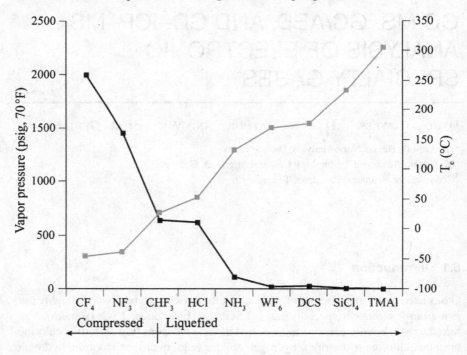

Figure 6.1   Vapor pressures and critical temperatures for selected specialty gases.

Inevitably, gas chromatography/mass spectrometry (GC/MS), gas chromatography–atomic emission detection (GC/AED), and gas chromatography–inductively coupled plasma–mass spectrometry (GC–ICP–MS) detection can come into play as tools for measurement of the impurities in both ESGs and bulk gases. These technologies have both qualitative and quantitative capabilities. GC/MS, GC/AED, and GC–ICP–MS all share important common and complementary features. Obviously, all three techniques use a GC as a front end, and it is this feature that does what one might call the heavy lifting. Much of the early work using mass spectrometry and some ICP–MS work were done by direct analysis, but review of the literature demonstrates coupling, with gas chromatography following shortly [1]. The problem with direct analysis is loss of information or speciation of impurities at low concentrations. The three detection techniques also use some form of ionization prior to final detection. Finally, the detectors used are all "smart" in their own way. In many cases, the mass spectrometer in GC/MS not only provides clues to the structure of the compound measured using conventional electron impact ionization or some form of chemical ionization, but also provides key molecular weight information. Due to its nature, the mass spectrometer can provide its own diagnostic information regarding atmospheric leaks, detector aging, and so on. The AED has the ability to identify the elemental composition of a given chromatographic peak (i.e., whether it contains carbon, hydrogen, chlorine, or many other elements of the periodic table). Carefully calibrated, it can be used to measure these ratios accurately such that an empirical formula can be determined. The ICP–MS as a detector can measure many, but not all, of the elements that the AED can. It is a true atomic or elemental detector. Due to its very high ionization efficiency, in most cases it can make these measurements several orders of magnitude lower than those of other techniques. Both GC/MS and GC–ICP–MS measurements can also provide reasonable isotopic data.

6.2 GC/MS

GC/MS is a mature technology that has evolved over the past five decades to occupy its current position as the preeminent workhorse for environmental and industrial testing where speciation, unequivocal identification, and quantification of trace-molecular contaminants are required [2]. GC/MS represents the coupling of two unrelated hardware systems: a GC to effect a temporal separation of a complex mixture, and a mass-selective detector (MSD) to provide a mass-based signature for each of the eluting compounds. The GC/MS provides two independent sets of data regarding the identity of an eluting molecule: its retention time on the GC column, and its mass fragmentation pattern. Depending on the application, the "MS" part of GC/MS can be selected from a number of mass-separation methods: quadrupole (QMS), time-of-flight (TOF), magnetic sector, $v \times B$ velocity filters, ion traps, and tandem mass detectors that combine multiple stages of mass selection (e.g., MS–MS). In this chapter we focus on GC/MS incorporating a single-stage quadrupole mass filter and an electron impact ionization (EI) source. This configuration is widely accepted in environmental and industrial laboratories, due to its combination of manufacturability, reasonable price, acceptable resolution and mass range, and small footprint. Modern GC/MS systems incorporating a QMS can be operated by nonexperts once the methods have been established.

There are, of course, many detectors that can be utilized for quantification of volatile impurities in specialty gases. However, GC/MS possesses the following advantages over other, non-MS detectors:

1. Unambiguous identification of contaminants, based on elution time and mass-fragmentation pattern. Competing "universal" detectors produce impurity peaks but provide no additional data, aside from elution time, to assign them. This key advantage for troubleshooting, analysis of reclaimed gas, and GC method development cannot be overstated.

2. A single GC/MS can serve as a gold-standard method development tool from which cheaper, non-MS production methods can be spawned.

3. The MS detector can resolve, by mass, co-eluting contaminants, which may be difficult to resolve by GC alone, without resorting to oven cooling.

A very detailed explanation of GC/MS is beyond the scope of this chapter. For more detailed discussion, various textbooks and tutorials are available online and in print [2].

6.2.1 Gas Chromatography

A gas chromatograph is a highly sophisticated oven whose temperature can be ramped rapidly and accurately under computer control, along with onboard sample gas and carrier gas introduction hardware, a gas-phase separation column, and a computer interface (Figure 6.2). The objectives of the GC system are to separate an injected specialty gas mixture and deliver some or all of its components to the MS detector in a preferred order. As discussed later, the success or failure of a GC setup for analysis of specialty gases hinges on judicious selection of the column(s) configuration, as well as correct material choices. ESGs are introduced to a GC via a gas sampling valve (GSV), which is a two-position multiport valve and sample loop, shown in Figure 6.3.* The body of the GSV is typically 316L stainless steel, although highly corrosion-resistant materials such as nickel, Hastelloy C-22, or other alloys, or even polymers, can be substituted. The rotor, fashioned from a fluoropolymer or fluoropolymer–graphite composite, has slots machined to allow passage of gas from one valve port to the next. For a two-position valve, the valve rotor must rotate smoothly between the two positions while maintaining a leaktight seal. As with the valve body, the rotor material for the targeted ESG must be selected with care, taking into account material compatibility, specified temperature range, permeability (with respect to air and moisture), and durability. The valve and sample loop inject a fixed volume of the sample gas into the carrier gas stream, which then carries the sample plug to the GC column for separation. For gases that are highly reactive with respect to oxygen and moisture, it may be necessary to enclose the GSV in a homemade or commercial enclosure, which is then actively purged with a dry gas such as nitrogen. Without this, reaction products can generate particulate and corrosion issues that result in leakage of air around the valve rotor. For specialty gas analysis, the GC separation method

---

*If a GC is not equipped with a GSV, but instead with a split-splitless injector, gas samples can be injected manually using a gas tight syringe. The GSV approach is preferred for reasons of reproducibility and to minimize exposure of reactive gases to air and moisture.

is most critical. Whereas much of the operation of modern mass spectrometers has been automated, gas chromatographic aspects still demand considerable attention by the analyst.

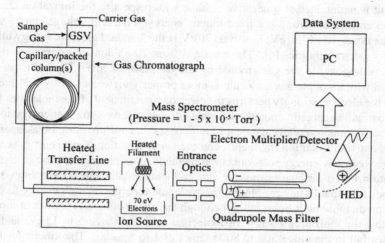

Figure 6.2 Principal components of a GC/MS system.

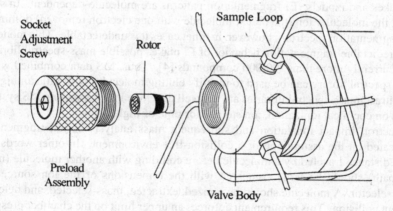

Figure 6.3 Gas sampling valve showing order placement of rotor and external socket adjustment screw. Courtesy of Valco Instruments Co., Inc. [3].

6.2.2 Mass Spectrometry

Electron Impact Ionization (EI) All versions of mass spectrometry are comprised of an initial ionization step followed by manipulation of the ions by electric and/or magnetic fields. The ionization step is crucial. It converts uncharged molecules that are practically unguidable into charged molecules or ion fragments that can be steered, sorted, focused, and even stored by electromagnetic fields. In an EI source, a wire filament typically of rhenium, tungsten, or iridium, with an electric current through it is heated, and electrons are produced by thermionic emission. The filament is maintained at a negative voltage with respect to the ionization chamber to repel emitted electrons at a fixed kinetic energy. Filament voltage can be varied over the range 10 to 150 eV; however, 70 eV is the standard electron energy utilized in mass-spectral libraries [4]. The electrons from 70-eV ionization traverse the ion source orthogonal to the gas emitted by the GC transfer line. The collision of an electron with a sample gas molecule is more properly viewed as a "fly-by" event, in which the electron strongly perturbs the electromagnetic field of the molecule. If this perturbation is energetic enough, the molecule can receive sufficient energy to eject an electron, undergo rearrangement to a more stable conformation, break apart into neutral and molecular ion fragments, and even emit radiation. On an energy scale, EI falls between low-energy soft-ionization methods (e.g., photoionization or chemical ionization based on charge transfer and collisional stabilization) and high-energy methods (plasma discharge). In EI, the molecule receives enough energy to cleave along predictable fault lines, but not enough energy to randomly shatter or atomize. Note that the ionization energy for molecules is in the range 5 to 15 eV, and bond energies fall in the range 300 to 800 kJ/mol (3.1 to 8.3 eV). The kinetic energy of 70-eV electrons far exceeds that required to ionize and break bonds in a molecule. In EI, 70-eV electrons produce similar fragmentation patterns for mass analyzers of all makes and models. EI fragmentation patterns are molecule-dependent. In some cases, the molecular ion (the parent molecule with one electron removed) dominates the fragmentation spectrum; however, in other cases it is undetectable. This moderate and predictable fragmentation behavior of EI makes possible mass-spectral libraries with currently more than 200,000 compounds [4]. Thus, MS data combined with a mass-spectral library can be used to identify eluting molecules with little ambiguity. In contrast, mass-spectral libraries are not well developed for competing MS systems based on chemical ionization, ion storage, or time of flight.

Electron impact ionization and subsequent mass analysis of ion fragments is predicated on the assumption of a collision-free environment. In other words, the mean distance traveled by a molecule before colliding with another molecule (mean free path) should be large compared with the dimensions of the ion source and mass selector. A molecule should be ionized, extracted, mass-selected, and detected without collision. This requirement enforces an upper limit on the chamber pressure. For a nitrogen molecule, the mean free path is approximately 5 cm/mtorr. For a typical mass analyzer chamber pressure of 10^{-4} to 10^{-5} torr, the mean free path of a nitrogen molecule is 50 to 500 cm. Even if the pressure at the outlet of the transfer line in the ion source reaches 10^{-3} torr, the mean free path is still 5 cm, long enough for molecules to ionize, fragment, and exit the ion source largely free of collisions.

In EI mode the sensitivity of a MS detector can vary greatly. Factors include the ionization potential (IP) of a given compound, background interference, acquisition speed, mass range selected for scanning, and column size (capacity). Using ion monitoring for hydrocarbons, GC/MS can be much more sensitive than a flame ionization detector (FID); however, it would pale in comparison to an electron capture detector (ECD) for a compound amenable to ECD.

Quadrupole Mass Spectrometer QMSs utilize a quadrupole mass filter, which is an arrangement of four rods, with either round or hyberbolic surfaces aligned precisely along the z-axis (Figure 6.2). Mass selection is accomplished only with electric fields; magnets are not utilized. A combination of direct-current (DC) and radio-frequency (RF) voltages is applied to the rods and ramped following a computer-controlled algorithm. Positive ions passing through the mass filter execute different trajectories, depending on their mass. The quadrupole acts as a bandpass filter, allowing only ions within a narrow mass range to pass on to the detector stage. Nonselected ions are lost to the quadrupole rods, on the entrance or exit optics, or are simply pumped away. Mass-filtered ions are deflected off-axis, accelerated to several kilovolts, and detected by an electron multiplier. Instruments available today typically have several auto-tune options that are available to the analyst. Selection is dependent on analytical requirements of type of analysis targeted, such as EPA methods, maximum sensitivity over a desired mass range, or mimicking of magnetic mass spectra.

Scan Mode Versus SIM Mode QMSs are generally operated in one of two modes: scan mode or selected ion monitoring (SIM) mode. In scan mode, the voltages on the quadrupole elements and ion optics are ramped continuously over a specified mass range. Depending on the mass range selected and intrinsic scan speed of the analyzer, a complete mass scan covering the entire selected mass range may take 10 to 100 ms. Depending on the scan speed and anticipated peak width, two or more individual scans are co-averaged and only the mean result is recorded. Mass scanning and data recording continue as impurity peaks elute from the GC column. At the conclusion of the run, the total ion current (TIC) chromatogram can be plotted. This is a plot of the summed ion intensity as a function of time. The GC/MS TIC chromatogram is comparable to signal versus time plots produced by "universal" detectors, such as flame ionization (FID), thermal conductivity (TCD), or helium ionization detector (HID). However, for each time increment in the TIC chromatogram (e.g., every 0.1 to 0.5 second), a complete mass spectrum is stored and can be plotted to see the distribution of ions contributing to the TIC at that moment. Scan-mode runs will detect any compound delivered by the GC, as long as that compound produces ion fragments within the mass range selected and assuming that the concentration is above the detection limit (DL). Scan-mode GC/MS analyses are particularly valuable when the impurities are not known beforehand.

In SIM mode, scans are produced by specifying a list of ion masses to be monitored, and programming the MS to "hop" rapidly from mass to mass throughout the GC run. The advantage of the SIM mode is a large improvement in signal-to-noise ratio due to the MS accumulating signal only for the target masses. SIM mode is used to quantify known impurities with the highest achievable sensitivity. Modern GC/MS systems can also be programmed to run using a combination SIM/scan mode [5].

The enhanced sensitivity of SIM versus scan modes can be demonstrated by this example: Analysis of a 10-ppm argon standard using SIM for mass-to-charge ration

(m/z) of 40 with three acquisitions per second yields a DL of 15 ppb. Scanning from m/z 30 to 60 yields a DL of 105 ppb. Increasing the scan range to m/z 30 to 260 yields a DL of 264 ppb.

Resolution and Mass Range   Quadrupole-based mass analyzers provide moderate mass resolution in a compact package. The resolution is a function of the time spent by ions in the filter which is dictated by the length of the filter, as well as the ion beam energy. An Agilent 5973 MSD has a specified mass range of 1.6 to 800 amu and achieves unit mass resolution over the entire range. The mass resolution can be improved slightly at the cost of reduced sensitivity. In practice, the MSD requires special tuning, and/or alteration of the hardware, for detection of very low-mass ion fragments (e.g., detection of hydrogen).

Chromatography for GC/MS of Specialty Gases   GC/MS is the method of choice for a wide range of applications, including environmental monitoring, drug testing, and contamination investigation. In these applications, GC/MS is often employed to resolve and quantify complex mixtures. For example, *U.S. EPA Water Method 524.2 Rev. 4.1* specifies more than 70 volatile compounds [6]. Complex mixtures produce rich, "inverted hairbrush" chromatograms of closely eluting and overlapping peaks. Fortunately, the impurity profiles of specialty gases are far simpler. This is true because specialty gases are manufactured via chemical reactions involving only two or three chemical precursors, each of which is relatively pure to begin with. In addition, the manufacturing process includes one or more stages of purification. Impurities in specialty gases may originate from the source reagents, the manufacturing process, interactions with the packaging material of the gas produced, or from spontaneous degradation of the gas itself. For example, diborane gas mixtures degrade due to self-reaction of diborane molecules to form higher boranes. Due to the relative simplicity of the specialty gas GC/MS chromatograms, as well as to the fact that EI cleaves molecules but does not atomize them, the unit mass resolution afforded by typical QMS analyzers is more than sufficient to identify and speciate most volatile impurities in the gas. Far more expensive and space consuming, magnetic sector, TOF, and ion trap analyzers are overengineered for this application.

Impurities in specialty gases can be broken down into the following main categories:

1. Air components (i.e., hydrogen, nitrogen, oxygen, argon, carbon dioxide, carbon monoxide, and methane)

2. Gas-specific impurities (idiosyncratic impurities related to the specific manufacturing method of the gas, or to by-products of reaction of the gas with moisture or oxygen)

3. Moisture

4. Metals

GC/MS typically addresses the first two categories. Metals and moisture (water vapor) are handled separately. A GC method for an ESG must be tailored to separate the air components and gas-specific impurities from the ESG itself, as well as to take into account the ESG's specific material compatibility requirements. For liquefied gases (see examples in Figure 6.1), a combination of heat and pressure reduction must be employed to deliver the sample gas to the GC at a temperature well above its dew-point temperature. The boiling-point temperature and polarity of an ESG and its volatile impurities dictate the elution order on a GC column. A decision must be made regarding the separation strategy. Typical approaches include:

1. Single column without back-flush: The gas sample is separated on a single column, and the run is complete when all components have eluted. This approach is best suited to "well-behaved" specialty gases: relatively unreactive, low to moderate polarity, and low to moderate boiling-point temperature. In this case, all the components of the injected gas sample are delivered to the MS ion source. As in all GC/MS methods, the ion source filament and electron multiplier detector voltage should be turned off temporarily during elution of matrix peaks.

2. Two column: In this approach a precolumn separates components based on their polarity, and the composite air peak is separated on a second column (typically, molecular sieve MS 5A). A switching valve can be interposed between the columns to make non-air peaks bypass the second column and pass directly to the detector. Alternatively, simply by altering the valve timing, only the precolumn needs to be employed in the separation. This approach was taken for the GC analysis of trifluoromethane in Figure 6.4.

3. Back-flushing: For late-eluting (high-boiling or polar) specialty gases, analysis time can be reduced substantially by waiting for the elution of lower-boiling impurities, and then reversing the flow of carrier gas in the main column and venting the high-boiling component(s) to the scrubber, without allowing them to make their way slowly through the column. An additional benefit of this strategy is that the high-boiling components do not enter, and potentially pollute, the ion source of the MS.

4. Heart-cut: A heart-cut GC method can deliver both low- and high-boiling components to the MS while selectively discarding the undesired matrix peak. Alternatively, this strategy can be employed to discard all components except for a selected group in the middle of the elution order.

Many GC column vendors provide examples of these strategies. Valco Instruments Company, Inc., in particular, provides an excellent summary of basic column/valve arrangements [3].

GC Column Material Selection   ESGs cover a broad range of reactivities, polarities, and boiling-point temperatures. In addition, the list of target impurities may vary with the end user. Therefore, it is not practical to recommend a particular column material or GC configuration for an ESG. Further, specific methods developed by individual industrial gas plants and labs are typically held confidential. However,

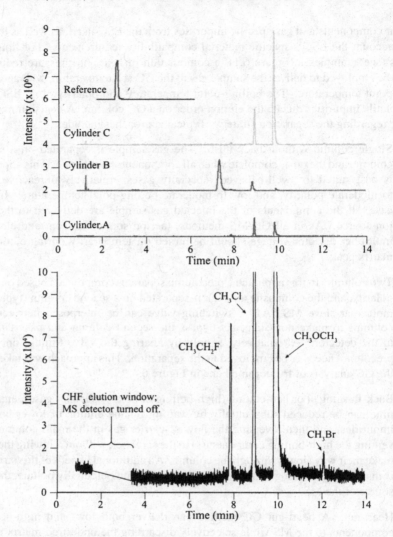

Figure 6.4   GC/MS analysis of trace contaminants in trifluoromethane cylinders.

in general, low-bleed columns (suitable for MS) with the minimum necessary film thickness and resolving power to effect the desired separation should be chosen. Fortunately, one can glean useful information through literature and Web-based searches. Column selection is discussed further in Chapter 8 as well as Subsection 6.4.8 of this chapter. Care must be taken with regard to column selection, as reactions can potentially occur between the sample matrix gas and the stationary phase. Such reactions can generate species that could be identified mistakenly as being in the sample, but are not actually present.

Capillary Versus Packed GC Columns   Either capillary or packed columns can be used for GC/MS analysis of specialty gases. Packed columns offer a wider variety of materials, as well as higher loading capacities, which may translate to increased dynamic range. Capillary columns provide much higher separation efficiency (i.e., sharper peaks in a shorter analysis time) and also generally have a lot less bleed than that of packed columns, due to immobilization of the stationary phases by cross-linking. Capillary and packed columns are generally not interchangeable, and the choice is usually dictated by the existing hardware. For example, a GC/MS configured for capillary columns may not accommodate the much higher carrier gas flow rates of packed columns. Small diffusion pumps found in benchtop GC/MS systems have flow-rate upper limits of around 2 mL/min; small turbo pumps can tolerate up to approximately 4 mL/min. However, neither of these small pumps will accommodate conventional packed columns, which typically operate with carrier gas flow-rates exceeding 10 mL/min. Porous-layer open tubular (PLOT) capillary columns, which combine traditional GC packed column materials with the advantages of capillary columns, are highly recommended. Packed, capillary, or micro-packed columns are other options that enable the use of low flow rates.

Use of Isotopes for GC/MS Calibration   Isotopically enriched chemicals are widely employed in liquid chemical GC/MS analyses, most often as internal standards. The utility of the isotope is that it mimics the behavior of the target analyte under the exact conditions of the analysis while standing apart in the GC/MS specific ion chromatogram, due to the mass offset of some or all of its mass fragments. Hundreds of chemicals are available which have been substituted, or enriched, by stable isotopes. The most common isotope chemicals substitute the elements hydrogen, nitrogen, carbon, and oxygen: ^2H for ^1H, ^{13}C for ^{12}C, ^{15}N for ^{14}N, and ^{17}O or ^{18}O for ^{16}O. A single added neutron separates these pairs. Poor analytical reproducibility, due to loss of analytes in sample preparation, imperfect vaporization or splitting in a split–splitless injector, or thermal decomposition, can be corrected by spiking the sample with an isotopically doped version of the target analyte.

Application of isotopes to specialty gas analysis is not commonly practiced, mainly because it is usually unnecessary and, in any case, is difficult to accomplish. Specifically: (1) ESGs are sampled and analyzed directly, without intervening sample preparation, thus reducing the opportunities for analyte loss; (2) injection of a gas sample via a GSV/sample loop is an inherently more reproducible process than are the injection, vaporization, and splitting associated with GC/MS of liquid samples; and (3) gas calibration standards are bulky and expensive compared with liquid-phase reagents. If it is necessary to introduce an internal standard in a gas-phase ESGs analysis, one option is the dual-sample-loop configuration using a 10-port GSV. This is discussed in more detail later.

Example Applications   Trifluoromethane (or Halocarbon 23) is a compressed fluorocarbon gas utilized in the electronics industry for plasma etching and chamber cleaning. The critical temperature of trifluoromethane is 25.6 °C, and therefore a liquid phase may or may not be present in the cylinder, depending on the cylinder temperature. Specified impurities include moisture, acids, air components (nitrogen,

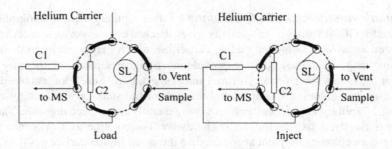

Figure 6.5   Injection valve with sample loop (SL), configured for two-column sequence reversal; load setup on left, inject setup on right. Column C1 separates composite air peak from carbon dioxide and the ESGs matrix gas. Column C2 breaks down the composite air peak into its constituent molecules. In this example, carbon dioxide arrives at the detector first, followed by the air components. The arrival time of the ESG is gas-dependent.

oxygen, argon, carbon monoxide, carbon dioxide), and other halocarbons. The typical purity of electronics-grade trifluoromethane is around 99.999 % (i.e., the overall volatile impurity budget is 10 ppmv). Three trifluoromethane cylinders were supplied by different companies to identify the impurities and their concentrations. For this analysis, the two-column GC configuration discussed in Section 6.2.2 was used (Figure 6.5).* The chromatograms obtained for the cylinders, along with a chromatogram obtained for a ppm-level trifluoromethane gas standard, can be seen in Figure 6.4. The bottom chromatogram zooms in on cylinder A to detail the region from 1.7 to 3.5 min, the elution window of the trifluoromethane matrix, during which the ion source and MS detector were turned off. Figure 6.4 demonstrates one key advantage of GC/MS for specialty gases: It affords unequivocal identification of impurities, some of which could not have been predicted. Although many of the impurities observed are plausible based on our knowledge of the trifluoromethane manufacturing process, we could not have confidently assigned the peaks based on their elution times alone. The mass fragmentation data embedded in each peak, combined with library matching software, assigned each of the peaks with high confidence. Although a universal-type detector such as a HID could have been used instead of the MS to detect the eluting species, it would not have been possible to assign identification. For a well-characterized gas stream, for example in a specialty gas production plant, a GC/MS could be used initially to identify all the potential impurities, and a less expensive low-maintenance universal detector could be swapped in later.

Resolution of Co-eluting Compounds   Molecular sieve columns have traditionally been used to separate the so-called "permanent" gases: nitrogen, argon, oxygen, methane, and carbon monoxide. A weakness of these columns has been their inabil-

---

*Note that in this case the injection valve was actuated and reset more rapidly than usual, so that the gas sample injected made only a single pass through column 1 before eluting directly to the MS.

ity to achieve adequate separation of argon and oxygen, even at room temperature. Analysts have therefore resorted to cooling the GC oven to subambient temperatures; others have given up and simply report a combined result for "Ar + O." To be sure, column manufacturers now offer improved molecular sieve columns that do achieve baseline separation; however, the end user is still saddled with a constraint on oven temperature and may lose valuable analysis time waiting for the oven to cool down between runs. GC/MS solves this problem neatly by taking advantage of the differing EI fragments of argon and oxygen: Argon produces a dominant positive ion at 40 amu, and oxygen produces ions at 32 and 16 amu. Therefore, simply by running the MS in SIM mode during the elution window for argon/oxygen, these molecules can be separated perfectly. Figure 6.6 illustrates the unresolved argon/oxygen peak (top) and the ion-extracted signals (middle and bottom). Although this example is somewhat simplistic, the same logic can be applied to more complex specialty gas analyses troubled by difficult separations.

Use of GC/MS for Rare Gas Impurity Quantification   In this application, GC/MS was used to quantify traces of krypton and xenon in argon gas to sub-ppbv concentrations. End users of high-purity argon gas for ICP–MS require low concentrations of krypton and xenon to avoid isobaric interferences with some elements [7–9]. Although argon is not a specialty gas, this example highlights a perhaps subtle advantage of GC/MS compared with competing universal-type detectors: background purity confirmation. The problem in this case is that GCs typically use helium as the carrier gas. The helium carrier gas is purified upstream of the GC using solid adsorbents or a heated purifier or getter. Some contaminants cannot be removed by this approach. Specifically, the other rare gases—argon, krypton, and xenon—are not removed from helium. Even worse, because the detector responds to all impurities, it cannot demonstrate gas purity in an absolute sense but only in a relative sense. The end user of the purified helium has no way to determine the absolute concentrations of krypton and xenon, or of any other contaminant for that matter. Therefore, if the helium carrier gas contains 10 ppbv of krypton impurity, the analysis will necessarily under-report krypton by the same amount. The MS detector, however, can be tuned to a specific ion fragment of krypton, and an absolute background of krypton in the purified helium can be measured and quantified by calibration. Figure 6.7 displays a chromatogram for a standard mixture of 5 ppmv each of krypton and xenon, along with 50 ppmv of neon, which has been further diluted approximately 100-fold in an argon matrix. For krypton and xenon in argon, reporting limits of less than 0.5 ppbv each were achieved.

6.3 GC/AED

Atomic emission detection (AED) is based on a helium microwave-induced plasma that converts species to their elemental state; the emission from this process is followed by an optical measurement. The first commercial instruments were introduced in the mid-1970s by Applied Chromatography Systems, Ltd. At that time they were

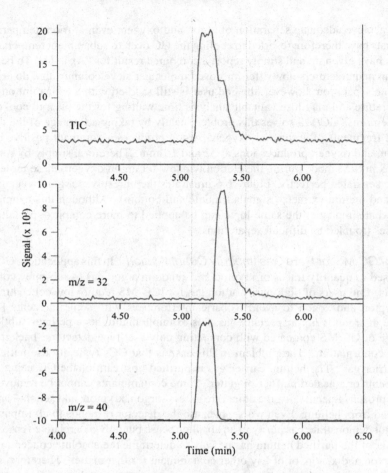

Figure 6.6   GC/MS scans demonstrating mass separation of co-eluting argon and oxygen peaks.

supplied without a data system. Lack of application publications and interest at that time brought production to a halt. Development continued and later Hewlett-Packard (now Agilent Technologies, Inc.) introduced the HP 5921A and later the G2350A, increasing interest and popularity [10,11]. The AED available today is manufactured by Joint Analytical Systems GmbH.

The AED utilizes a microwave-induced helium plasma. Compounds eluting from the chromatographic column enter the plasma, emitting light that is characteristic of each element present [12]. The light from the plasma discharge tube is sent through a grating system, where it is separated so that the intensity at various component wavelengths can be measured using a photodiode array detector (PDA) [13]. The computer-controlled spectrometer provides automatic wavelength calibration and measurements of intensity, wavelength, and linewidth. The PDA detects emissions

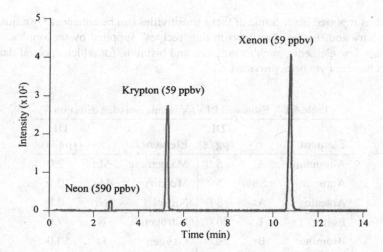

Figure 6.7 SIM-mode chromatogram demonstrating separation and sensitivity for selected rare gases. GC/MS: Agilent 6890/5973; capillary columns in series: (1) 30 m × 0.53 mm ID Poraplot Q, (2) 30 m × 0.53 mm ID MS5A PLOT; temperature: 35 °C, 5 min to 100 °C at 40 °C/min, hold 9 min; selected ion mode: 20, 84 and 129 amu.

at selected combinations of diodes, which are recorded to form the signal due to the analyte. It also collects emissions from another combination of diodes to form the background portion of the chromatogram. The analyte and background signals are subtracted to obtain the final chromatogram for that species. Figure 6.8 illustrates the plasma source and spectrometer in detail. Construction of the plasma cavity and attendant pneumatics allow for venting of the solvent or matrix gas peak. This is a critical feature since the small size and power of the plasma limit how much sample can be introduced without fouling the plasma tube.

The AED is capable of measuring any element whose IP is below that of helium (24.6 eV), making it a universal detector. It can be considered a trace detector for many elements and possibly an ultratrace detector for a few. For instance, it is as sensitive to sulfur species as are other sulfur-specific detectors [14], and is roughly an order of magnitude more sensitive for carbon than is an FID. When measuring sulfur species, it is less prone to signal quenching caused by the other portions of the molecule as is sometimes encountered with other sulfur-specific detectors. When measuring carbon, unlike an FID, a carbon–hydrogen bond is not necessary for a response. Therefore, the AED response to equal molar amounts of carbon monoxide, methane, and carbon dioxide is the same, whereas the response of ethane to methane would be double, on a molar basis. The downside to this is that carbon monoxide as a matrix sample would not be transparent to the detector in an analysis determining light hydrocarbon impurities. Table 6.1 lists estimated DLs for some elements that can be measured using AED. It should be noted that these are estimates; some empirical data indicate that compounds such as germanium are actually more sensitive than the results reported here. Some of these sensitivities can be enhanced by adjustment of auxiliary and reactant gas flows in the "recipes" supplied by the vendor. There are also a few elements, such as tungsten and bismuth, for which spectral data and recipes have not yet been provided.

Table 6.1   Estimated DLs for atomic emission detection

Element		DL (pg/s)	Element		DL (pg/s)
Aluminum	Al	5.0	Manganese	Mn	2.0
Antimony	Sb	5	Mercury	Hg	0.5
Arsenic	As	3.0	Nickel	Ni	0.8
Boron	B	20.0	Nitrogen	N	7.0
Bromine	Br	20.0	Oxygen	O	50.0
Carbon	C	0.5	Palladium	Pd	5.0
Chlorine	Cl	15.0	Phosphorus	P	1.0
Chromium	Cr	7.0	Selenium	Se	4.0
Deuterium	D	7.0	Silicon	Si	1.0
Fluorine	F	20.0	Sulfur	S	1.0
Gallium	Ga	200.0	Tellurium	Te	10.0
Germanium	Ge	1.0	Tin	Sn	1.0
Hydrogen	H	0.5	Titanium	Ti	1.0
Iron	Fe	0.05	Vanadium	V	4.0
Lead	Pb	2.0			

In addition to monitoring the wavelength of a given element, the AED can acquire and store spectral data (similar in concept to mass spectrometry) over a 26-nm range. Atomic emission detection is an invaluable tool in confirming the identification of compounds tentatively identified by mass spectrometry, but it is better suited as a quantitative rather than as a qualitative tool. Its great power is in its ability to quantify both known and unknown compounds based on the equal atom-for-atom response of any particular component.

Since the AED is an optical detector, the emission lines are not only relatively complex, but response can be nonlinear; spectral lines can overlap and have both positive and negative interference from species not being measured. The vendor's software takes this into account with tools to remove this interference, primarily with respect to interference from carbon, as shown in Figure 6.9. The default parameters work well for most signals, including nitrogen, sulfur, carbon, and silicon channels, and are usually not an issue with most gas analyses. A good example of the specificity is demonstrated in Figure 6.10. The chromatogram shows a product gas containing approximately 3 % diborane (B_2H_6) with a balance gas of approximately

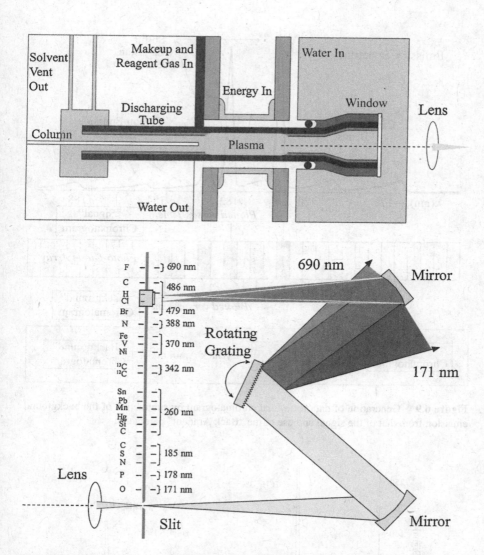

Figure 6.8 Microwave plasma source and spectrometer: The light from the plasma discharge tube is sent through a grating system, where it is separated so that the intensity at various component wavelengths can be measured using a photodiode array detector. The computer-controlled spectrometer provides automatic wavelength calibration and measurements of intensity, wavelength, and linewidth. The PDA detects emissions at selected combinations of diodes, which are recorded to form the signal due to the analyte. It also collects emissions from another combination of diodes to form the background portion of the chromatogram. The analyte and background signals are subtracted to obtain the final chromatogram for that species. Courtesy of Agilent Technologies.

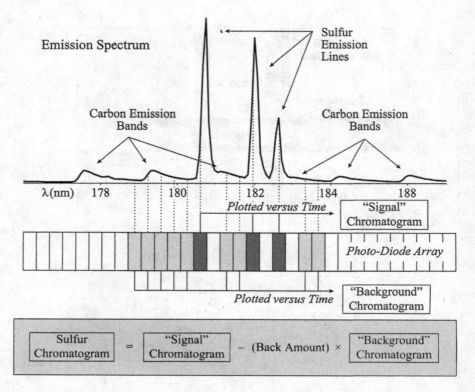

Figure 6.9   Generation of data for a final chromatogram by subtraction of the background emission from that of the signal and use of the "Back Amount" parameter.

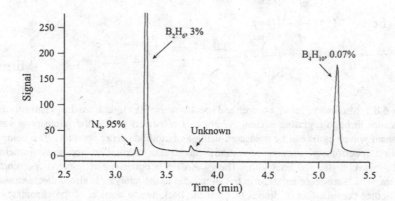

Figure 6.10   GC/AED chromatogram of a mixture of 3 vol% diborane and balance nitrogen obtained using the wavelength for boron.

95 % nitrogen. The response from nitrogen corresponds to less than 100 ppm as boron using the boron wavelength for detection. Use of a universal detector for this analysis would have required a column with far greater separation ability to isolate the target analyte from the nitrogen matrix. In cases where the concentration of interfering species is very high or where less commonly used channels are employed, removal of background interference may require some user intervention. For example, carbon-containing components can result in a low positive response in the germanium channel and lead to erroneous results when performing analysis at trace levels. Therefore, in such cases, the user may need to manually adjust the "Back Amount" parameter in the instrument software until the interference is completely minimized.

The AED plasma is a closed source that operates slightly above ambient pressure, which is why it is able to measure elements such as nitrogen, oxygen, and carbon that would not be possible with an open source such as ICP–MS. However, at approximately 38 µL, the plasma volume is very small, and although the plasma is energetic, it is not particularly robust compared to ICP. Thus, the matrix gas or high analyte levels can easily coke or deposit in the plasma tube. This also means that care must be taken that trace analytes, particularly with metallic or reactive species, are transported to the plasma tube without deposition. The easiest technique for protecting the plasma tube is to use the built-in vent switch to divert the matrix to vent during its elution time. Another technique commonly used when determining low-boiling contaminants is to use two columns with back-flush of the first column to vent as seen in the flow paths in Figure 6.11. In the diagram, the sample is loaded in the upper diagram, and in the lower diagram, it is injected. As the compounds of interest elute from column 1 onto column 2, the valve is returned to the load or back-flush position. The chromatogram of carbon monoxide, methane, and carbon dioxide in Figure 6.12 shows a typical application for any gas with an elution time later than that of carbon dioxide on the first column. By changing the back-flush time or columns, this arrangement can be configured to measure any low boiler in a higher boiler matrix, keeping the matrix gas off the analytical column.

AED is one of the few detectors that exhibits true specificity for fluorine. This makes it well suited for confirmational analysis of impurities in refrigerants, nitrogen trifluoride, and other fluorine-containing products. An example of this is the quantitative analysis of a hexafluorobutadiene (C_4F_6) reaction mixture containing fluorinated and brominated species (Figure 6.13). Note that the carbon signal could have been used as well. If the species of interest is known to contain carbon, measurement using the carbon wavelength frequently can provide the greatest sensitivity. With careful calibration, stoichiometric ratios of elemental composition can be determined. This is particularly useful for identifying compounds for which there are no reference infrared or mass spectra.

One of the issues when analyzing ESGs is that the matrix gas of the sample can have a significant effect on the retention time of component peaks due to changes in stationary-phase selectivity. This is observed with other GC techniques and is not unique to GC/AED. In hydride and corrosive gases, components elute earlier than in inert gas matrices, so users need to bear this in mind when setting valve switching times. The chromatogram in Figure 6.14 shows over a 2-minute difference between

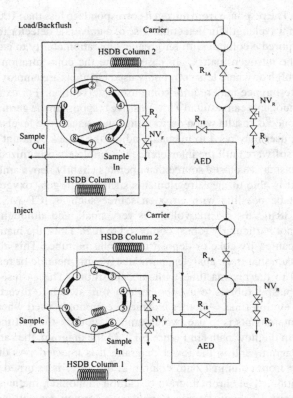

Figure 6.11 Two columns with back-flush to the vent of the precolumn. This flow path protects the analytical column and detector from matrix gas.

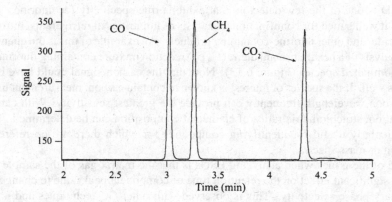

Figure 6.12 10 ppm standard using a 30 m × 0.32 mm ID porous polymer PLOT column; DL approximately 10 ppb.

injection of 100 ppb germane in phosphine (top) versus helium (bottom) under the same GC/AED conditions. A technique for dealing with this is use of a dual-loop 10-port valve plumbed to inject a sample simultaneously with an independent gas standard (Figure 6.15). Not only does this technique provide verification of one's ability to measure a contaminant in the sample matrix, it also allows for standard addition and further virtual dilution of the standard by using a smaller standard loop than the sample loop. When not used for standard additions, inert gas can be used to blank it out.

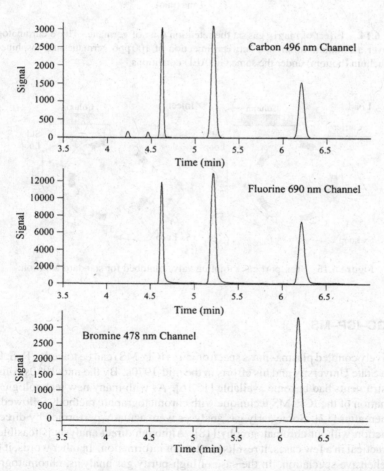

Figure 6.13 GC/AED analysis of a hexafluorobutadiene reaction mixture using the carbon, fluorine, and bromine channels.

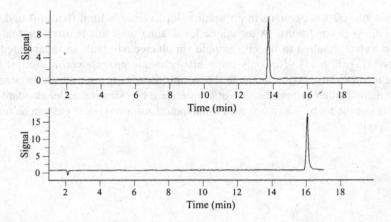

Figure 6.14   Effect of matrix gas on the retention time of germane. These chromatograms show over a 2-minute difference between injection of 100 ppb germane in phosphine (top) versus helium (bottom) under the same GC/AED conditions.

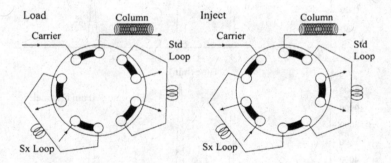

Figure 6.15   Ten-port gas sampling valve plumbed for standard addition.

6.4   GC–ICP–MS

Inductively coupled plasma–mass spectrometry (ICP–MS) can be traced to R. S. Houk at Iowa State University and his efforts in the mid-1970s. By the mid-1980s commercial instruments had become available [15,16]. As with many new technologies, the hyphenation of the ICP–MS technique with chromatographic methods followed very soon thereafter [17]. Some early gas analyses were actually performed by direct gas introduction without chromatography [18]. Although direct analysis is feasible and even practical in a few cases, it results in a loss of information. In other words, it lacks comprehensive speciation. In the case of high-purity gas analysis, chromatography allows for the temporal removal of the interfering matrix from detection of the target analyte(s).

For most elements, ICP–MS as a stand-alone instrument has DLs equal to or better than those of graphite furnace atomic absorption (GFAA), ICP–optical emission spectrometry (ICP–OES), or microwave-induced plasma-AED.

Three significant types of mass spectrometers are used for ICP–MS: TOF, magnetic sector, and quadrupole instruments. Each type of instrument has particularly unique characteristics that make it more or less appropriate for its intended use. TOF instruments operate at very high data acquisition rates and thus are well suited to measurements of very rapid transient signals in a flowing stream such as laser ablation, chromatography, or fast reaction processes. Since its data acquisition is nearly simultaneous, it is also well suited for accurate isotopic measurements, second only to double-focusing mass spectrometers equipped with multicollector detection systems. The drawback to current TOF instruments is that they are still somewhat less sensitive than quadrupole and magnetic sector instruments. Magnetic sector instruments, particularly double-focusing instruments, have characteristics that are nearly opposite those of TOFs. Magnetic sector instruments have very high resolving power, approximately 10,000 amu at high-resolution settings, close to 100 % ion transmission at low-resolution settings. This capability is tempered by the fact that as higher resolution is required, the ion transmission can drop significantly. For example, distinguishing $^{75}As^+$ and $^{40}Ar^{35}Cl^+$ requires a resolution of 7725, dropping the ion transmission to 2 % [19,20]. Using multicollector detectors or Mattauch–Herzog geometry, magnetic sector instruments are the choice for exacting isotopic measurements. Although they are no match for TOF or quadrupole instruments in data acquisition speed, modern magnetic instruments can scan with sufficient speed for chromatographic data acquisition by limiting the masses collected. Perhaps the reason there are fewer magnetic instruments than quadrupoles is cost—typically, they are three times the cost of most quadrupole-based instruments. The most commonly used ICP–MS instruments today are based on quadrupole technology. These instruments can acquire data more rapidly than can magnetic instruments, but the data acquisition is actually sequential, moving from one mass to the next by rapid scanning. The time required to do this can cause sufficient spectral skewing such that they are not suitable for highly accurate isotopic measurement. The sensitivity of quadrupoles is better than that of TOF instruments, but since there is some inefficiency in ion transmission and more background noise than in magnetic instruments, they are less sensitive than are magnetic sector instruments. There are numerous excellent reference books describing these instruments, and how they operate in more detail as well as the myriad ancillary equipment and techniques available in what has now become a mature technology [16,19,21,22]. Since quadrupoles are the predominant type used both in general and as a chromatographic detector, and since our concern is the application of GC–ICP–MS to gas analysis, we will our discussion primarily to quadrupole technology and the techniques and challenges encountered with its use as a chromatographic detector.

6.4.1 Basic Operation

All GC–ICP–MS systems are composed of the major components shown in Figure 6.16. The sample introduction and gas chromatograph appear to be redundant since they are both a type of introduction, but as will be seen later, it can be useful to have both acting simultaneously. The RF generator provides power between 500 and 1700 W to an induction coil surrounding a torch through which there is a flow of argon gas. As this power is applied, an alternating current oscillates within the field. The oscillations set up magnetic and electrical fields, and when a spark is applied, electrons are stripped from some of the argon atoms. These electrons are then trapped in and accelerated in closed circular paths in this field. This inductive coupling is what is then called an inductively coupled plasma (ICP). Further collisions with neutral atoms establish a chain reaction that causes the plasma to reach temperatures between 6000 and 10,000 K. It is the plasma that transforms the analytes of interest from their molecular form to ions predominantly of a monoelemental nature. The ions formed from the plasma are separated in the mass spectrometer and measured by the detector. All of these devices are controlled using the manufacturer's or vendor's software on the computer.

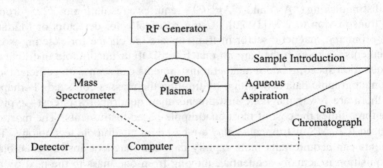

Figure 6.16 GC–ICP–MS components.

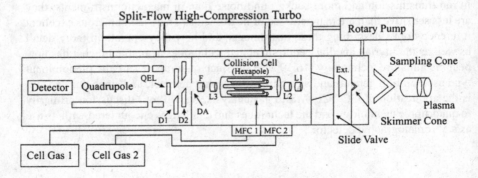

Figure 6.17 Quadrupole ICP–MS instrument.

In order to transfer ions from the high-temperature plasma to the mass spectrometer, a water-cooled interface composed of two or more nickel or platinum cones is used to step the pressure down from atmospheric to the operating pressure of the mass spectrometer, approximately 0.5×10^{-7} mbar (Figure 6.17). The first cone or sampling cone is at atmospheric pressure. Between the sampling cone and the second or skimmer cone, a roughing pump maintains a pressure of approximately 2.2 mbar. On the other side of the skimmer cone the ions travel into the ion optics or, if so equipped, into the collision/reaction cell. Appropriate voltages are applied to present the ions to the quadrupole mass filter. The quadrupole mass filter consists of four precisely machined rods. By applying DC and RF potentials to opposing pairs of rods that are 180° out of phase, an electromagnetic field is created. At a given field strength, only a single unique ion of specific mass-to-charge ratio will be able to traverse along the axis of the rod assembly and be detected by an electron multiplier, sending the signal back to the computer. By changing the potentials rapidly and repeatedly, temporal scans or spectra are achieved. If the instrument is equipped with a collision/reaction cell, the ions formed in the plasma can be manipulated further to enhance their detection or minimize interfering ions by introducing small amounts of reagent gases into the collision/reaction cell region with mass flow controllers prior to their entrance into the mass filter.

Measurement Capabilities of GC–ICP–MS   ICP–MS can measure most of the elements in the periodic table down to the nominal DLs listed in Table 6.2. The high-energy plasma essentially breaks down the analyte to its elemental ions. Once ionized, its elemental spectrum can be recorded. Unlike other elemental analyzers, ICP–MS can additionally determine the isotopic ratios of elements having more than one isotope. Because the ionization efficiency is one or more orders of magnitude greater than that of a conventional organic EI mass spectrometry, the sensitivities for most elements are quite impressive. Since unlike the atomic emission plasma, the plasma used is an open source, there are some limitations to elements that can be measured. Elements present in the atmosphere (nitrogen and oxygen) produce background signals too high for trace measurements. Carbon can be measured, but its response is also poor due to its high IP (11.26 eV) and ambient levels of carbon dioxide. Elements such as fluorine and neon have IPs higher than that of the plasma (Ar IP = 15.79 eV), which prevents them from being measured. Finally, there is the problem of isobaric interferences such as $^{28}NN^+$ with $^{28}Si^+$, $^{56}ArO^+$ with $^{56}Fe^+$, and $^{32}OO^+$ with $^{32}S^+$. These can be avoided using a high-resolution magnetic sector instrument. However, a quadrupole, having only unit resolution, requires other strategies, such as cool plasma or collision/reaction cell technology (CCT).

6.4.2 Gas Chromatography

When considering the analysis of ESGs, we find that many of these gases have a metallic or semimetallic nature or other atoms present that are amenable and sensitive to ICP–MS detection. These include volatile metal and metalloid hydrides, metal carbonyls, metallic fluorides, silanes, chlorosilanes, alkylsilanes, germanes,

Table 6.2 Nominal DLs and IPs for elemental analysis by ICP–MS

Element		DL (ppb w/w)	IP 1[a] (eV)	Element		DL (ppb w/w)	IP 1[a] (eV)
Aluminum	Al	0.01	5.98	Neodumium	Nd	0.001	5.51
Antimony	Sb	0.001	8.64	Nickel	Ni	0.004[b,c]	7.63
Arsenic	As	0.002[c]	9.81	Niobium	Nb	0.001	6.88
Barium	Ba	0.001	5.21	Osmium	Os	0.002	8.50
Beryllium	Be	0.002	9.32	Palladium	Pd	0.002	8.33
Bismuth	Bi	0.001	7.29	Phosphorus	P	0.5[d]	10.48
Boron	B	0.04	8.30	Platinum	Pt	0.001	9.00
Bromine	Br	0.04[c]	11.84	Potassium	K	0.005[b,c]	4.34
Cadmium	Cd	0.001	8.99	Praseodymium	Pr	0.001	5.46
Calcium	Ca	0.005[b,c]	6.11	Rhenium	Re	0.001	7.87
Cerium	Ce	0.001	5.60	Rhodium	Rh	0.001	7.46
Cesium	Cs	0.001	3.89	Rubidium	Rb	0.001	4.18
Chlorine	Cl	5–10	13.01	Ruthenium	Ru	0.002	7.36
Chromium	Cr	0.002[b,c]	6.76	Samarium	Sm	0.001	5.60
Cobalt	Co	0.001	7.86	Scandium	Sc	0.01	6.54
Copper	Cu	0.002[d]	7.72	Selenium	Se	0.01[c]	9.75
Dysprosium	Dy	0.001	6.80	Silicon	Si	5[d]	8.15
Erbium	Er	0.001	6.08	Silver	Ag	0.001	7.57
Europium	Eu	0.001	5.67	Sodium	Na	0.005[b]	5.14
Gadolinium	Gd	0.001	6.16	Strontium	Sr	0.001	5.69
Gallium	Ga	0.002	6.00	Sulfur	S	0.5[c]	10.38
Germanium	Ge	0.007	7.88	Tantalum	Ta	0.001	7.88
Gold	Au	0.001	9.22	Tellurium	Te	0.01	9.01
Hafnium	Hf	0.001	7.00	Terbium	Tb	0.001	5.98

[a] First ionization potential.
[b] Cool plasma.
[c] Collision or reaction cell.
[d] Collision or reaction cell may be useful.

alkylgermanes, alkylstannanes, and phosphorus and sulfur compounds. Before ICP–MS became practical, popular, and reasonably affordable, many different detectors, such as electron capture, photo-ionization, electrolytic, chemiluminescent, electron impact mass spectrometer, atomic emission, and others, might be required to speciate and quantify these compounds. Other than AED, ICP–MS provides nearly universal elemental detection combined with high specificity in a single instrument, generally higher sensitivity than that of most selective detectors, and intelligence in the form of isotopic information. Today, GC–ICP–MS is used for an extremely diverse range of analyses, including organometallic gases in landfills [23,24], flame retardants [25, 26], environmental analysis [27,28], petroleum and petrochemical characterization [29,30], and photovoltaics and ESGs [31,32], to name just a few applications. Gas chromatographic analysis of specialty and electronic gases is a very small subset of these applications. Although the types of samples encountered in product gas analyses are generally not very complex chromatographically, the challenge is the measurement of a target analyte of very low concentration in what is usually a difficult matrix. A further challenge is that unlike air or water samples, there is rarely an opportunity to enhance DLs by concentration techniques.

6.4.3 Interfacing

Some manufacturers provide a physical interface to couple the GC to the ICP–MS. Although not trivial, an interface is not very difficult to fabricate if not provided [33–36]. The configuration we have found useful is described by Figure 6.18. The column is fed through a tee fitting into the annular space of a 1/16-in SilcoSteel tube that is inside a heated transfer line. The end of the column extends just to the end of this line. Heated makeup argon gas is delivered at the tee. A second tee between the argon preheater and the makeup tee is connected to a vacuum pump. The vacuum is useful when analyzing high-purity reactive gases for trace impurities. By automating the valve to be opened during elution of the matrix gas, a vacuum pulls the effluent from the end of the column before it reaches the plasma torch, keeping the torch relatively clean of deposits. Due to the robust nature of the plasma and the low mass of carbon, this is not necessary for hydrocarbon samples, but it is important in keeping residual background levels down when analyzing gases such as phosphine or arsine for impurities. The torch also has an extra leg with a spray chamber and nebulizer attached. The spray chamber and nebulizer would ordinarily be the most common way to introduce an aqueous or organic liquid sample and is not particularly necessary for it to be attached for gas work. In fact, most analysts performing GC–ICP–MS do not use such a configuration and operate in "dry" plasma mode. However, there are some important advantages to aspirating an aqueous solution while simultaneously operating the GC; it makes instrument tuning easier, the plasma is somewhat more stable, but most important, it provides for a technique that can be used for calibration when specific gas standards are unavailable. This last feature is discussed later.

The robust nature of the plasma and the relative low flows used in gas chromatography allow for virtually any carrier gas (i.e., hydrogen, helium, nitrogen, or argon). The makeup gas is not restricted to argon, and the use of other makeup gases can be used to enhance the sensitivity of some species [37].

6.4.4 Tuning and Optimization of the ICP–MS Detector for Gas Analysis

Tuning the ICP–MS for best performance is highly dependent on the target analyte. The two main issues affecting performance or sensitivity are the IP of the analyte

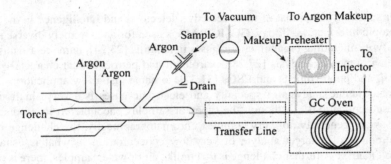

Figure 6.18   Physical interface of a GC to an ICP–MS.

and interferences. The addition of a small amount of nitrogen to the plasma gas has been reported to increase the overall sensitivity of many elements [37,38]. Methane addition has also been reported to increase the sensitivity for arsenic, germanium, and selenium [39].

Using GC for sample introduction into the ICP–MS has an advantage versus conventional aqueous aspiration. Other than the matrix gas, there are few, if any, chemical interferences. As an example, the $^{40}Ar^{35}Cl^+$ interference with $^{75}As^+$ is of little concern. However, one needs to be concerned with isobaric spectral interferences. These are due primarily to ions formed with argon, nitrogen, oxygen, hydrogen, and carbon. These issues are generally dealt with by the addition of various gases to the collision or reaction cell. Xenon is a typical gas used to minimize interferences and can be used to optimize the response for sulfur [40]. Oxygen can also be used for some analytes to form AO^+ (A = analyte), shifting the analyte away from the interference. Many other gases can be used, such as hydrogen, helium, ammonia, nitrous oxide, nitric oxide, carbon dioxide, methyl fluoride, sulfur hexafluoride, and carbon disulfide [41]. These gases can be used in what can generally be called either kinetic energy discrimination (KED) mode or non-KED mode. The precise mechanism is somewhat dependent on the manufacturer's design. However, simply put, KED mode uses the cell gas to suppress the interference at a greater rate than the analyte of interest, while the non-KED mode uses the cell gas to react with the analyte of interest faster than the interference, forming a new ion shifted away from the interference. The choice of gas to use and whether KED or non-KED mode is used is dependent on the relative ion–molecule reaction rates of the analyte and interference.

When operating a wet plasma, the tuning can be accomplished in virtually the same fashion as if no GC were present. The only difference would be to lower the nebulizer flow to account for the added makeup gas that is being supplied by the GC to punch through the plasma. When operating a dry plasma, it is not uncommon to tune or optimize using xenon that might be present in the argon plasma gas. The alternative would be to leak a small amount of gas containing the species of interest into the column effluent for tuning.

There is insufficient space here to cover all optimization strategies for all the elements of interest in gas analysis, but some examples of collision cell use and interferences typically encountered are discussed later in the chapter.

6.4.5   Detector Linearity and Data Acquisition

Measurements of high-purity gases for contaminants at extremely low concentrations is perhaps the best attribute of using ICP–MS as a GC detector. However, GC–ICP–MS can also be used to quantify complex mixtures based on its atomic response. To do so with confidence requires that the detector be linear. Current instrumentation uses a dual-stage discrete dynode detector. This type of detector incorporates both pulse counting and analog modes in one detector. At low concentrations the detector operates in the pulse mode; once sufficient signal is encountered, the circuitry switches to analog mode, extending the dynamic range. A cross-calibration solution of 100 ppb is typically used to set up the pulse count and analog calibration correlation. This easily allows for measurements over six orders of magnitude. Figure 6.19 describes the linearity of a calibration made for germane.

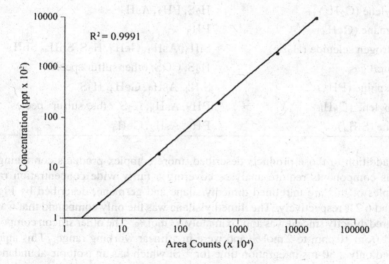

Figure 6.19   Linearity study of germane (190 ppt to 1 ppm).

When acquiring chromatographic data with a quadrupole ICP–MS it is advisable to decide what elements or ions to measure. Due to the speed limitations, full scans will severely limit the sensitivity of the instrument. Although the instrument has great sensitivity for elemental analysis, this is usually defined based on constant aspiration of a liquid sample over a very long time period. When performing chromatographic analysis at ppb levels and lower, consideration must be given to expected chromatographic peak width of the analyte and desired sensitivity. When using medium-sized or megabore capillary columns, total dwell times for all ions might range from 50 to 500 ms. Dwell times beyond this risk loss of data at peak apexes.

Consideration should also be given to confirmational analysis of isotopes. Isotopic measurement can be used not only for confirmational analysis, but also to extend the dynamic range of measurement for compounds in high concentration.

6.4.6   Analytes and Matrices

The gases typically encountered that require GC–ICP–MS analysis and the target contaminants are described in Table 6.3.

Table 6.3   Gases typically encountered that require GC–ICP–MS analysis and the target contaminants

Matrix Gas	Contaminant Gas
Acetylene (C_2H_2)	PH_3, H_2S, AsH_3
Arsine (AsH_3)	SiH_4, GeH_4, PH_3, H_2Se, H_2S
Carbon monoxide mixtures (CO_x)	$Ni(CO)_4$, $Fe(CO)_5$, $Cr(CO)_6$, $Mo(CO)_6$
Ethylene (C_2H_4)	H_2S, PH_3, AsH_3
Germane (GeH_4)	PH_3
Hydrogen selenide (H_2Se)	SiH_4, AsH_3, GeH_4, H_2S, SnH_4, SbH_3
Natural gas	H_2S, COS, other sulfur species
Phosphine (PH_3)	SiH_4, AsH_3, GeH_4, H_2S
Propylene (C_3H_6)	PH_3, AsH_3, H_2S, other sulfur species
Silane (SiH_4)	PH_3, AsH_3, GeH_4

In addition to those products described, more complex products containing numerous components require analyses covering a fairly wide concentration range. Examples of this are unrefined dimethylsilane and germane, described by Figures 6.20 and 6.21, respectively. The dimethylsilane was the only compound that was not measured directly since it was approximately 63 mol %. The other silicon compounds ranged from 10 ppm to 2 mol % and were in a linear working range. This analysis required only a 50-ms integration time for ^{29}Si which has an isotopic abundance of 4.68 %. What is notable in the germane chromatogram is use of the vacuum vent to minimize the background level for subsequent speciation. In both of these cases we are still measuring relatively high concentrations. If we consider an element that has a low IP and no interferences, we see the full power of GC–ICP–MS as a technique. An example of this is the analysis of germane in electronic-grade arsine. For some electronic devices germane can be problematic even at single-digit ppb levels, this is just at the practical limit of GC/AED. Germanium is an n-type impurity of concern in gallium arsenide (GaAs) and aluminum gallium arsenide (AlGaAs) epilayers and detrimental to high-speed transistors such as high electron mobility transistors (HEMTs) and field-effect transistors (FETs). Using GC–ICP–MS, germane can be

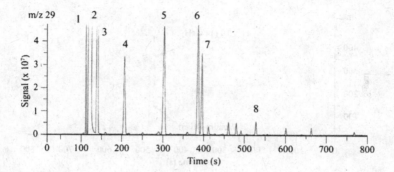

Figure 6.20 Unrefined dimethylsilane analysis: (1) methylsilane, (2) dimethylsilane, (3) trimethylsilane, (4) trimethylsilanol, (5) 1,1,3,3-tetramethyldisilane, (6) tetramethyldisilane, (7) dimethyl((methylsilyl)methyl)silane, (8) hexamethyltrisiloxane. Column: 60 m × 0.32 mm ID × 1.5 μm Restek RTXVolatiles; carrier: hydrogen at 12 psig; initial temperature: 35 °C; initial time; 4 min; ramp: 20 °C; final temperature: 200 °C.

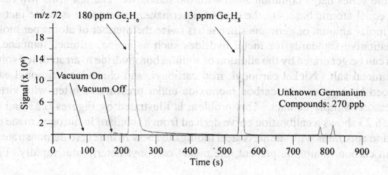

Figure 6.21 Unrefined germane analysis: digermane 180 ppm, trigermane 13 ppm, unknown germanium compounds 270 ppb. Column: 100 m × 0.53 mm ID × 5.0 μm Agilent DB-1; carrier: argon at 25 psig; initial temperature: 40 °C; initial time: 3 min; ramp: 15 °C/min; final temperature: 200 °C; sample size: 50 μL.

detected at nearly three orders of magnitude lower than current specifications. This is illustrated in Figure 6.22.

6.4.7 Standards and Calibration

Standards for compounds such as silane, phosphine, arsine, and germane are available from several specialty vendors. This is not necessarily the case for other metallic hydrides or metal carbonyls. It is virtually impossible to acquire stable gas-phase standards for metal carbonyls. Measurement of substituted species is not a problem if any standard containing the target element is available. For example, in the germane and dimethylsilane analyses, a single compound was used to calibrate an entire series

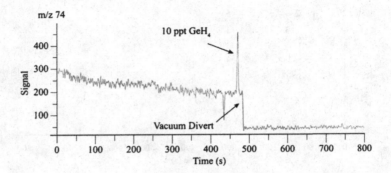

Figure 6.22   Trace germane in arsine product. Column: 200 m × 0.53 mm ID × 5.0 μm Agilent DB-1; carrier: argon at about 6 mL/min; temperature: 35 °C isothermal; sample size: 400 μL.

since the series had a common atom with the standards. The ICP–MS will respond on an equal atomic basis. In the case of digermane, its response is twice that of an equal molar amount of germane since it has twice the number of atoms per mole.

Qualitative standards for metal hydrides such as arsine, stibine, stannane, and others can be generated by the addition of sodium borohydride to an acidified solution of the metal salt. Nickel carbonyl, iron carbonyl, and chromium carbonyl can be produced by contact with carbon monoxide under pressure or often with virtually no pressure, simply contact. This problem is illustrated by Figures 6.23 and 6.24. Figure 6.23 shows a calibration curve derived from a standard in nitrogen made using a dilution apparatus with stainless steel tubing: the bow in the curve demonstrates that without carbon monoxide present, the nickel carbonyl deteriorates rapidly. Figure

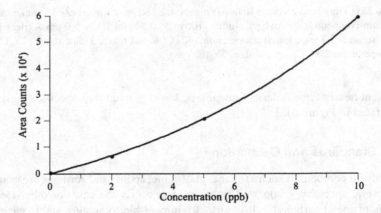

Figure 6.23   Calibration curve derived from a standard in nitrogen made using a dilution apparatus with stainless steel tubing. The bow in the curve demonstrates that without carbon monoxide present, the nickel carbonyl deteriorates rapidly.

6.24 describes an increasing concentration of nickel carbonyl on replicate injections as it is passing through a 1/4-in-long by 1/16-in-ID stainless fitting. When analyzing carbon monoxide, it has been found that nickel carbonyl can be formed almost instantly as the gas passes through stainless steel. This is such a problem that the chromatographic system employed for this analysis uses all poly(ether ether ketone) (PEEK) tubing, valves, and fittings to avoid its formation.

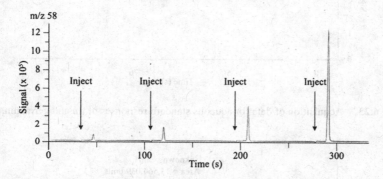

Figure 6.24   Nickel carbonyl formation in hydrogen–carbon monoxide mixture. Continuous flow through stainless steel fitting, demonstrating increasing carbonyl formation.

The problem of quantitating compounds for which gas standards are unavailable is solved by use of the novel torch design shown in Figure 6.18. The extra leg allows simultaneous aspiration of an aqueous solution and chromatographic interfacing. This feature has allowed the development of a cross-calibration technique between aqueous-based and gas standards for determining the concentration of gases for which no standards are available. As an example, known molar amounts of a reference element and the target element are aspirated through a conventional nebulizer and spray chamber. This step establishes a relative molar response multiplication factor. Once this factor is established, a known gas standard is introduced chromatographically. The area of the known and unknown are compared, and the unknown target compound concentration is calculated by using the multiplication factor established in the first step, as illustrated in Figures 6.25 and 6.26.

In our example we first determine the multiplication factor by aspirating known molar concentrations of tin (^{120}Sn):

$$\frac{4,990,000}{833} = 5990$$

and germanium (^{74}Ge):

$$\frac{2,810,000}{1351} = 2080$$

$$X \text{ RRF}(^{120}\text{Sn}) = \frac{2080}{5990} = 0.347$$

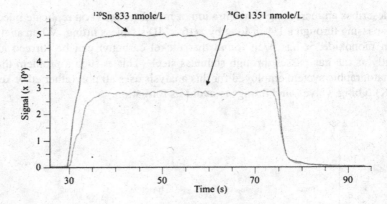

Figure 6.25   Acquisition of data for aqueous standard responses of tin and germanium.

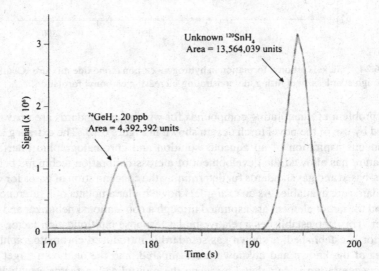

Figure 6.26   Acquisition of a known germane standard and an unknown stannane sample.

This is followed by gas chromatographic analysis of a known amount of germane and the unknown amount of stannane, measuring the areas and completing the algebra:

$$13{,}564{,}099 \times 0.347 \; [\times RRF(^{120}Sn)] \times 20 \; (^{74}GeH_4) \; 4{,}392{,}392 = 21$$

A study was made to determine the efficacy of this technique with available and known standards. Results are presented in Table 6.4. The worst two cases are arsine (60.9 %) and phosphine (44 %) at the 200-ppb level. This would ordinarily not be considered very good performance, but it should be understood that these are fairly reactive species and that the gas standards may be just as suspect as the calibration technique. Indeed, the fact that all concentrations found are low using this cross-calibration technique suggests that the actual gas standard values may be lower than those reported by the vendor.

Table 6.4   Test results for aqueous water/gas cross-calibration

Component	Concentration Expected (ppb)	Concentration Found (ppb)	Recovery (%)	Error (+/-%)
Arsine	200	121.7	60.9	−39.2
Carbonyl sulfide	160	119.3	74.6	−25.4
Germane	200	158.2	79.1	−20.9
Phosphine	200	88.0	44.0	−56.0

Ideally, it is desirable to have standards no more than an order of magnitude greater than the DL and to be able to make a standard addition or "spike" to the sample matrix. Vendors who manufacture gas standards usually are not inclined to supply standards with concentrations much below 1 ppm. Permeation devices are available for only a few of the target compounds and have a limited calibration range. Gas dilution systems are the most practical solution and are available commercially. These systems can usually span three orders of magnitude. These systems can be based on mass flow controllers or pressure/fixed restrictor devices. This hardware is described in more detail in Chapter 9.

6.4.8   Columns

In contrast to GC/MS and GC/AED, GC–ICP–MS uses very high argon flow rates near ambient pressure and has virtually no constraints with respect to the flows from the GC column. The consequence of this is that large-megabore (0.53 mm ID) capillary columns are ideally suited for trace analyses of labile materials at very low levels. They have high capacity, are highly inert, and are usually able to make the desired separation. Packed columns made of steel and having a significant amount of surface area should be avoided, due to the potential for loss of labile analytes. In circumstances such as speciation where ultimate sensitivity is not required, columns with IDs of 0.25 and 0.32 mm are satisfactory. This is usually the case when quantitation of samples that have been "road-mapped" by conventional GC/MS is required, as in the case of the disilane analysis.

Boiling-Point Columns   It would be convenient and quite simple if we only required a single column for all the analyses we perform using GC–ICP–MS, even if only for the electronic gases. That is not the case, but we have found that the large-bore (0.53 mm ID) capillary, thick-film (5.0 μm) boiling-point column comes very close to being the universal column for many applications. Due to its inertness (thick film) it can be used for a wide variety of contaminants in various matrices. The largest range of applications is for the various hydride gases.

Figure 8.8 describes the separation and retention times of several hydrides on a 300-m methyl silicon column. This study was part of a project investigating the presence of possible hydrides in hydrogen selenide product. A 300-m column is used rarely and only in the most demanding situations. A 100-m column is obviously easier to manage, but 200 m is required for a number of hydride separations. A good example of this is the analysis of hydrogen sulfide in bulk phosphine shown in Figure 6.27. This is a particularly difficult analysis since hydrogen sulfide elutes shortly after phosphine and is separated only by one mass unit; thus, there is a lot of mass "splash-over." A 300-m column may have worked a bit better in this case. A shorter 30-m phenyl methyl silicone column works well for the metal carbonyls. Figure 6.28 illustrates an analysis of carbon monoxide containing nickel, iron, and chromium carbonyls. Note the reasonably good match for the isotopic abundance ratios in the chromium measurement.

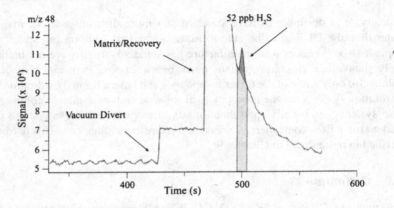

Figure 6.27 Analysis of hydrogen sulfide in bulk phosphine. Column: 200 m × 0.53 mm ID × 5.0 μm Agilent DB-1; carrier: argon at about 6 mL/min psig; temperature: 35 °C isothermal.

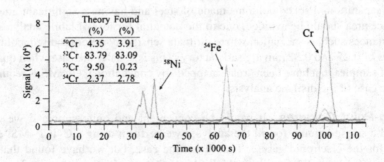

Figure 6.28 Analysis of carbon monoxide containing nickel, iron, and chromium carbonyl. Column: 30 m × 0.53 mm ID × 1.5 μm RTX-5; temperature: 60 °C; carrier: argon at 8 psig.

PLOT Columns   A PLOT column that has been found to be useful is the $60\,m \times 0.53\,mm$ ID $\times 6.0\,\mu m$ Chrompack CP-SilicaPLOT column. It reverses the elution order of hydrogen sulfide and carbonyl sulfide from that of a boiling-point column. It has also been useful for the analysis of sulfur species in silicon tetrafluoride. This is because the silicon tetrafluoride either reacts or is irreversibly absorbed on the column without deterioration of subsequent analyses* of the sulfur species. It is under investigation for providing a better separation of hydrogen sulfide in bulk phosphine, a difficult determination.

Porous Polymer Columns   Porous polymers have separation characteristics that make them very attractive. However, impurities in the polymer and their absorptive nature make them somewhat suspect when analyzing ppb- and ppt-level materials, as some of the species exhibit tailing and loss. A good test for any column is usually its performance for the analysis of hydrogen sulfide at low concentrations.

Other Columns   The GasPro column has been used occasionally for difficult separations, but since it is available only in 0.32-mm bores, its use for ultratrace analysis is limited.

6.4.9 Interferences

Interferences encountered in ICP–MS are generally of an isobaric nature and include isotope overlaps and ion product overlaps. There are numerous elemental isobaric interferences that must be accounted for when analyzing complex aqueous or liquid samples. When analyzing gases, many of these isobaric interferences are eliminated because the GC presents a very "clean" sample to the ICP–MS, notwithstanding the matrix gas, which is usually minimized by separation or venting. Table 6.5 addresses those that are due to atmospheric contamination. Most of the interferences are still relatively minor, and those interferences generally contribute only to an offset baseline in the worst case. The most problematic are those that interfere with silicon, phosphorus, and sulfur measurement. Since a quadrupole instrument has insufficient resolution to separate exact masses, a collision or reaction cell can be used to overcome interferences to some degree. $^{28}Si^+$ has a low IP but it is affected significantly by interference from NN^+. Because it is a relatively light element, its transmission through the mass filter is also less efficient. By adding a small amount of a hydrogen or hydrogen/helium mixture to the collision cell, the NN^+ ion is depressed. The $^{28}Si^+$ is also reduced somewhat but to a lesser extent and will still maintain reasonable signal strength. $^{15}N^{16}O^+$ or $^{14}N^{16}O^1H^+$ can cause a high background level for monoisotopic P^+. Caruso has described the use of additional nitrogen in the plasma combined with helium in the collision cell to enhance phosphorus sensitivity [37]. Sulfur also has a very high IP; this, combined with significant OO^+ interference, makes it problematic. Use of a cool plasma described by Wilbur and Soffey is one solution [42]. Mason et al. have has used a

---

* As of the date of writing.

mixture of gases, including xenon in the collision cell to minimize OO^+ formation and enhance S^+ transmission [40]. The technique used in this lab is addition of a small amount of oxygen to the collision cell [43]. The S^+ species will react with oxygen to form SO^+ ($m/z = 48$), shifting it away from the interference. This technique also maintains the isotopic integrity for higher-level sulfur measurements and works equally well for phosphorus measurement. Figure 6.29 describes the analysis of a fuel gas with low levels of hydrogen sulfide and carbonyl sulfide. Oxygen was used in the collision cell to form SO^+ ($m/z = 48$).

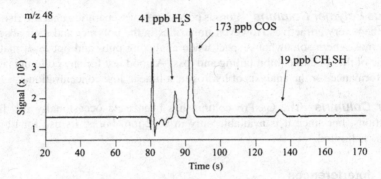

Figure 6.29   Fuel gas analysis for trace sulfur species. Column: 100 m × 0.53 mm ID × 5.0 μm Agilent DB-1; carrier: argon at 24 psig; temperature: 35 °C isothermal.

6.4.10   Detection Limits

Detection limits for ICP–MS are available from instrument manufacturers, standard vendors, numerous literature sources, and empirical experience. These DLs are reported in many cases as single-digit ppt levels or lower. These detection levels require some qualification. The amount of time spent signal averaging is crucial. If one is measuring a single element at a single mass by signal averaging for 5 seconds, the detection level will be significantly lower than analyzing 10 elements signal-averaging 40 ms each. Other factors, such as the cleanliness of the environment, the quality of water and acid used, the sample matrix, and the mode of operation are also determining factors. When comparing aqueous or liquid DLs to chromatographic DLs, one must remember that the aforementioned are usually weight per weight values, and gas analysis is usually considered on a molar basis, thus is not a direct translation. The major issue with gas analysis DLs compared to liquid analysis DLs is that there are significant time constraints when it comes to signal averaging. There is a compromise that must be struck with the column resolution, sampling rate, and best or required DL. If one uses thin-film small-bore columns, the signal height versus noise will be good, but the overall sample capacity will be limited. The opposite is true using packed columns; the capacity is high but the peaks will generally be broader. Megabore-type columns (0.53 mm ID) seem to be a good compromise with 500- to 800-ms signal averaging, providing good results.

Table 6.5   Routinely achievable GC–ICP–MS DLs using megabore (0.53 mm ID) capillary columns

Compound	Formula	IP 1[a] (eV)	m/z	Atmospheric Interferent	Typical DL (ppbv)
Silane	SiH_4	8.15	28	NN^+	2–5
Phosphine	PH_3	10.48	31, 47 CCT	$^{15}N^{16}O^+$, $^{16}O^1H^{14}N^+$	2–5
Germane	GeH_4	7.88	74		<0.01
Arsine	AsH_3	9.81	75		<0.02
Hydrogen selenide	H_2Se	9.78	80	$ArAr^+$	0.8
Stannane	SnH_4	7.34	120		<0.01
Stibine	SbH_3	8.64	121		<0.02
Hydrogen sulfide	H_2S	10.36	32, 48 CCT	OO^+	2–5
Nickel carbonyl	$Ni(CO)_4$	7.63	58	$Ar^{18}O^+$	<0.02
Iron carbonyl	$Fe(CO)_5$	7.87	56	ArO^+	<0.02
Chromium carbonyl	$Cr(CO)_6$	6.76	52	ArC^+, $^{36}ArO^+$	<0.02
Molybdenum carbonyl	$Mo(CO)_6$	7.09	98		<0.01
Organochlorides	C_xCl	13.01	35	$^{18}OH^{16}O^+$, $^{16}OH^{18}O^+$	50–100
Organobromides	C_xBr	11.84	79	$H^{78}Kr^+$	0.5–1.0
Organoiodides	C_xI	10.45	127		–
Mercury	Hg	10.43	202		–
Alkylleads	Pb	7.42	208		–

[a] First ionization potential.

Since the chromatographic signal is transient, the most accurate description of sensitivity should be picograms per second (pg/s). However, analysts typically think of measurements as a molar concentration such as ppmv or ppbv. This presupposes a typical frame of reference for the volume of sample typically used in gas chromatography. Table 6.5 describes the DL that GC–ICP–MS can achieve routinely using megabore (0.53 mm ID) capillary columns. These DLs are based on measuring a single component in a noncompromising matrix using a 400-μL sample size. These values also assume that the retention time is optimal with respect to peak width and height. DLs can vary based on tuning, electron multiplier life, and system cleanliness. Other factors affecting DLs are IP, isobaric interference, and to some degree the atomic mass. A higher atomic mass can lower the background noise such that the DL will be lower than the limit that might be predicted due to its IP, and higher masses tend to be transmitted more efficiently through the mass filter.

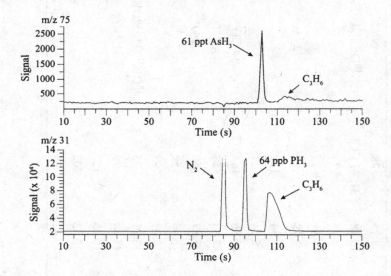

Figure 6.30   Analysis of propylene for arsine (upper) and phosphine (lower) impurities. Column: 100 m × 0.53 mm ID × 5.0 μm Agilent DB-1; carrier: argon at 20 psig; temperature: 35 °C isothermal.

Two examples of DLs with relatively simple matrices are illustrated in Figure 6.30. This describes analyses of arsine and phosphine in polymer-grade propylene. Both are catalyst poisons detrimental to downstream processes. Note the negligible response from the matrix in the arsine chromatogram. This analysis was performed without the use of the collision/reaction cell technology for phosphorus so that arsine and phosphine could both be measured using the same tune parameters, which is why there is considerable signal from nitrogen in the chromatogram. Arsenic has a lower IP (9.81 eV) than phosphorus (10.48 eV), less background interference, and is a heavier element. All these factors enhance the detector's sensitivity to arsine relative to phosphine.

6.5 Conclusions

GC/MS is a relatively inexpensive, information-rich technique for gas analysis. Modern instrumentation has become very reliable and has excellent sensitivity for trace analysis. However, it does require individual compound calibration.

GC/AED has the ability to measure numerous elemental species with very good sensitivity and provide spectral confirmation. It allows for compound-independent calibration (CIC). Care must be taken to prevent matrix gases from contaminating the detector. It is a moderately expensive technology.

GC–ICP–MS analysis is with few exceptions an extremely sensitive technique for the analysis of trace contaminants in both high-purity and complex mixture–type gases. Properly calibrated, it is linear over six orders of magnitude. It is particularly suited to measurement of trace and ultratrace impurities in products not amenable to concentration techniques. It is element- and isotope-specific and has CIC capability. It is typically the most expensive of the three technologies.

Additional references discussing GC with MS, AED, and ICP–MS detection can be found in other publications [44–49].

REFERENCES

1. Bouyssiere, B., Szpunar, J., & Lobinski, R. (2002). Gas chromatography with inductively coupled plasma mass spectrometric detection in speciation analysis. *Spectrochimica Acta B: Atomic Spectroscopy, 57,* 5, 805–828.

2. Hübschmann, H.-J. (2001). *Handbook of GC/MS: Fundamentals and Applications.* New York: Wiley-VCH.

3. Valco Instruments Co. Inc. (2012). *Home.* Retrieved from http://www.vici.com/index.php.

4. National Institute of Standards and Technology (2011). *NIST/EPA/NIH Mass Spectral Library 2011 (Update).* New York: Wiley-VCH.

5. Agilent Technologies, Inc. (2012). *KB001864: How is synchronous SIM/Scan performed?* Agilent Technologies, Inc. Retrieved from http://www.chem.agilent.com/en-US/Support/FAQs/MS/Software/MS-ChemStation/Pages/KB001864.aspx.

6. National Exposure Research Laboratory, Office of Research and Development, U.S. Environmental Protection Agency (1995). *Method 524.2. Measurement of Purgeable Organic Compounds in Water by Capillary Column Gas Chromatography/Mass Spectrometry (Rev 4.1.),* J. W. Munch (Ed.). Cincinnati, OH; U.S. EPA.

7. Duan, Y., & Jacksier, T. (2009). *Removal of krypton and xenon impurities from argon by MOF adsorbent.* International Patent No. WO 2011/053820 A1.

8. Wilbur, S. M., & Soffey, E. (2000). *Analysis of Arsenic, Selenium, and Antimony in Seawater by Continuous-Flow Hydride ICP–MS with ISIS.* Agilent Technologies Application Note 5980-0243E.

9. U.S. Environmental Protection Agency (2010). *Inorganic Superfund Methods (ISM01.3) Statement of Work (SOW).* Exhibit D, Part B, analytical methods for inductively coupled plasma–mass spectrometer, Section 4.1, Iso-

baric elemental interferences. Washington, DC: U.S. EPA. Retrieved from http://www.epa.gov/superfund/programs/clp/download/ism/ism12d.pdf.

10. Quimby, B. D., Uden, P. C., & Barnes, R. M. (1978). Atmospheric pressure helium microwave detection system for gas chromatography. *Analytical Chemistry, 50*, 14, 2112–2118.

11. Sullivan, J. J., & Quimby, B. D. (1989). Detection of C, H, N, and O in capillary gas chromatography by atomic emission. *Journal of High Resolution Chromatography, 12*, 5, 282–286.

12. Quimby, B. D., & Sullivan, J. J. (1990). Evaluation of a microwave cavity, discharge tube and gas flow system for combined gas chromatography–atomic emission detection. *Analytical Chemistry, 62*, 1027–1034.

13. Sullivan, J. J., & Quimby, B. D. (1990). Characterization of a computerized photodiode array spectrometer for gas chromatography–atomic emission spectrometry. *Analytical Chemistry, 62*, 1034–1043.

14. Eckert-Tilotta, S. E., Hawthorne, S. B., & Miller, D. J. (1992). Comparison of commercially available atomic emission and chemiluminescence detectors for sulfur-selective gas chromatographic detection. *Journal of Chromatography A, 591*, 313–323.

15. Houk, R. S., Fassel, V. A., Flesch, G. D., Svec, H. J., Gray, A. L., & Taylor, C. E. (1980). Inductively coupled argon plasma as an ion source for mass spectrometric determination of trace elements. *Analytical Chemistry, 52*, 14, 2283–2289.

16. Taylor, H. E. (2001). *Inductively Coupled Plasma–Mass Spectrometry: Practices and Techniques*. San Diego, CA: Academic Press, pp. 2–7.

17. Chong, N. S., & Houk, R. S. (1987). Inductively coupled plasma–mass spectrometry for elemental analysis and isotope ratio determinations in individual organic compounds separated by gas chromatography. *Applied Spectroscopy, 41*, 1, 66–74.

18. Hutton, R. C., Bridenne, M., Coffre, E., Marot, Y., & Simondet, F. (1990). Investigations into the direct analysis of semiconductor grade gases by inductively coupled plasma mass spectrometry. *Journal of Analytical Atomic Spectrometry, 5*, 463–466.

19. Thomas, R. (2008). *Practical Guide to ICP–MS: A Tutorial for Beginners*. Boca Raton, FL: CRC Press.

20. Busch, K. L. (2009). Higher resolution mass analysis in inductively coupled plasma-mass spectrometry. *Spectroscopy, 24*, 1, 30–37.

21. Jarvis, K. E., Gray, A. L., & Houk, R. S. (1992). *Handbook of Inductively Coupled Plasma Mass Spectrometry*. London: Blackie & Son.

22. Nelms, S. M. (2005). *ICP–Mass Spectrometry Handbook*. Oxford, UK: Blackwell Publishing.

23. Feldmann, J., Koch, I., & Cullen, W. R. (1998). Complementary use of capillary gas chromatography–mass spectrometry (ion trap) and gas chromatography–inductively coupled plasma mass spectrometry for the speciation of volatile antimony, tin and bismuth compounds in landfill and fermentation gases. *Analyst, 123*, 815–820.

24. Maillefer, S., Lehr, C. R., & Cullen, W. R. (2003). The analysis of volatile trace compounds in landfill gases, compost heaps and forest air. *Applied Organometallic Chemistry, 17*, 154–160.

25. Wilbur, S., & Soffey, E. (2004). Detecting "new PCBs" using GC–ICPMS: challenges of PBDE analysis. *Agilent ICP–MS Journal, 18*, 2–3.

26. Kucklick, J. R., & Davis, W. C. (2008). The determination of polybrominated diphenyl ether congeners by gas chromatography inductively coupled plasma mass spectrometry. *Journal of Analytical Atomic Spectrometry, 23*, 12, 1557–1696.

27. DeSmaele, T., Vercauteren, J., Moens, L., Dams, R., & Sandra, P. (2001). Capillary GC–ICP–MS: unsurpassed sensitivities for metal speciation in environmental samples. *American Laboratory, 33*, 17, 39–40.

28. Meija, J., Montes-Bayon, M., Le Duc, D. L., Terry, N., & Caruso, J. A. (2002). Simultaneous monitoring of volatile selenium and sulfur species from Se-accumulating plants (wild type and genetically modified) by GC/MS and GC/ICP–MS using solid-phase microextraction for sample introduction. *Analytical Chemistry, 74*, 22, 5837–5844.

29. Geiger, W. M., McSheehy, S., & Nash, M. J. (2010). Application of ICP–MS as a multi-element detector for sulfur and metal hydride impurities in hydrocarbon matrices. *Journal of Chromatographic Science, 45*, 10, 677–682.

30. Ellis, J., Rechsteiner, C., Moir, M., & Wilbur, S. (2011). Determination of volatile nickel and vanadium species in crude oil and crude oil fractions by gas chromatography coupled to inductively coupled plasma mass spectrometry. *Journal of Analytical Atomic Spectrometry, 26*, 1674–1678.

31. Meyer, C., & Geiger, W. M. (1997). The chromatographic analysis of trace atmospheric gases. In J. D. Hogan (Ed.), *Specialty Gas Analysis*. New York: Wiley-VCH, pp. 43–79.

32. Geiger, W. M., & Raynor, M. W. (2008). ICP–MS: A universally sensitive GC detection method for specialty and electronic gas analysis. *Spectroscopy, 23*, 11, 34–42.

33. Kim, A. W., Foulkes, M. E., Ebdon, L., Hill, S. J., Patience, R. L., Barwise, A. G., & Rowland, S. J. (1992). Construction of a capillary gas chromatography inductively coupled plasma mass spectrometry transfer line and application of the technique to the analysis of alkyllead species in fuel. *Journal of Analytical Atomic Spectrometry, 7*, 1147–1149.

34. Pretorius, W. G., Ebdon, L., & Rowland, S. J. (1993). Development of a high-temperature gas chromatography–inductively coupled plasma mass spectrometry interface for the determination of metalloporphyrins. *Journal of Chromatography, 646*, 369–375.

35. Bayon, M. M., Camblor, M. G., Alonso, J. I. G., & Sanz-Medel, A. (1999). An alternative GC–ICP–MS interface design for trace element speciation. *Journal of Analytical Atomic Spectrometry, 14*, 1317–1322.

36. Glindemann, D., Ilgen, G., Herrmann, R., & Gollan, T. (2002). Advanced GC/ICP–MS design for high-boiling analyte speciation and large volume solvent injection. *Journal of Analytical Atomic Spectrometry, 17*, 1386–1389.

37. Vonderheide, A. P., Meija, J., Montes-Bayon, M., & Caruso, J. A. (2003). Use of optional gas and collision cell for enhanced sensitivity of the organophosphorus pesticides by GC–ICP–MS. *Journal of Analytical Atomic Spectrometry, 18*, 1097–1102.

38. Hu, Z., Liu, Y., Gao, S., Hu, S., Dietiker, R., & Geèunther, D. (2008). A local aerosol extraction strategy for the determination of the aerosol composition in laser ablation inductively coupled plasma mass spectrometry. *Journal of Analytical Atomic Spectrometry, 23*, 9, 1192–1203.

39. Park, G. (2003). Transition states of homologous SN reactions of all-(e)-polyenolate derivatives. *Bulletin of the Korean Chemical Society, 24*, 3, 265–266.

40. Mason, P. R. D., Kaspers, K., & Van, B. M. J. (1999). Determination of sulfur isotope ratios and concentrations in water samples using ICP–MS incorporating hexapole ion optics. *Journal of Analytical Atomic Spectrometry, 14*, 7, 1067–1074.

41. D'Ilio, S., Violante, N., Majorani, C., & Petrucci, F. (2011). Dynamic reaction cell ICP–MS for determination of total As, Cr, Se and V in complex matrices: Still a challenge? A review. *Analytica Chimica Acta, 698*, 6–13.

42. Wilbur, S. M., & Soffey, E. (2003). *Quantification and Characterization of Sulfur in Low Sulfur Reformulated Gasolines by GC–ICP–MS*. Agilent Technologies Application Note 5988-9880EN.

43. Bandura, D. R., Baranov, V. I., & Tanner, S. D. (2002). Detection of ultratrace phosphorous and sulfur by quadrupole ICP–MS with dynamic reaction cell. *Analytical Chemistry, 74*, 7, 1497–1502.

44. Crompton, T. R. (1982). *Gas Chromatography of Organometallic Compounds*. New York: Plenum Press.

45. McDonald, J. C. (1985). *Inorganic Chromatographic Analysis*. New York: John Wiley & Sons.

46. Hill, H. H., & McMinn, D. G. (1992). *Detectors for Capillary Chromatography*. New York: John Wiley & Sons.

47. Sievers, R. E. (1995). *Selective Detectors*. New York: John Wiley & Sons.

48. Scott, R. P. W. (1996). *Chromatographic Detectors*. New York: Marcel Dekker, Inc.

49. Hogan, J. D. (Ed.). (1997). *Specialty Gas Analysis*. New York: Wiley-VCH.

CHAPTER 7

TRACE WATER VAPOR ANALYSIS IN SPECIALTY GASES: SENSOR AND SPECTROSCOPIC APPROACHES

MARK W. RAYNOR,[1] KRIS A. BERTNESS,[2] KEVIN C. COSSEL,[3] FLORIAN ADLER,[3] AND JUN YE[3]

[1]Matheson, Advanced Technology Center, Longmont, Colorado
[2]National Institute of Standards and Technology, Boulder, Colorado
[3]JILA, National Institute of Standards and Technology, and University of Colorado, Department of Physics, Boulder, Colorado

7.1 Introduction

The analysis of water vapor impurity (often also referred to as *moisture*) is important in a number of specialty gas applications. However, the main driver for the development and advancement of trace moisture analysis techniques has been the microelectronics industry. The International Technology Roadmap for Semiconductors (ITRS) details the gas purity requirements for various wafer fabrication processes on their website (www.itrs.net). Oxygenated contaminants such as water vapor in the materials used and in the wafer environment are primary causes of defects and process variations that compromise yield. Because even trace levels of water in the process gases can seriously decrease device performance, analytical techniques must be capable of detecting water vapor from ppmv (µmol/mol) down to the low and even sub-ppbv (nmol/mol) range.

Measurement of water vapor at trace levels is quite challenging not only due to the adsorptive nature of the water molecule on metal and other surfaces, but also because of the range of gas matrices (including inert, oxide, halide, hydride, corrosive, hydrocarbon, and halocarbon gases) that can potentially interfere with the measurement process. There is no single approach that meets all analysis requirements, and hence a range of techniques have been investigated over the years [1]. These include:

(a) Gas chromatography (GC)

(b) Mass spectrometry with electron impact and atmospheric pressure ionization sources (EI–MS, APIMS)

(c) Ion mobility spectrometry (IMS)

(d) Sensor-based methods such as chilled-mirror hygrometry, oscillating quartz crystal microbalance (QCM), capacitance cell, and electrolytic cell

(e) Spectroscopic methods such as Fourier transform infrared (FTIR) spectroscopy, tunable diode laser absorption spectroscopy (TDLAS), cavity ring-down spectroscopy (CRDS), intracavity laser spectroscopy (ILS), and cavity-enhanced frequency-comb spectroscopy (CE–DFCS)

Some approaches are better suited than others to water vapor analysis. For example, gas chromatography and mass spectrometry, methods that are more commonly used for analyzing other gas-phase impurities, are not employed routinely for quantitative water vapor measurements in many specialty gases. This is due mainly to issues such as elevated water vapor background concentrations in the instrumentation, loss of water vapor due to adsorption, sampling difficulties, matrix gas interference, poor sensitivity, or need for frequent calibration. Reaction GC, where, by reaction with calcium carbide, water in the sample is converted to acetylene, a species that is easier to chromatograph, is an option for ppmv-level water vapor detection. However, a reported variance of ± 1 ppmv for water vapor challenges in helium and ammonia precludes its use for sub-ppmv analysis [2]. Mass spectrometry–based methods can be used but suffer from water vapor background and drift issues. Further, the sensitivity of electron impact ionization sources is limited to around 0.5 ppmv water vapor levels unless selected matrix gases such as helium or hydrogen are used [1]. Atmospheric pressure ionization mass spectrometry (discussed in more detail in Chapter 5) and ion mobility spectrometry offer sub-ppbv detection capability for water vapor in bulk gases and are employed primarily for gases such as argon, nitrogen, helium, and hydrogen [3–6]. However, neither of these techniques is commonly applied in specialty gases. In many cases the high ionization potential of water prevents ionization by charge transfer. Therefore, APIMS methods in gases such as ammonia, silane, germane, and hydrogen chloride require development of specialized ionization mechanisms and cluster analysis by specialized personnel in order to measure water vapor [7–10]. Further, corrosion of source components, formation of solid decomposition products, and the potential for contamination also complicate analysis of many reactive hydride and corrosive gases.

In this chapter we therefore focus on technologies that not only provide high sensitivity for water vapor but can also be applied to a wide range of specialty gases. They can be broadly categorized into sensor- and spectroscopic-based approaches. Aspects related to water standards and calibration are also discussed.

7.2   Primary Standards for Water Vapor Measurement

The primary standard for water vapor measurement in gases is the standard hygrometer, which has been developed to achieve an expanded uncertainty of 0.1 % for water vapor in inert gases at concentrations of 250 µg/g or above. (The expanded uncertainty, u, is defined as the range that covers two standard deviations, σ, about the mean of a normal distribution of measurements $u = k\sigma$ with $k = 2$, for a 95 % confidence interval.) The standard hygrometer maintained at the National Institute of Standards and Technology (NIST) has recently been improved to achieve this level and also to measure lower concentrations (down to 250 µg/g) with somewhat higher uncertainty (1 %) [11]. This hygrometer measures water vapor by separating it from the carrier gas using desiccant material and cold traps and then measuring both the change in weight of the traps and the total mass of dried carrier gas that flowed during the sampling period. The centerpiece of this instrument is a set of prover tubes that collect the dry gas and measure its volume, temperature, and pressure with high accuracy (see Figure 7.1). The ideal gas law is then applied to calculate the total mass of the gas collected. A complete description of the system and uncertainty analysis is given by Meyer et al. [11].

Although the NIST standard gravimetric hygrometer can achieve exceptional accuracy, as with many primary standards, its design emphasizes accuracy over convenience. The system is not readily transportable and requires long measurement periods and control of ambient conditions to maintain accuracy. Transfer of the measurement to secondary methods is accomplished by using the standard hygrometer to characterize primary humidity generators, which are then used to calibrate hygrometers for field use, such as chilled-mirror hygrometers, electrolytic hygrometers, and cavity ring-down spectroscopy systems. NIST maintains two such primary humidity generators, the hybrid humidity generator (HHG) [12] and the low-frost-point generator (LFPG) [13], as a part of their calibration services for portable hygrometers. Both of these systems generate saturated water vapor in equilibrium with a carrier gas (typically, purified air or nitrogen) at a constant temperature. The concentration of water vapor at the system outlet can be varied either by adjusting the gas pressure after leaving the saturation chamber or by mixing in additional dry gas, or both. The HHG shown in Figure 7.2 has a range of water vapor mole fractions from 1 ppmv to 0.3 % (frost-point/dew-point temperatures from −76 to 40 °C). Compared to the NIST standard hygrometer described above, the two systems agreed within their combined expanded ($k = 2$) uncertainty of 0.22 %. As its name suggests, the LFPG was developed to calibrate instrumentation in even lower water vapor concentration ranges, from 3 ppmv down to 3 ppbv, corresponding to frost-point temperatures from −10 to −100 °C. Volumetric flow rates of up to 4 slpm can be accommodated. The relative expanded ($k = 2$) uncertainty for low-water-vapor mole fractions (3 nmol/mol) is only 0.8 % [14].

The LFPG also forms the basis of a new transfer calibration standard based on permeation tube humidity generators. These generators flow liquid water through a tube with a stable permeation rate into a gas stream in a temperature-stabilized chamber from which the humidified gas is extracted and then mixed with additional

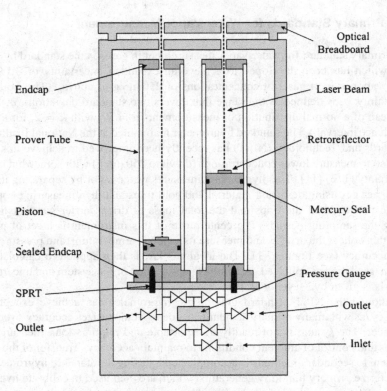

Figure 7.1   Prover tubes from a second-generation NIST standard hygrometer. Each prover tube is about 1 m in length and 14 cm in diameter.

dry gas if desired. The low cost and high stability of these systems makes them popular for the calibration of various commercial detectors. The exact permeation rate of a given tube, however, is difficult to predict from its geometry and average materials properties. Using a nulling technique whereby the humidity output of the primary standard LFPG is set equal to that of a permeation tube generator under test, errors related to humidity analyzer zero, nonlinearity, and absolute response can be eliminated. By use of this technique, the permeation rate of a typical tube was found to be stable to within 7 % over five calibrations spanning a nine-month period, varying less than 1 ppbv over the total gas flow range of the permeation tube generator [15]. This stability compares favorably with 10 to 30 % errors for permeation tube moisture generators calibrated with less reliable methods.

A portable humidity source alternative to a permeation tube is the water diffusion vial. These vials contain a reservoir connected to a long narrow capillary tube that restricts the flow of water vapor. The mass diffusion of water out of the vial depends on the vial temperature, total gas pressure, and capillary cross-sectional area, along with diffusion and density parameters for water vapor. The mass diffusion rate can be calculated within approximately 10 % for a given diffusion vial, or the vial

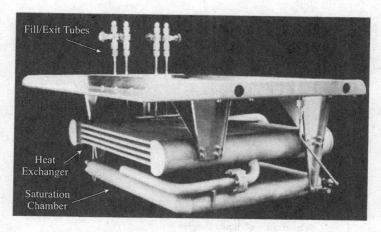

Figure 7.2   Final saturation chamber of a NIST hybrid humidity generator. During actual operation, this portion of the system is submerged in a temperature-controlled bath.

can be maintained at constant pressure and temperature for several days or weeks and its change in mass measured with a precision balance. Mass flow rates of 1×10^{-6} g/min are typical. When the diffusion vial is placed in a stream of dry gas, the water vapor concentration is a simple function of the mass diffusion rate, total pressure, temperature, and carrier gas flow rate.

7.2.1 Sampling for Instrument Calibration and Gas Analysis

A generic manifold for generating and delivering water vapor standards to analytical instrumentation as well as for the analysis of specialty gas samples for water vapor is shown in Figure 7.3. For calibration, known concentrations of water vapor in dry nitrogen are first generated in diffusion or permeation systems as described above and diluted dynamically [16]. This entails spiking the moist stream into a stream of the dried sample matrix gas under set conditions of flow rate, temperature, and pressure to generate standards at different concentrations. Dry matrix gas is typically required as a reference or diluent for most instruments and is obtained by passing the sample gas through a suitable purifier for the gas concerned. Since the water vapor added to the dry matrix gas is in a nitrogen or inert carrier stream, the amount of carrier gas is typically kept below 10 % to minimize any effects of the matrix gas composition on the instrument response. This is particularly important for spectroscopic methods, as discussed later in the chapter.

As water is adsorptive, the surface area and dead-volume regions within the sampling manifold are minimized in the design. This approach ensures that the water vapor concentration rapidly equilibrates in the gas delivered to the analyzer. Regulators, sampling lines, and the analyzer are also thermally controlled (e.g., 60 °C) to hasten equilibration and to prevent ambient temperature fluctuations from affecting the readings. An electronic pressure controller is most often installed downstream of

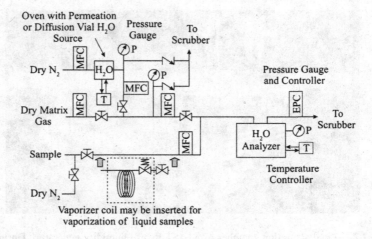

Figure 7.3 Generic manifold for instrument calibration and sample gas analysis.

the measurement point so that pressure can be accurately controlled in the analyzer. Manifold materials require careful selection, as the addition of water vapor/nitrogen to some reactive and acid gases can result in corrosion and particulate generation. Electropolished 316 stainless steel components and lines are most commonly used. However, other alloys, such as Hastelloy, are preferred for corrosive gases such as hydrogen bromide. Additional information, including the effects of reactions between process gases and tubing walls, may be found in the review by Funke et al. [1].

Water vapor in the sample gas is measured by flowing gas from the cylinder directly to the calibrated analyzer. In the case of compressed liquefied gases, liquid-phase sample can also be withdrawn from the cylinder and vaporized in a vaporizer coil prior to introduction to the analyzer (Figure 7.3). In both cases, sample gas is delivered to the analyzer under the same conditions of temperature and pressure as those used for calibration.

7.3 Sensor Technologies

There are a number of conventional sensor technologies used for water vapor detection that rely on adsorption or condensation of water vapor from the gas sample onto solid materials. These are typically robust, convenient, fairly inexpensive devices that have a small footprint and are widely used in inert or nonreactive gas matrices. Four main sensor technologies are discussed here.

7.3.1 Electrolytic Cells

Electrolytic hygrometers operate on the basis that water molecules can be electrolyzed into molecular oxygen and hydrogen by application of a voltage greater than the thermodynamic decomposition voltage of water (2 V). A typical electrolytic hygrometer comprises a sensing cell coated with a thin film of phosphorus pentoxide electrolyte, which is highly hygroscopic and absorbs water from the sample gas stream forming $P_2O_5(H_2O)_n$. The cell consists of a hollow glass tube with two inert platinum or rhodium electrodes spirally wound around the inside wall and in contact with the phosphorus pentoxide electrolyte layer (Figure 7.4). The gas sample flows through the tube at a set flow rate (100 mL/min) and contacts the phosphorus pentoxide electrolyte. By applying an electrical potential to the electrodes, the water in the film is electrolyzed. Once equilibrium is reached, the rate at which water molecules enter the cell will exactly match the rate at which molecules are electrolyzed. Each water molecule produces the flow of two electrons through the external circuit as shown below, and the steady-state current created by these electrons can be related directly to the water concentration using Faraday's law and the sample flow rate.

$$H_2O + e^- \rightarrow \frac{1}{2} H_2 + OH^-$$

$$OH^- \rightarrow \frac{1}{2} O_2 + \frac{1}{2} H_2 + e^-$$

The theory of operation of phosphorus pentoxide electrolytic hygrometers has been discussed in more detail by McAndrew [17] and Wiederhold [18].

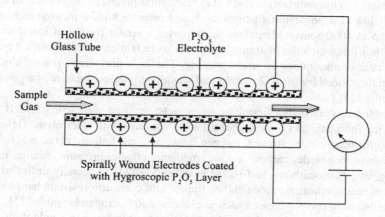

Figure 7.4 P_2O_5 electrolytic sensor.

Phosphorus pentoxide electrolytic cells were invented by Keidel [19] and developed by the DuPont Company. The technology is commercially available today from companies such as Meeco and GOW–MAC Instruments.* Since phosphorus pentoxide cells are used in the electrolysis of water vapor, the method is often considered to be a primary measurement method. However, cell-specific effects and different modes of operation that result in measurement errors are observed, and therefore certification against a water vapor standard is recommended on a periodic basis. Typical electrolytic hygrometers can detect water vapor in the range from a few ppbv to 2000 ppmv with an accuracy of ±5 %, provided that the flow rate of the sample gas through the cell is properly controlled. Water vapor levels higher than 2000 ppmv should be avoided, as the electrolyte will become completely hydrated and form liquid orthophosphoric acid that can leave the cell. At low water vapor concentrations, loss in sensitivity and long equilibration and response times can be experienced, especially when the sensor has been exposed to very dry gas for long periods. This occurs as the electrolyte becomes progressively drier. The problem has been minimized through the addition of glycols and other proprietary components to the electrolyte [20] or by the cyclic addition of controlled amounts of water vapor in combination with flow modulation to keep the layer sufficiently wet [21]. However, even so, electrolytic phosphorus pentoxide hygrometers still respond relatively slowly to step changes and are more similar to capacitive sensors in this regard [22]. In recent years rhodium has been used for the electrodes rather than platinum, as platinum acts as a catalyst in a recombination reaction between chemisorbed hydrogen on the platinum and the oxygen product from the electrolysis. Electrolysis of the recombined water results in erroneously high water vapor measurements in oxygen and hydrogen matrices. The use of rhodium eliminates this recombination reaction. In addition, rhodium minimizes the tendency of metallic ions to migrate under the influence of the potential applied, which results in the formation of metallic compounds and the eventual bridging of the electrodes. Modern developments continue with respect to miniaturization and novel fabrication methods. Meeco manufactures a small low-cost phosphorus pentoxide–based monitor with a measurement range from 0.5 to 1000 ppmv. Manickam has reported a sensor for use at low flow rates (5 mL/min) in which the platinum electrodes were fabricated either with a pulsed-laser ablation technique or by screen printing. The thick-film screen-printed electrode version functioned for over 12 months and was able to detect moisture in argon down to 1 ppmv [23].

Phosphorus pentoxide electrolytic hygrometers can be used with all inert gases and many inorganic and organic gases that do not react with electrolyte. Gases that are compatible include nitrogen, oxygen, hydrogen, air, helium, neon, argon, krypton, carbon monoxide, carbon dioxide, methane, ethane, propane, butane, natural gas, Freons, fluorocarbons, and sulfur hexafluoride [24]. Specially designed units operated in amperometric mode and equipped with corrosion-resistant hardware are compatible with corrosive gases such as chlorine and hydrogen bromide [25]. However, measurements in corrosive gases are only possible down to single-digit ppmv

*Manufacturers and product names are given solely for completeness. These specific citations neither constitute an endorsement of the product nor imply that similar products from other companies would be less suitable.

levels, as only a fraction of the incoming water is electrolyzed, and electrolysis is diffusion limited. Gases that should be avoided include amines, ammonia, and certain unsaturated hydrocarbons (alkynes, alkadienes, and alkenes higher than propylene), which can polymerize and form liquid or solid residues that may block the cell.

7.3.2 Oscillating Quartz Crystal Microbalances

Oscillating quartz crystal microbalances (QCMs) are piezoelectric quartz crystals that have been coated with a thin film of a hygroscopic material which absorbs water vapor selectively from the sample gas passing over it. As the water vapor concentration increases, the water absorbed increases the weight of the film and decreases the oscillation frequency of the QCM, according to [26]:

$$\Delta f = -Cm\,\Delta m \qquad (7.1)$$

where Δf is the change of the crystal resonance frequency resulting from a change in mass per unit surface area (Δm) and the constant Cm is a property of the crystal. Frequency-change measurements are made relative to a reference crystal that is not exposed to the gas stream. This compensates for temperature drifts and other external factors that similarly affect both crystals and enables high-sensitivity measurements down to low- and sub-ppbv levels. By alternating the flow of sample and dry reference gas over the QCM for short intervals (e.g., 30 seconds), the frequency difference due to water absorption during each cycle can be determined:

$$\Delta f = (f_{\text{sample gas}} - f_{\text{reference crystal}}) - (f_{\text{dry gas}} - f_{\text{reference crystal}}) \qquad (7.2)$$

Because the frequency change is proportional to the gas-phase water concentration, a dedicated microprocessor collects the frequencies and calculates the concentrations using a polynomial expression obtained through prior calibration. During measurements, the period during which the sensor is exposed to sample gas flow is not sufficient to equilibrate the coating with water. However, the amount of water accumulated in a specific sampling interval is proportional to the water vapor concentration and can be used as a measure of the actual concentration during each sampling cycle (Figure 7.5a and b). The theory of QCM technology has been discussed and reviewed in more detail by O'Sullivan and Guilbault [27]. Instrument design and practical aspects of the technology for gas analysis are also covered in some detail by Wiederhold [24] and by Funke et al. [1].

Various hygroscopic coating materials that absorb water preferentially have been investigated for QCM analyzers, including zeolites, metal oxides, salts, and polymers [27–30]. Commercial QCM technologies typically use proprietary polymer coatings. Early units were introduced by DuPont Company and Shimadzu Corporation (Tokyo, Japan). Today, instrumentation manufacturers include Ametek Process Instruments (Newark, DE) and Mitchell Instruments (Ely, Cambridgeshire, U.K.). Figure 7.5, shows an Ametek 5800 QCM hygrometer, including a flow schematic with main components, the process measurement cycle and performance for ppbv-level water vapor detection in nitrogen. The QCM in this unit is an AT-cut quartz crystal operated with a thickness-shear mode of oscillation at 9 MHz and temperature controlled at 60 °C. Filtered sample and reference gases at 200 mL/min are passed over the sensor sequentially in each measurement cycle.

A number of factors that to be considered when making measurements with QCM hygrometers. First, as measurements are made relative to a reference gas stream, a high-efficiency purifier (typically, a molecular sieve or other suitable material) must be used to generate a "dry" reference gas with a water vapor concentration well below the detection limit of the sensor. Second, as the amount of water absorbed on the sensor is dependent on temperature and pressure, control of these parameters within the analyzer is vital for accurate measurements, especially at low concentrations. Other factors to account for include differences in flow rate, line length, and gas composition between sample and reference streams. In the case of the latter, different sample and reference gas matrices can be used in some QCM systems by applying a correction factor that accounts for factors such as gas viscosity [17]. However, more accurate and sensitive results are typically obtained when the same sample and reference gas matrix are used. Users should also be aware that the absorption of water vapor by the QCM sensor is dependent on the material properties of the polymer film. Although sensors are quite rugged, these properties can change over time or can be affected by exposure to different sample gases. Therefore, periodic calibration of the sensor, as well as detection limit confirmation is required. Commercial systems typically include a permeation-based moisture generator that is integrated into the instrument sampling system and allow users to verify calibrations using a single concentration point (ca. 1 ppmv level). However, multipoint calibration and permeation tube replacement are advisable at least on an annual basis.

Detection limits for QCM instrumentation in specialty gases are typically in the low to single-digit ppbv range, with a dynamic range between 0 and approximately 2000 ppmv (Figure 7.5c). However, detection limits in the 150 pptv range have also been reported recently for specialized units used in inert matrix gases [31]. Response time is important for process monitoring applications. The speed of response to water vapor changes is relatively rapid compared to that of other sensor-based methods. Wet-up response periods of 80 % to a step change in less than 10 minutes have been reported at the 10-ppbv level. In other dry-down tests, for an external water vapor change from 50 to 23 ppbv (Figure 7.5d) the sensor reached 95 % of the final value in just over an hour. QCM hygrometers are compatible with various gas matrices, including inert gases, air, oxygen, hydrogen, nitric oxide, carbon monoxide, carbon dioxide, hydrocarbons (such as methane, ethylene, and propene), fluorocarbons (such as tetrafluoromethane or hexafluoroethane), nitrogen trifluoride, and sulfur hexafluoride. In addition, some reactive gases, such as silane and arsine, can also be analyzed [32]. Figure 7.6 shows a time plot of water vapor spiked into dry arsine from 22.6 ppbv to 1.8 ppmv and the corresponding graph of measured versus added water vapor using a QCM hygrometer. Due to the toxicity or flammable/pyrophoric nature of some reactive gases, the analyzer must be specially prepared to ensure material compatibility and leak-free operation of switching valves over time in the unit.

The technology is not suitable for use in corrosive gases or with gases such as ammonia that can strip off the polymer layer or otherwise degrade the sensor. In

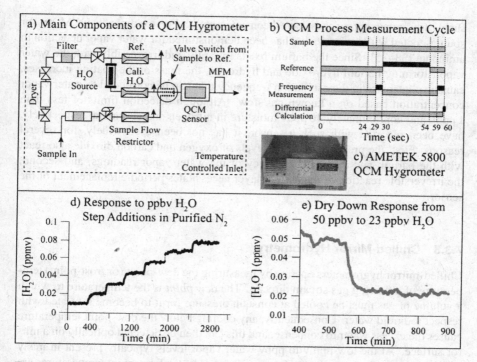

Figure 7.5 Main components, process measurement cycle, and wet-up and dry-down responses of a QCM hygrometer.

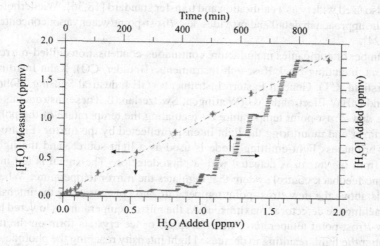

Figure 7.6 Measurement of water vapor in arsine using a QCM hygrometer (Ametek model 5800). Time plot shows water vapor addition to dry arsine (upper x and right y axes). The graph shows added versus measured water vapor in the concentration range 22.6 ppbv to 1.8 ppmv (lower x axis and left y axis).

such cases other approaches have been attempted. For example, a QCM sensor sputter-coated with barium metal has been demonstrated for water vapor detection in ammonia [33–35]. Since the barium reacts instantly and irreversibly with the water vapor, forming barium hydroxide and hydrogen, the mass of the coating increases, causing a decrease in the oscillation frequency that can be correlated to water vapor concentration based on a known gas flow. Although detection limits of less than 1 ppbv and approximately 10 ppbv moisture in nitrogen and ammonia, respectively, have been reported with such a sensor, it has not been used widely, for several reasons. First, the presence of ppmv levels of oxygen and carbon dioxide also react with the barium and affect the accuracy of the water vapor readings, and second, the irreversible reaction of the barium layer necessitates periodic replacement of the sensor.

7.3.3 Chilled-Mirror Hygrometry

Chilled-mirror hygrometers operate by measuring the dew-point (or frost-point) temperature of a flowing gas stream directly. The *dew point* is the temperature to which a volume of gas must be cooled at constant pressure for it to become saturated with respect to liquid water. Consequently, any cooling below the dew-point temperature causes the excess water to condense, and this can then be detected optically on a mirror surface. At the low-ppmv to ppbv water vapor levels typically present in many specialty gases, the excess water forms as ice, and hence frost-point temperatures are measured. Since the dew- or frost-point temperature is a fundamental thermodynamic property, chilled-mirror hygrometry is considered an absolute measurement method and is also used widely as a calibration and transfer standard [18,36]. Wiederhold discusses the approach in detail and provides dew-/frost-point/water vapor concentration tables [24].

A number of companies manufacture continuous-condensation chilled-mirror hygrometers, including Buck Research Instruments (Boulder, CO), Kahn Instruments (Wethersfield, CT), General Eastern Instruments (GE Industrial Sensing, Bellerica, MA), and MBW Electronik AG (Wettingen, Switzerland). These instruments measure the dew-/frost-point temperature by regulating the temperature of the polished metal mirror and monitoring the light intensity reflected by the mirror (Figure 7.7). A high-brightness light-emitting diode is used as a light source, and the light reflected from the mirror is detected with a photodetector. The signal is fed into an electronic feedback control system that regulates the mirror temperature. When the mirror is above the dew-/frost-point temperature (no condensation) the intensity of light reaching the detector is maximized. As the mirror temperature is lowered below the dew-/frost-point temperature, water droplets or ice crystals form on the mirror and scatter the light, resulting in decreased light intensity reaching the photodetector. The frost layer thickness is typically displayed as a balance voltage. With a properly designed feedback control system, the mirror temperature is regulated such that the rate of condensation exactly equals the rate of evaporation, and the ice layer thickness remains constant. Under this condition the dew/frost on the mirror is in equilibrium with water vapor in the gas, and the surface temperature of the mirror is precisely at the dew/frost point of the flowing gas stream.

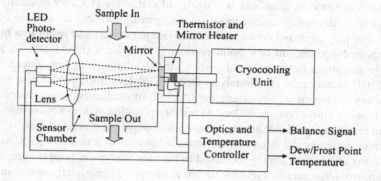

Figure 7.7 Principal components of a conventional chilled-mirror hygrometer.

The detection limit of a chilled-mirror hygrometer is based on the lowest mirror temperature that can be achieved by the cooling system or that can be reached before any matrix gas condensation occurs on the mirror (in the case of some specialty gases). Thermoelectric or Peltier cooling units used in many instruments cool the mirror down to −85 °C (240 ppbv water in nitrogen). Other, more costly cryo-coolers are capable of cooling to temperatures below −96 °C (30 ppbv water in nitrogen). The Kleemenko cycle cryo-cooler in the Buck Research CR-3 unit reportedly cools to −120 °C, enabling detection below 1 ppbv. Care must be taken not to cool the mirror below the temperature at which the matrix gas condenses, as the ability to measure the dew point is lost at that point. Most hygrometers will allow a minimum mirror set-point temperature to be entered into the control unit of the instrument. Funke et al. discuss the case for analysis of phosphine and demonstrates water vapor detection at the 60 ppbv level [1]. In that work the mirror was kept above −86 °C to prevent phosphine condensation (boiling point −87.7 °C at 1 atm).

At sub-ppmv water vapor concentrations, several hours may be required before a suitable condensation layer is established. To hasten the process, a water vapor source can be pulsed into the flowing sample stream for a short period. Once a condensation layer is present, chilled-mirror hygrometers respond within minutes to changes in water vapor concentrations, although low-level readings may still take several hours to fully stabilize. As with other analysis methods, the sampling-manifold design, sensor chamber design, and sampling conditions may also affect equilibration times.

Several factors can affect the accuracy of measurements. First, it is important to minimize the error in the measurement of the mirror surface temperature at the point where condensation occurs. Therefore, high-end instruments employ precision NIST-traceable platinum resistance thermistors, embedded into the mirror surfaces, which have standard uncertainties below 0.1 °C. Second, the presence of other contaminants can affect readings adversely. Particulate or volatile impurities other than water can collect or condense on a cold mirror surface and reduce mirror reflectivity. Therefore, a common practice is to open the feedback control loop periodically and heat the mirror to a new "clean" state, then readjust the balance of the optical circuit to compensate for the reduced mirror reflectance. Even with rebalancing, some reactive gases, such as chlorine, ammonia, and oxides of sulfur, tend to react gradually with or dissolve in the water layer deposit on the mirror, forming an acid or caustic product that may decrease the vapor pressure (Raoult effect) and result in a higher-than-expected equilibrium dew-point reading [37]. Users should also be aware that at temperatures between 0 and $-40\,°C$, the dew layer may not convert to ice immediately or may exist indefinitely as supercooled dew. Supercooled dew tends to occur when the mirror is very clean, due to a lack of nucleation sites or because of certain types of mirror contamination. As misinterpretation of dew- versus frost-point data can result in significant errors, some chilled-mirror hygrometers have a "force frost" mode whereby the mirror temperature is lowered well below $-30\,°C$ to form a frost layer and is then slowly increased to the true frost-point temperature. Other chilled-mirror hygrometers cycle the mirror temperature periodically to measure the dew point [24]. Cycling chilled-mirror hygrometers repeatedly lowers the mirror temperature at a precisely controlled rate until dew is detected, and then the mirror is heated to evaporate the dew before a continuous layer is formed. Typically, the measurement cycle is performed in 20 seconds. With this approach the buildup of contaminants on the mirror is minimized and measurement of the dew-point rather than the frost-point temperature is ensured. However, cycling parameters must be set carefully, and the instrument must be calibrated against a NIST-traceable hygrometer to ensure accurate measurements. Consequently, the method is employed more for industrial applications rather than as a laboratory calibration standard [18].

Chilled-mirror measurements can be made in a variety of nonreactive and reactive gases, including air, nitrogen, argon, helium, hydrogen, oxygen, hydrogen chloride, hydrogen bromide, and phosphine. When analyzing corrosive gases, corrosion-resistant sensor chamber components must be used to prolong the lifetime of the analyzer. For example, sensor housings are usually fabricated from stainless steel and the replaceable part of the mirror is made from an inert metal such as platinum or rhodium. Other custom modifications or materials related to the particular gas being analyzed may be required. As dew-/frost-point curves can be strongly matrix gas–dependent, curves generated in one gas are not generally applicable in another [38]. For example, interactions between water and corrosive gases such as hydrogen chloride and hydrogen bromide gases significantly increase the frost points by approximately 30 and $51.6\,°C$, respectively, compared to nitrogen at the 1 ppmv moisture level (Figure 7.8). In comparison with phosphine, there is significantly less molecular interaction, resulting in only a minor positive shift of $4.8\,°C$ at the 1-ppmv level. Therefore, empirically derived calibration curves should be determined for each matrix gas of interest. Furthermore, chilled-mirror hygrometry is not applicable to all gases. For example, trace water vapor detection is not possible in gases such as ammonia and chlorine, which will condense on the mirror at temperatures corresponding to relatively high water vapor concentrations. However, in cases where it can be applied, the method offers significant advantages over other approaches, particularly with regard to its simplicity, accuracy, and fundamental measurement principle.

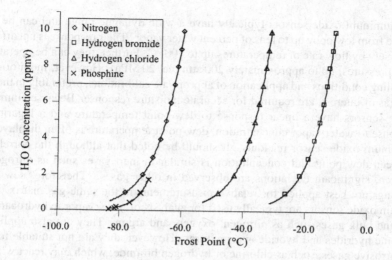

Figure 7.8 Plots of frost-point temperature versus moisture concentration in nitrogen, hydrogen chloride, hydrogen bromide, and phosphine matrix gases at 760 torr (101 kPa).

7.3.4 Capacitance-Based Sensors

Capacitance- or impedance-based sensors for water vapor analysis consist of a hygroscopic material that also functions as a dielectric and is situated between the two electrodes of a capacitor. Sensors based on aluminum oxide, silicon oxide, and porous silicon are commercially available. Materials for humidity sensors have been reviewed by Chen and Lu [39]. The most common capacitance hygrometer consists of a high-purity aluminum electrode that has been anodized to provide a porous oxide layer. A very thin gold coating is deposited over these structures to form the second electrode. During analysis, water vapor is transported rapidly through the gold layer and equilibrates in the aluminum oxide pore walls, affecting the dielectric constant of the material and as a result, the capacitance of the unit. The capacitance can be calculated according to

$$C = \varepsilon \varepsilon_0 \frac{A}{d} \tag{7.3}$$

where ε is the dielectric constant of the material, ε_0 is the vacuum permittivity, A is the overlapping area of the electrodes, and d is the distance between the electrodes. The large difference between the dielectric of the base material [$\varepsilon(Al_2O_3) = 9$] and water [$\varepsilon(H_2O) = 80.4$ at 20 °C] causes detectable changes in the capacitance, even though only small amounts of water adsorb on the dielectric from the gas phase. Resistive components of the hygroscopic layer are also affected by water adsorption, and hence commercial instrumentation typically measures the impedance (resistive, capacitive, and inductive component in an oscillating circuit) of the sensor and correlates changes with the appropriate calibration data.

Aluminum oxide sensors typically have a wide dynamic range and can be calibrated from low ppbv up to tens of percent water vapor. Their response is not affected adversely by flow rate or temperatures up to 100 °C. Further, they can be operated at high pressures, up to approximately 200 atm (ca. 20 MPa). However, monitoring of sampling conditions and application of appropriate calibrations, preferably in the matrix gas of interest, are required for accurate moisture response. Because aluminum oxide sensors have a linear response to dew-point temperature and a logarithmic response to water vapor concentration, dew-point temperature is often displayed on aluminum oxide sensor readouts. It should be noted that although the correlation between dew point and concentration is similar in inert gases such as nitrogen or oxygen, significant deviations are observed in other gases. Therefore, dew-point readings are best applied for relative measurements in the same gas matrix. Aluminum oxide sensors are typically used for analysis of water vapor in hydrocarbons, air, and bulk gases such as nitrogen, oxygen, and argon. They are also applicable in some hydrides and hydride–gas mixtures. However, they are not suitable for use in corrosive gases, such as chlorine or hydrogen bromide, which may react with the sensor materials.

Although these sensors offer advantages of low price, rugged design, and applicability to inline measurements at varying temperatures and pressures, they do require frequent calibration (every six months). Further, aluminum oxide sensors have traditionally suffered from slow response as well as drift (decreasing sensitivity) and dormancy issues [40] related to the intrinsic adsorption–desorption properties of aluminum oxide. Mehrhoff, for example, reported that after 11 days, the response of five aluminum oxide sensors decreased to 25 to 50 % of the true moisture level [41]. The decrease in sensitivity observed with time has been attributed to loss of sites for water adsorption, entrapment of water within the pores, and changes to the pore structure of the material [42,43]. To minimize such issues, recently developed aluminum oxide sensors have been fabricated with an integrated heating element so that the sensor temperature can be cycled periodically (Figure 7.9). In this approach, a temperature pulse is first applied to dry down the sensor, after which the sensor is allowed to cool down. The water concentration measurement is made based on the wet-up slope and resulting change in impedance [44].

Tests on one such sensor (Hygrotrace, GE Sensing, Billerica, MA) have shown good detection sensitivity (1.8 ppbv), accuracy (correlation coefficient of 0.9985 for measured versus actual water vapor concentration), detection limit (ca. 10 ppbv in nitrogen), acceptable wet-up and dry-down periods in the range 13 to 108 min (to reach 95 % of the final value), and good stability over 15 hours [45]. Thermally cycling the sensor is effective for low water vapor concentrations up to the maximum concentration specified (ca. 100 ppbv). However, once the heating step does not dry-down the sensor reproducibly, the accuracy of measurements decreases progressively as the water vapor concentration increases.

Silicon-based sensors are similar to aluminum oxide sensors, except that the dielectric material is either silicon oxide or porous silicon manufactured using silicon chip technologies. They have a wide dynamic range, from low ppbv to percent levels and operate independent of flow and at a pressure from vacuum to approximately 200 atm

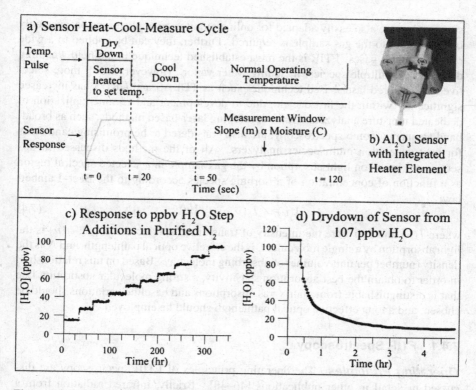

a) Sensor Heat-Cool-Measure Cycle

Temp. Pulse →

Dry Down

Sensor heated to set temp. | Cool Down | Normal Operating Temperature

Sensor Response →

Measurement Window
Slope (m) α Moisture (C)

t = 0   t = 20   t = 50   t = 120
Time (sec)

b) Al_2O_3 Sensor with Integrated Heater Element

c) Response to ppbv H_2O Step Additions in Purified N_2

[H₂O] (ppbv) axis: 0, 20, 40, 60, 80, 100
Time (hr) axis: 0, 100, 200, 300

d) Drydown of Sensor from 107 ppbv H_2O

[H₂O] (ppbv) axis: 0, 20, 40, 60, 80, 100, 120
Time (hr) axis: 0, 1, 2, 3, 4

Figure 7.9 Operating principle and performance of a Hygrotrace aluminum oxide sensor in nitrogen: (a) time plot showing the heat–cool–measure step sequence of the sensor; (b) sensor chip with an integrated heating element; (c) response of sensor to water vapor step changes in the concentration range 15 to 93 ppbv; and (d) dry-down response at low ppb levels. The sensor took 108 min to reach 95 % of the final value from 107 to 10 ppbv.

(ca. 20 MPa). Although they are reported to drift less and have a faster response time than that of aluminum oxide sensors, they also require periodic calibration. Further, as their response is very temperature dependent, silicon-based sensors are typically thermostated. The sensor in the MAnalytical Microview hygrometer, for example, is held at 46 °C. It has an integrated heater that can also be used to increase the sensor temperature to 130 °C to desorb water when necessary and for rapid dry-down of a sensor from a wet condition. Silicon-based sensors are used for detection of water vapor in gases such as ethylene, natural gas, sulfur hexafluoride, hydrogen, air, and bulk gases.

7.4 Spectroscopic Methods

Spectroscopic methods have many advantages for the analysis of trace water vapor in process gases. These techniques not only offer rapid response times and high sensitivities but are also easily adapted for online monitoring applications because only optical access to the gas sample is required. Further, they can be applied to a wide range of matrix gases. FTIR is the most established technique in the field, enabling detection of multiple species, including water vapor. However, use of more selective near-infrared laser-based techniques such as TDLAS and CRDS has increased significantly within the last decade, due to development and commercialization of dedicated moisture analyzers. Other developing laser-based methods, such as broadband frequency-comb spectroscopy, are also considered to be promising candidates for next-generation multispecies analyzers. All of the methods discussed in this section are based on light absorption by the gas sample in a selected spectral region as a function of concentration of absorbing species according to the Beer–Lambert law:

$$I(\nu) = I_o(\nu)e^{-\alpha(\nu)LN} \tag{7.4}$$

where $I(\nu)$ and $I_o(\nu)$ are the intensity of transmitted and incident light, $\alpha(\nu)$ is the light absorption by a single molecule, L is the effective optical pathlength, and N is the density (number per unit volume) of absorbing molecules. Based on this relationship, in order to obtain the best absorption sensitivity, a strong molecular absorption line that is distinguishable from matrix gas absorptions and baseline variations should be chosen, and a long effective optical pathlength should be employed.

7.4.1   FTIR Spectroscopy

Operating Principles   The operating principles of FTIR spectroscopy are discussed in detail in other publications [46–48]. Briefly, infrared radiation from a blackbody radiator is modulated in a Michelson interferometer (consisting of a fixed mirror, a moving mirror, and a beamsplitter), passed through a cell containing the gas to be analyzed, and then directed to an infrared light–sensitive detector (Figure 7.10a). The resulting interferogram, which is a plot of infrared intensity versus pathlength difference, is Fourier transformed to produce the infrared spectrum of absorbance versus wavenumber. FTIR spectroscopy is a single-beam method. Therefore, a background spectrum of the dried cell (and instrument bench) is first acquired as a reference, after which a spectrum of the sample is collected and referenced to the background spectrum to obtain the final absorbance spectrum. Because acquisition of a single spectrum takes less than a second, multiple spectra can be averaged in several minutes to increase the signal-to-noise ratio of each measurement. Quantitative results are obtained by relating the absorbance of moisture bands in sample spectra to those in calibration spectra of known standards, measured under the same conditions. Although FTIR spectroscopy is a mature technique, it should not be considered as a perfect "black-box" analyzer. Knowing where the limitations of performance arise and which components to optimize for trace analysis is critical to obtaining repeatable and accurate results [49]. These aspects are discussed further below.

FTIR Optical and Gas Sampling Components   Although a variety of infrared sources exist for FTIR spectroscopy, silicon carbide glow-bars are an appropriate choice for the mid-infrared region. By varying the current through the glow-bar, it is

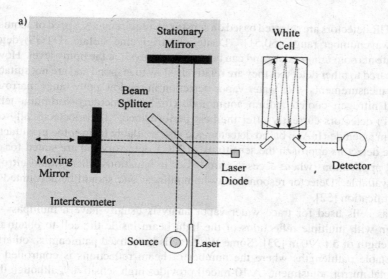

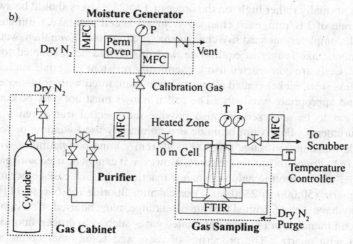

Figure 7.10   Schematics showing (a) main optical components and (b) gas-sampling components of an FTIR system for water vapor analysis.

possible to control the source temperature. Typical operation is at 1100 to 1200 °C, to ensure good sensitivity and lifetime. However, higher temperatures can be used for short durations to maximize the source intensity in the region 2200 to 4000 cm^{-1} when long path cells are employed. Other higher-temperature sources include zirconium oxide Nernst and molybdenum disilicide cermet sources. Potassium bromide is commonly used as a beamsplitter material. As potassium bromide is hygroscopic and will fog in the presence of atmospheric moisture, FTIR bench components must be kept under purge with dry nitrogen even when not in use.

FTIR detectors are selected based on sensitivity requirements, speed of acquisition, and wavenumber range [50,51]. Deuterated triglycine sulfate (DTGS) detectors operate at room temperature and can be used for analysis at the ppmv level. However, compared to other detectors they are relatively slow to respond and are not suitable for trace measurements. For water vapor detection in the low-ppbv range, narrowband liquid nitrogen–cooled indium antimonide (InSb) or mercury–cadmium–telluride (MCT) detectors currently offer the best performance. Thermoelectrically cooled indium arsenide (InAs)–based detectors are also available from some manufacturers. These detectors approach the sensitivity of MCT detectors and are suited for online FTIR applications where access is limited or in situations where liquid nitrogen is not available. Detector response may be nonlinear and should be accounted for in quantification [52].

Gas cells used for trace water vapor analysis usually have a multipass White design with multiple reflections of the light beam inside the cell to obtain a total pathlength of 5 to 20 m [53]. Some White cells have fixed pathlengths; others have adjustable pathlengths where the number of beam reflections is controlled by an external mirror adjustment. A 10-m cell provides high sensitivity, although the cell volume is normally rather high, on the order of 1 to 2 L. Users should be aware that at a flow rate of 1 L/min, each changeover in a 1-L cell will take 1 minute, and to flush a cell completely, at least five changeovers are required. Even then, water vapor measurements take longer to equilibrate, and this should be accounted for during sampling. Cells are constructed using chemically resistant materials such as glass, 316 stainless steel, nickel-coated alloys, or aluminum, fitted with gold-coated nickel mirrors and appropriate windows. The cell windows must not only be compatible with the gas to be analyzed but also provide the spectral transparency required for measurements. Potassium bromide is a commonly used window material, due to its wide transmission range (40,000 to 400 cm^{-1}) and compatibility with many nonreactive hydrides and corrosive gases. However, it cannot be used with gases such as chlorine, which result in anion-exchange reactions. In such cases other materials, such as quartz (50,000 to 2500 cm^{-1}) or calcium fluoride (66,666 to 1110 cm^{-1}), which may have a more limited transmission range, must be used. Calcium fluoride is also recommended for reactive fluorinated gases such as hydrogen fluoride, which will react with quartz. The properties of these and some other commonly used infrared transparent window materials are listed in Appendix B. O-ring materials can be metal or polymeric and should also be reviewed for gas compatibility, permeability, and outgassing in the case of the latter. Gas cells are thermally controlled and the cell temperature is chosen according to the sample gas properties. For example, some gases may decompose at elevated cell temperatures, while other gases, such as hydrogen fluoride, that hydrogen-bond have very temperature-sensitive infrared spectra [54,55].

The construction materials, design, and operating conditions of the manifold used to deliver the sample gas or water standards to the cell are similarly important to optimize (Figure 7.10b). Electropolished 316 stainless steel is widely used for manifold components and lines due to its compatibility with most gases. However, other alloys, such as Hastelloy, are used for some corrosive gases. Surface area and dead volume in the delivery manifold are minimized to prevent long dry-down times, and regulators and sampling lines are thermally controlled (e.g., 60 °C) to prevent ambient temperature fluctuations from affecting the readings. Typically, gases are sampled dynamically through the cell at 0.5 to 2 L/min, and the cell pressure is controlled using an electronic pressure controller.

A key concern when making trace water vapor measurements by FTIR is the infrared absorption by water vapor in the beam path within the bench but outside the cell. Therefore, care must be taken to purge atmospheric moisture out of the bench to a negligible or low and steady level using purified nitrogen or other appropriate dry purge gas, so that the background can always be effectively subtracted from the spectrum of the sample. Several FTIR manufacturers, including MKS Instruments (Andover, MA), Thermo Scientific (Madison, WI), MIDAC Corporation (Irvine, CA), and CIC Photonics, Inc. (Albuquerque, NM), supply instrumentation specifically designed for online gas analysis that have optimized containment of bench and cell transfer optics for efficient purge of water vapor. In some cases the bench components can also be heated to assist with dry-down. Most other general laboratory FTIR instruments, in addition to the regular internal bench purge, require modification by placing the entire FTIR optical bench inside a sealed enclosure (i.e., a plastic purge box). This must be purged independently with a purified gas stream at sufficient flow rates (ca. 20 L/min dry nitrogen). Other approaches investigated to minimize or account for background water vapor include bench evacuation or the use of a specially designed optical system that allows sequential measurement of water vapor in the bench alone or in the gas cell alone to reference out the water vapor in the bench [56–59]. Note that auto- or self-referencing methods that are used for background spectral removal from within the gas cell (discussed in more detail in Chapter 3) still require a low and steady-state level of water vapor in the bench purge [60].

FTIR Operating Parameters   Important parameters to optimize for trace water vapor measurement include spectral resolution, number of co-added spectra, scan time, and apodization function [47,48,61]. Several studies have shown that the best sensitivity is obtained using a spectral resolution of 2 to 4 cm^{-1} [56,57,62]. Higher resolution enables the absorption peaks of interest to be better resolved from those due to interfering molecules, but it results in an increased noise level and longer scan times. Lower-resolution measurements can be made more rapidly but suffer from a higher degree of spectral overlap, which can be a problematic with some sample gas matrices [62,63]. Because the signal-to-noise ratio of a spectrum increases with the square root of the number of spectral scans, the co-addition of a large number of spectra increases sensitivity. However, the increased sampling period can offset the benefit of lower noise levels. Consequently, for online water vapor analysis of flowing gas streams, a compromise between sensitivity and data acquisition rate must be reached. Typically, for low-ppbv detection of water vapor, each infrared data point is obtained from the co-addition of approximately 200 to 400 spectral scans collected in under 5 minutes.

Finally, the apodization function selected has an impact on spectra in terms of the spectral line width and the presence of side lobes. Clear differences in the spectra of water vapor that have been processed with different apodization functions are evident [61]. Therefore, the same function must be used when processing calibration and sample spectra. The use of the Norton–Beer weak function is recommended for high resolution or when good quantitative analysis is required [47].

Spectral Region Selection and Quantification Method In the absence of matrix gas or cell window absorbance spectral overlap, water vapor can be quantified in two mid-infrared regions where strong water absorptions occur: 1900 to 1300 and 3900 to 3600 cm^{-1}. When analyzing gases such as hydrogen, argon, or chlorine that don't have permanent dipole moments, either region can be used, since they are infrared transparent. However, for other gases, a region free from matrix gas absorptions must be located for water vapor detection. Sampling of neat (100 %) matrix gas in long-pathlength cells invariably results in strong and broad matrix gas absorptions, many of which are off scale. This is not immediately apparent from library spectra, which are often measured with short-pathlength cells or with the gas at low pressure or diluted with nitrogen.

For many gases, including the hydrogen halides and the carbon and nitrogen oxides, one or both regions are available for water vapor analysis [54–59,64–66]. An example of the analysis of a low-purity nitric oxide sample is shown in Figure 7.11. Clearly, care is required to avoid regions where the matrix gas or other impurities, such as carbon dioxide (3550 to 3750 cm^{-1}), absorb. Many hydride gases, such as arsine, phosphine, ammonia, silane, and disilane, also interfere with detection of low-level water vapor in many infrared regions. Trace measurements are only possible by selecting very narrow spectral windows in which the absorbance bands for water and the matrix gas are strong and weak, respectively, and by using regression-based data analysis [65,67]. In ammonia, water bands in the region 3682 to 3988 cm^{-1} have been used to measure water vapor down to 10 ppbv [65,67–69]. Similarly, in arsine, there are several bands in the region 3660 to 3760 cm^{-1} that can be used. In phosphine, water vapor can be detected in narrow windows in either region, but bands between 1500 and 1570 cm^{-1} have the least interference from the phosphine matrix [70].

Quantification using classical least squares (CLS) is preferred in cases where matrix gas interference occurs [71–73]. With this approach, a spectrum of water vapor in nitrogen at a known concentration is added spectrally to that of the dry sample matrix gas until a match to the sample spectrum is obtained. The actual concentration of water vapor in the sample is then calculated from the individual contributions of the two spectra. The CLS method not only improves the analysis precision but also enables multiple components to be analyzed simultaneously. Further data on the residual spectra provide feedback to the analyst on the presence of interfering impurities in the region used for quantification. The success of the CLS method is dependent on the quality of the spectra used for quantification. As the sampling conditions (cell pressure and temperature, matrix gas type, and FTIR spectral collection parameters) can significantly affect the spectral line shapes, the spectra of purified matrix gas, calibration standards, and sample should be measured under the same conditions. This is discussed further in Chapter 3. Typically, FTIR detection limits for water

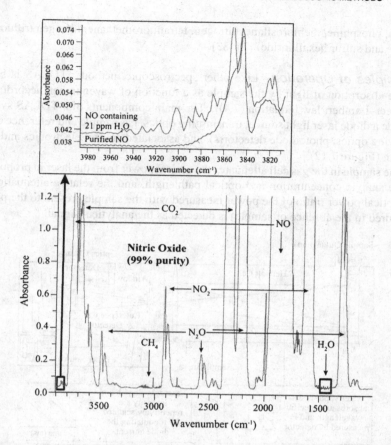

Figure 7.11 Infrared spectrum of nitric oxide (99 % purity) showing the presense of methane, carbon dioxide, nitrogen dioxide, nitrous oxide, and water vapor impurities. The upper insert illustrates infrared spectra of nitric oxide containing 21 ppm water and the same sample gas stream after passing through a purifier to remove the water.

vapor in many corrosive and hydride matrix gases are in the range 10 to 30 ppbv. To measure water vapor in reactive gases at single-digit and sub-ppbv levels in real time, more sensitive laser spectroscopy methods are required.

7.4.2 Tunable Diode Laser Absorption Spectroscopy

TDLAS is an established method for quantitative assessment of species in the gas phase. Its major advantages over other techniques are high selectivity, fast response, and ability to achieve low detection limits. These attributes are particularly important in microelectronic applications. Therefore, TDLAS has been used to detect water vapor in a variety of inert, corrosive, and reactive gases (including argon, helium, oxygen, nitrogen, hydrogen, hydrogen chloride, hydrogen bromide, chlorine, ammonia, phosphine, dichlorosilane, nitrogen, tetrafluoromethane, nitrogen trifluoride, silane, and sulfur hexafluoride) [74–83].

Principles of Operation   Like other spectroscopic methods, TDLAS is based on the absorption of light by the sample as a function of wavenumber, according to the Beer–Lambert law [equation (7.4)]. The main components of a TDLAS system include a diode laser light source, transmitting optics, gas sample and reference cells, receiving optics, photodiode detector(s), and associated control electronics and data system (Figure 7.12).

The sample in the gas cell attenuates the optical power from the laser in proportion to the analyte concentration and optical pathlength, and the relative attenuation of the optical power (ratio of the power measured with the sample present to the power measured in the absence of sample) is detected as the analytical signal.

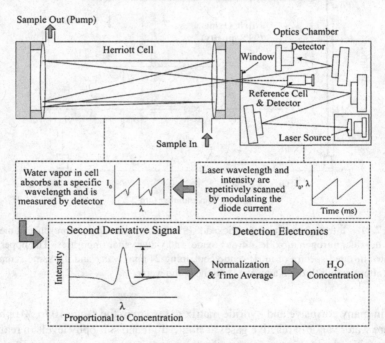

Figure 7.12   TDLAS schematic depicting the principle of operation and sensitivity enhancement using a $2f$ second-derivative absorption spectrum.

In TDLAS the amount of water vapor is measured by quantification of the absorbance of a single rotational line within a fundamental vibrational, a vibrational overtone, or a combination vibrational band of water vapor, and the selectivity of the technique stems from the precise overlap of the diode laser wavelength with the water absorption. The water absorption line is typically scanned by maintaining the laser at a constant temperature, sweeping the laser frequency with the drive current and measuring the decrease in laser power after passing through the sample. The ability to measure low concentrations of the analyte is limited by noise present in the measurement background. However, the sensitivity of the TDLAS measurement can be improved significantly by the use of wavelength-modulation spectroscopy (WMS), which reduces the contribution of excess laser noise to the total noise current [74]. Signal-to-noise improvements of three orders of magnitude compared with direct absorption spectroscopy have been reported [84]. In WMS, the laser wavelength is slowly tuned over the absorption line and modulated simultaneously by rapidly varying the drive current at a frequency f. The resulting absorption signal is then demodulated by a phase-sensitive detector at the modulation frequency or at the first or second harmonics of the modulation signal ($1f$ and $2f$, respectively). The line shapes produced at the first and second harmonics are approximately the first and second derivatives of the original line shape, respectively. Wavelength-dependent features such as sloping backgrounds contribute to the first-derivative profile. For example, with a constant slope, these appear as a constant offset in the $1f$ case. However, such features are removed in the second-derivative profile while the sharp variations due to the actual sample absorption remain intact (Figure 7.12). It should be noted that background spectral features such as étalon interference fringes that tend to drift in time are not removed in WMS. The second-derivative profile is directly proportional to concentration and can be modeled explicitly according to known spectroscopic parameters of laser-tuning rate, modulation depth of the laser wavelength, and pressure-induced line broadening of the specific absorption to interpret the strength of the moisture signal. Optimization of spectroscopic measurements with modulation techniques, such as frequency modulation and two-tone modulation, are discussed and reviewed in several publications [85–87].

Effects of Pressure Broadening and Matrix Gas

An important consideration with laser spectroscopic techniques is that the resulting line shape is dependent on the pressure-broadening effects of the matrix gas. At 760 torr (101 kPa), collisional broadening of absorption lines (due to intermolecular collisions between gas molecules) is the dominant line-broadening mechanism, giving rise to a nominally Lorentzian line shape with a width that is proportional to the total pressure. At low pressure (e.g., 10 torr or ca. 1.3 kPa), Doppler broadening is dominant (due to the Maxwell–Boltzmann distribution of molecular speeds), resulting in a Gaussian line shape. At intermediate pressures of 50 to 250 torr (7 to 35 kPa), a Voigt line shape, which is a convolution of Lorentzian and Gaussian features, is observed. Other important mechanisms influencing moisture line shapes are Dicke narrowing and speed-dependent broadening and shifting [88,89]. The systematic errors associated with pressure broadening are on the order of 10 % when a Voigt fit is used because the details of these linewidth effects are ignored, while the systematic errors associated with the total line strength (peak area) are less than 1 %. Detection in the intermediate pressure range is commonly employed to minimize interference from the matrix gas and to balance signal strength while keeping the line shape narrow enough that the full shape can be recorded within the wavelength scan of the laser. In different gas matrices there is a varying amount of broadening, so each gas must have an associated unique line measurement or calibration. Vorsa et al. reported measured pressure-broadening coefficients for the 1392.53-nm water line in hydrogen chloride, hydrogen bromide, chlorine, and oxygen, which are, respectively, 2.76, 2.48, 1.39, and 0.49 times that of the nitrogen pressure-broadening coefficient [90].

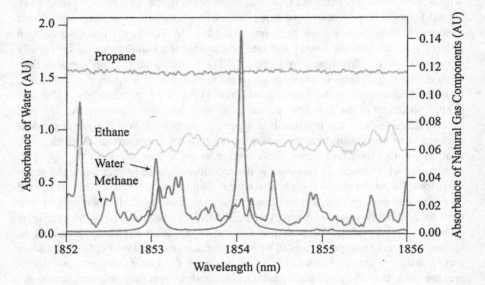

Figure 7.13 Spectra of methane, ethane, propane, and water vapor in the region 1852 to 1856 nm. Spectral overlap with water vapor exists for methane, but to a much lesser extent with ethane and propane. Reprinted with permission of ISA from [91].

Unlike broadband infrared sources, the emission of diode lasers provides a strong signal with an extremely narrow linewidth, on the order of $10^{-3}\,cm^{-1}$ or smaller. This greatly facilitates the resolution of the sample absorption from those of the matrix gas or other potentially interfering species at nearby wavelengths. However, even so, some gas matrices (in particular, hydrides such as ammonia, phosphine, arsine, and methane) absorb in many of the same wavelength regions as that of water vapor (Figure 7.13). Therefore, it is important not only to select a suitable water line in a region where matrix gas absorbance is minimized, but also to account for any matrix gas absorbance contribution to the water reading [76,91]. Double-beam acquisition and subtraction is one way to account effectively for matrix absorption [76]. With this approach, a reference cell containing a purified stream of the matrix gas is employed to remove absorbance lines from the gas matrix, cancel common-mode noise from the diode laser, and remove the contribution of atmospheric moisture absorption outside the sample cell. This, together with the second-derivative absorption signal to remove the impact of laser beam intensity variations induced by laser wavelength sweep, enables low-ppbv detection limits. Much of the work on dual-cell TDLAS has been reported by Wu et al. They developed a system for water vapor analysis in nitrogen, ammonia, and hydrogen chloride, based on an indium gallium arsenide phosphide (InGaAsP)–distributed feedback (DFB) diode laser emitting at 1380 nm.

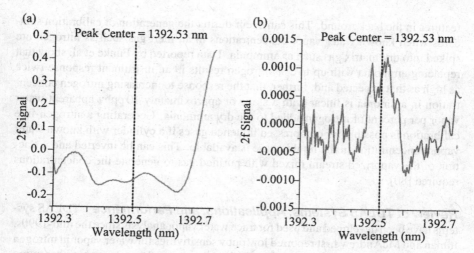

Figure 7.14   Absorption of water vapor in ammonia. (a) Superimposed spectra of 21.9 to 29.2 ppbv water vapor in ammonia, where spectra are dominated by ammonia absorption features, and (b) subtraction result from the spectra in (a) showing detection of water vapor at about 7 ppbv (note difference in the scales in the spectra). Reprinted with permission [78].

In the case of hydrogen chloride, water vapor was detected down to less than 20 ppbv in a 50-cm cell at 60 torr (8 kPa) using the line at 1380.64 nm [76]. For ammonia, the water vapor impurity could be detected down to 12 ppbv using a 92-cm pathlength cell and a laser, tuned from 1370.93 to 1370.99 nm at room temperature [75].

As the dual-cell arrangement is instrumentally complex, and it is difficult to maintain the necessary balance between the two beams over time, most TDLAS systems are equipped with only a single cell. In this case, if there are matrix gas absorption contributions at the wavelength used for water vapor measurement, then the spectrum of a purified gas stream must be generated, modeled, and used as a reference. This can be subtracted from that of the sample [78], as shown in the example for water vapor in ammonia in Figure 7.14. However, to make accurate measurements, the sample pressure must be tightly controlled at the pressure used to generate the reference. Outside this pressure range, water vapor readings can deviate significantly from the true water vapor content, as the line-fitting routine ceases to account accurately for the line-broadening effects. For example, increasing water vapor readings on a Delta F DF740 ammonia analyzer (hardware described in more detail below) are observed as the cell pressure decreases from the pressure recommended (70 ± 3 torr or 9.33 ± 0.40 kPa), and vice versa. If the pressure is lowered progressively by 10 torr (1.33 kPa) and then 20 torr (2.66 kPa), the respective water vapor reading increases by approximately 10 and 250 ppbv [80]. Therefore, use of an electronic pressure controller is recommended in such cases.

In addition to the importance of pressure control, TDLAS users should be aware that any change in gas matrix composition (even a minor change) can also significantly, affect water vapor readings due to line broadening of the ammonia spectral features in the background. This can occur during the generation of calibration standards when known water vapor concentrations in an inert gas such as nitrogen are spiked into dry matrix gas such as ammonia. Data reported by Funke et al. show that replacing ammonia with up to 5 % nitrogen results in an instrument response twice as high as that expected and, further, that the response to increasing nitrogen concentration in ammonia is linear with a slope of approximately 10 ppbv apparent water vapor per percent of nitrogen added to the dry ammonia. Generating a nitrogen-free calibration is possible for compressed liquefied gases if a cylinder with known water vapor concentration in the liquid phase is available. This can be inverted and a fraction of the vaporized stream mixed with purified gas to generate the concentrations required [80].

Review of TDLAS Systems, Applications, and Performance   TDLAS systems have been developed and used for trace water vapor analysis since the mid-1990s. Inman and McAndrew first reported low ppbv sensitivities for water vapor in nitrogen (1 ppbv) and hydrogen chloride gases (3 ppbv) by use of the water vapor absorption line at 1456.888 cm^{-1} [79]. Their TDLAS setup used second-derivative detection and signal averaging and was calibrated directly by introduction of a water vapor standard into a 30-cm Herriott cell with a 9-m optical pathlength. The cell was operated at a pressure of 20 to 40 torr (2.66 to 5.32 kPa) so that the signal due to water vapor in the cell could be distinguished from the broader absorption from water vapor in the optical path outside the cell. The ultimate detection limit of the system was 10 ppbv water vapor in dry nitrogen and 63 ppbv water vapor in dry hydrogen chloride. A similar TDLAS system was used for online monitoring of water vapor in gases (hydrogen chloride, hydrogen bromide, tetrafluoromethane, nitrogen trifluoride, ammonia) supplied to process tools used to make microelectronic devices [77,81].

An early commercially available TDLAS analyzer used for semiconductor gas applications was developed in prototype form by Southwest Sciences, Inc. This system is based on a hermetically sealed laser chamber containing an InGaAsP diode laser source emitting at 1392.5 nm and tuning optics, connected to approximately a 1-L stainless steel Herriott multipass cell (50-m pathlength) fitted with sapphire windows. Tests in nitrogen at 100 torr (13.3 kPa), performed with the NIST low-frost-point hygrometer [24] and using moisture challenges in the range 5 ppbv to 3 ppmv range demonstrated sub-ppbv noise levels, residual moisture in the low ppbv range, and a 3σ detection limit of 3 ppbv [74]. Commercial versions of this system developed by Delta F Corporation (now Servomex) with a more compact optical design to reduce outgassing and incorporation of a water reference to prevent the laser drifting off the moisture absorption line show comparable sensitivity [78,80]. The DF740 unit for moisture analysis in ammonia has approximately 2 to 4 ppbv sensitivity and a 3σ detection limit of 9 ppbv. The unit for use in hydrogen chloride (DF730) has a reported sensitivity of 0.25 ppbv and a detection limit of 1 ppbv water [82].

The initial success of early TDLAS analyzers has encouraged the development of other commercial systems for continuous real-time monitoring applications. Ametek

Process Instruments, GE Sensing and Inspection Technologies, and SpectraSensors, Inc. have all developed process analyzers for the detection of water vapor in natural gas and hydrocarbons in the range from several thousand ppmv down to single-digit ppmv levels [91,92]. The Ametek 5100 unit is based on a DFB diode laser operating with a wavelength centered at 1854 nm. Detection of water vapor in natural gas has been reported down to the 4 ppmv range using a 90-cm pathlength cell at 760 torr (101 kPa) and 21 °C and second-derivative measurements [91]. Siemens Applied Automation has reported a LDS6 TDLAS process analyzer for in-situ measurement of water vapor in chlorine streams [93]. Other instrumentation will undoubtedly be developed as analytical needs arise in the future.

TDLAS is being used increasingly for in-situ monitoring and process control in a variety of gas applications [94] because of its ppbv sensitivities for water (assuming that absorbance lines without spectral interference from the matrix gas are available) and its ability to handle corrosive or aggressive gas environments [76,81,82,93]. The technology is rugged provided that the materials used for the gas-wetted parts are selected correctly for the gas being analyzed. TDLAS is less likely than CRDS to suffer performance losses due to particulates that can coat mirror surfaces of long-pathlength gas cells. However, mirror surfaces must still withstand the corrosive nature of some gases, such as hydrogen chloride, hydrogen bromide, or chlorine. TDLAS is also capable of responding rapidly to changing water vapor concentrations. Although the response time of TDLAS systems will vary depending on cell volume, flow rate, and sampling system design, readings typically equilibrate rapidly. Equilibration to 90 % of the final reading for a roughly 150-ppbv water step change can be expected within 5 to 10 minutes, depending on whether the analyzer is being wet-up or dried-down [80,82].

7.4.3 Cavity Ring-Down Spectroscopy

Principles of Operation In CRDS, the time dependence of the decay of light intensity in an optical cavity is used to measure absorption losses associated with a gaseous impurity and thereby calculate the impurity concentration. The gas under test flows through a resonant optical cavity illuminated with light absorbed by the impurity, in this case water vapor. A schematic of the basic elements is given in Figure 7.15. When the laser wavelength is in resonance with the cavity, a strong optical field builds inside the cavity, and a small amount of that light passes through the right-side mirror (sometimes called the "output coupler") and is measured by the photodetector. When the incident light is turned off with the optical switch, the optical intensity in the cavity decays with a characteristic time constant τ. This time constant is determined by the total optical losses in the cavity, including the mirror transmittance and any absorption by the host gas and impurities. The cavity lifetime is determined by fitting the detector output with a simple exponential decay.

The number density of absorbing molecules, N, in the optical beam path inside the resonant cavity is given by the simple relationship

$$N = [c\sigma(\nu)]^{-1} [\tau(\nu)^{-1} - \tau_0(\nu)^{-1}] \qquad (7.5)$$

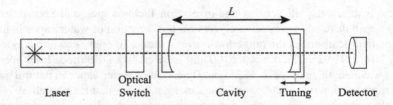

Figure 7.15   Basic elements of a CRDS system. The resonant cavity has length, L, defined by the two cavity mirrors.

where c is the speed of light, ν is the frequency of the laser light, σ is the absorption cross section for each molecule, τ is the measured lifetime, and τ_0 is the lifetime when no impurity absorption is present. In an ideal cavity, $\tau_0 = L/[c(1-R)]$, where R is the geometrical mean of the reflectivities of the cavity mirrors (i.e., the square root of the product of the two reflectivities).

The absorption cross section is a function of the laser frequency, and for an isolated transition it can be modeled in terms of the line shape parameters (e.g., frequency, line intensity, broadening parameter, Doppler width). Most laser linewidths (typically, 1 MHz) are orders of magnitude smaller than the width of a pressure-broadened gas absorption line (typically, 1 to 3 GHz), and the width of the optical cavity resonance is usually even smaller still (ca. 1 kHz for a high-finesse cavity). Thus, application of equation (7.5) requires knowledge of the line shape for the gas absorption line and the exact position of ν relative to the center of the absorption line, ν_0. For research measurements, the usual approach is to scan ν over a large range to achieve "zero" absorption at the ends and then fit the entire gas absorption peak to extract total absorption line strength, S, and estimate the proper background absorption. This process is most conveniently executed by first converting the τ values into absorbance, $\alpha(\nu)$, using $\alpha = (c\tau)^{-1}$. The most common line shape used for fitting gas absorption lines is the Voigt function, a convolution of a Lorentzian function, representing pressure broadening from intermolecular collisions, and a Gaussian function, representing the Doppler width of the transition for a particular temperature. Because the Gaussian width can be calculated from the Doppler broadening, this parameter can be held constant during the Voigt fit. For data acquired in air or nitrogen, the Lorentzian width can be determined from the AGAM value for water in the high-resolution transmission molecular absorption (HITRAN) database and also held constant. More complicated line shape functions are recommended for the highest accuracy, as discussed in Section 7.4.2.

An example of such a fit for water absorption in nitrogen gas is given in Figure 7.16. While the height and width of the absorption peak will change with environmental conditions, most notably total gas pressure, the integrated area under the line shape will remain constant. The Voigt line shape is parameterized in different ways depending on the analysis software used, but the fitting parameters are used to determine an integrated peak area A that has units of frequency (or wavenumber) divided by length, 4.57×10^{-5} MHz/cm in the example. The number density of water molecules can

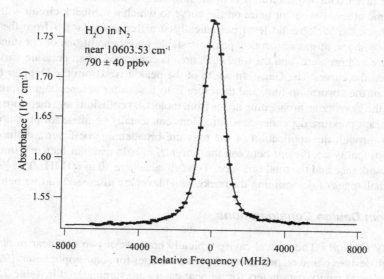

Figure 7.16 Curve fitting of water vapor absorption line in nitrogen carrier gas as measured by CRDS for a total cavity pressure of 13.3 kPa (100 torr) at room temperature. The cavity length is 73.5 cm. The instrument used to acquire these data is described in more detail by Lehman et al. [95].

then be extracted as $N = A/S$. The absorption line intensity, S, is typically reported in units of $cm^{-1} \, cm^2 \, molec^{-1}$, for example, in the HITRAN database [96]. For water vapor, there are several references [97,98] covering limited frequency ranges that include newer data than are generally reported in HITRAN. Converting the units of the published value [98] of $S = 5.899 \times 10^{-22} \, cm^{-1} \, cm^2 \, molec^{-1}$ with $1 \, cm^{-1} = 30 \, GHz$ yields $S = 1.77 \times 10^{-17} \, MHz \, cm^2 \, molec^{-1}$, and from the example curve we calculate a water molecule density of $2.58 \times 10^{12} \, molec \, cm^{-3}$. The uncertainty in peak area returned by the fitting program provides an estimate for number density uncertainty of $3 \times 10^9 \, molec \, cm^{-3}$. The water vapor mole fraction in the gas is given by the ratio of number density of water molecules to the total number density of gas molecules in the cavity, N_{tot}, which in turn is generally calculated from the cavity gas pressure P and temperature T using the ideal gas law ($PV = nRT$), where R is the gas constant $8.31 \, J \, mol^{-1} \, K^{-1}$. Multiplying the result by Avogadro's number (6.02×10^{23}) will convert moles to molecules, and for the example above, we arrive at a volume mole fraction for water vapor of $790 \pm 40 \, nmol \, mol^{-1}$ (ppbv). The larger relative uncertainty in the concentration compared with the water molecule number density arises because the chamber pressure and temperature vary under laboratory conditions, which in turn generates 5 % uncertainty in the total number density of molecules in the cavity.

For faster data acquisition, it is often sufficient to measure the peak absorbance (or some other convenient point on the curve to which a feedback circuit will lock) and check the background level periodically at a frequency far off from the peak. The line shape of gas absorption peaks is stable under conditions of constant temperature and pressure, and the total peak area is conserved when pressure variations broaden the curve. The Gaussian width of the peak arises from the Doppler broadening of the absorption line, and therefore is to first order independent of pressure, while the Lorentzian broadening arises from molecular collisions and therefore shows significant pressure dependence. Corrections can usually be made for pressure variations through the application of the pressure-broadening coefficient, defined as a proportionality coefficient between the Lorentzian half-width at half-maximum for the absorbance and the total pressure. Typical values are 30 to 60 MHz/kPa [90,95]. The challenges of determining the background level are discussed further below.

System Design Considerations

Cavity Design The optical cavity typically consists of two collinear mirrors, although designs based on prisms also have advantages for some applications [99,100]. The relevant cavity parameters for a linear cavity are summarized in Table 7.1. Although in principle cavity mirrors can be flat, greater stability and optical coupling can be obtained with mirrors that have a slight radius of curvature. A common configuration is one in which the cavity length is approximately 0.75 times the radius of curvature of the mirrors. Optical cavity design for CRDS is covered in detail by Busch et al. [101]. Mirrors with very high reflectance can now be obtained for the near infrared, based on careful evaporation of dielectric stacks. In fact, optimization of high-reflectivity mirror coatings was the first application of CRDS [102]. Low-

Table 7.1 Characteristics of a linear high-finesse optical cavity, with typical values based on cavity length of 75 cm and mirrors with reflectivity of 0.99993 %

Parameter	Equation	Typical Values	Comments
Free spectral range, ν_{SFR}	$\nu_{SFR} = c/2L$	200 MHz	Spacing of longitudinal cavity modes
Finesse, f	$f = \pi R^{1/2}/(1 - R)$	105,000	
Effective length, L_{eff}	$L_{eff} = 2Lf/\pi$	51 km	Length of a single path with the same total absorption
Background, τ_0	$\tau_0 = L/[c(1 - R)]$	83 μs	
Resonance width, $\Delta\nu$	$\Delta\nu = c/(2Lf)$	1.9 kHz	Full width at half maximum

loss mirrors produce a long τ_0 and therefore enhance the minimum sensitivity of the CRDS system. Because the detector sensitivity and response period will limit the ability to measure high impurity concentrations (with high absorption and short τ), the dynamic range of the impurity measurement is also enhanced by high-reflectivity mirrors. The cavity resonance width becomes narrower, however, as the mirror reflectivity increases, which increases the demands on laser frequency stability and mechanical stability for the cavity. Using cavity supports with minimal thermal expansion (Invar metal), limiting thermal variations, and isolating the system from vibration all lead to greater stability. Most high-finesse CRDS cavities are also equipped with a feedback circuit that either tunes the laser frequency to match the cavity or slightly adjusts the cavity length with a piezoelectric actuator on the mirror or cavity end supports and stabilizes the cavity related to a frequency reference such as a frequency-stabilized helium–neon laser [103]. Cavity optics should also be designed to minimize coupling to parasitic cavities formed by collinear mirrors, detector windows, front and back surfaces of mirrors, and so on. Thermal expansion and other drift in these components can generate variations in τ_0 over time and add to uncertainty in absolute concentration measurements [104].

Light Sources and Detectors   The basic requirements for a light source are short-term stability and a tuning range over at least the molecular absorption line of interest. Three fundamental vibrational bands of water, designated as ν_1, ν_2, and ν_3, and corresponding to wavelengths of 2.73, 6.27 and 2.66 μm, respectively, comprise thousands of rotational–vibrational transitions that have the potential to yield high sensitivity and selective measurements. The strongest transitions in these bands have room-temperature intensities over 10^{-18} cm^{-1} cm^2 molec^{-1}. However, the absence of laser light sources and sensitive detectors in this range has meant that most work has been carried out in the overtone bands near 0.9 and 1.4 μm (10,600 and 7000 cm^{-1}, respectively), where intensities are orders of magnitude weaker than those of the fundamental bands. Tunable laser sources for these bands include distributed feedback Bragg reflector diode lasers and external cavity diode lasers. The stronger molecular lines at longer wavelengths are becoming increasingly accessible with the development of optical parametric oscillators (OPOs) that combine laser sources operating in the near infrared (ca. 1 μm) and nonlinear optics to generate coherent light with overlap to the 1.8- and 3.2-μm water bands. Quantum cascade lasers are now commercially available with wavelengths greater than 4 μm, and these have been used successfully in CRDS systems [105,106]. Distributed feedback laser diodes have also been used in the near infrared to improve the compactness of CRDS systems [107]. In general, these coherent light sources are continuous wave (CW), which simplifies analysis of the ring-down signal, but single-mode cavity excitation can be achieved with pulsed sources by working with relatively short ring-down cavities [108,109]. A recent development in laser technology has been the frequency-comb laser, and its application to cavity ring-down spectroscopy and cavity-enhanced absorption spectroscopy is discussed elsewhere in this chapter.

Detectors must be chosen to have a response time well below that of the minimum τ to be measured. A high-finesse cavity can have an empty-cavity time constant

τ_0 of hundreds of microseconds, which is readily measured with semiconductor photodiodes provided that the detector area is kept small. Of the bands discussed above, only those near $0.9\,\mu m$ are detectable with silicon photodiodes; InGaAs, InAs, InSb, and HgCdTe detectors are employed most commonly for the longer wavelengths. As the bandgap of the photodiodes decreases, thermal noise and shunt resistance become more significant, requiring design trade-offs between sensitivity and response period relative to the specific target molecules and gas. Detectors also must operate with DC coupling over a dynamic range of three or four orders of magnitude to allow accurate τ measurement. If the detector signal is digitized, the electronics must also be capable of high resolution to cover the full dynamic range of the detector (analog-to-digital signal with 12-bit or higher resolution) [110].

Optical Components The optical switch in Figure 7.15 is typically an acousto-optic modulator (AOM) oriented so that the first-order diffracted beam is directed to the CRDS cavity. AOMs have fast switching times (<100 ns) for small spot sizes and therefore do not affect the measurement of the decay time constant. A semiconductor optical amplifier can perform a similar function [111]. Other methods, such as detuning the laser, can also be employed as a switch, but often any laser adjustment increases undesirable chirp in the laser frequency during ring-down measurements. When the cavity is in resonance with the laser light wavelength, the cavity throughput is high as measured on the detector. The detector signal with suitable processing can serve as a trigger signal to turn off the cavity illumination via the optical switch. The same trigger signal can also start acquisition of the ring-down signal.

An optical Faraday isolator is usually inserted between the laser and the cavity to avoid reflection of the light from the front cavity mirror back into the laser source. This reflection is quite strong when the cavity is off-resonance and can interfere with laser stability. If the laser does not produce a clean Gaussian beam, coupling to only a single TEM00 mode of the cavity will be difficult or impossible. A spatial filter or single-mode optical fiber can be added to the laser path to prevent excitation of higher-order cavity modes. A video camera or beam profiler that responds to the laser frequency is very useful for monitoring the mode patterns in the cavity during alignment. Finally, it will be necessary to measure the frequency of the light accurately. For frequency-stabilized cavities, it is often sufficient to tune the laser through successive longitudinal modes of the cavity and use the known position of the gas absorption line to place these changes on an absolute wavelength or wavenumber scale. Interferometric wavemeters are now commercially available with 0.2 ppm resolution in the near infrared, or 0.06 GHz for the laser frequency in the example spectrum, which is sufficient for acquisition of full line spectra. Although not necessary to the measurement, a Pound–Drever–Hall locking feedback system to the laser [112] will improve control over the acquisition of ring-down data. This lock mechanism requires additional optics to modulate the beam to produce an error signal for feedback on the cavity, and electronics to suspend feedback during retuning of the laser [113,114].

Review of CRDS Systems, Applications, and Performance The primary appeal of CRDS as a method to measure water vapor contamination in gases is that unlike most other methods based on optical absorption, CRDS does not rely on accurate measurement of light intensities. CRDS systems therefore do not require periodic calibration to correct for degradation of optical components or drifts in light source intensity and photodetector sensitivity. This method can also achieve less than 1 ppbv sensitivity and short measurement periods (<1 s). Although more sophisticated systems can occupy a large optical bench, the basic elements can be selected so as to fit inside a benchtop instrument enclosure. Water vapor analysis systems based on CRDS are now commercially available.

To a large degree, the history of CRDS development tracks the development of laser light sources. The method was developed initially with pulsed dye lasers. These sources produce a pulse train with a low duty cycle (1 to 10 Hz) operating at multiple frequencies due to the broad gain response of the dye relative to the laser cavity mode spacing. This method was used successfully to measure transition strengths for forbidden transitions in molecular oxygen, with an estimated absorbance sensitivity of 10^{-8} cm^{-1} [115]. Absorbance sensitivity on the order of 2×10^{-9} to 7×10^{-10} cm^{-1} was reported for the measurement of weak absorption lines in hydrogen cyanide [116]. The short duration of the pulse relative to the round-trip time period of the circulating beam meant that the cavity operated in a regime where light could be coupled into the resonator regardless of the wavelength detuning of the probe laser relative to the cavity resonances. Using frequency-domain analyses of the intracavity field, Hodges et al. [117], Lehman and Romanini [118], and Looney et al. [109] showed that for short pulse lengths ("long cavities"), there will be complications caused by mode beating effects and multiexponential decays in ring-down signals. For long pulse lengths ("short cavities") single-mode excitation of the ring-down cavity results, leading to exponential decay signals [108]. For pulsed excitation, the laser bandwidth can yield significant systematic error in the spectra measured and hence in the concentration deduced for the light-absorbing species [119]. In subsequent years, these multimode effects in CRDS were largely mitigated by the use of narrowband CW probe lasers. As these tunable sources became more widely available, CW–CRDS became more widespread, in part due to the simpler form of the underlying theory and data analysis. Detection limits on the order of 10^{-9} cm^{-1} were demonstrated in the measurement of weak acetylene overtones in the vicinity of 570 nm [120]. Heterodyne detection techniques were also shown to offer high sensitivity, on the order of 10^{-10} to 10^{-13} cm^{-1}, by overcoming technical noise sources, such as laser instabilities, acoustic noise, and electromagnetic noise from laser power supplies [110,121]. Most recently, frequency-comb lasers have also been employed as CRDS light sources in a very broad wavelength range [122]; this topic is discussed Section 7.4.4. An informative and more complete discussion of the historical development of CRDS may be found in other publications [123,124].

Advances in CRDS instrumentation are typically demonstrated for dilute analytes in vacuum or inert gases such as nitrogen. New practical considerations arose as the method was developed for commercial use to detect water vapor as an impurity in host gases (or matrix gases). Many of these gases have complicated spectra with a

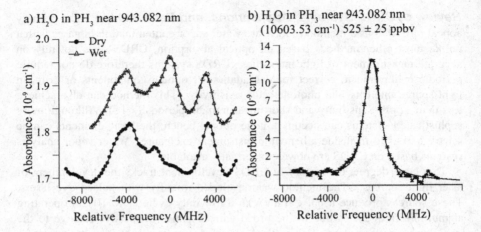

a) H_2O in PH_3 near 943.082 nm

b) H_2O in PH_3 near 943.082 nm
(10603.53 cm⁻¹) 525 ± 25 ppbv

Figure 7.17   CRDS spectra for water vapor in phosphine gas. (a) Absolute absorbance scale showing spectra with and without water vapor present. The frequency scale zero corresponds to the position of the water absorption peak; other peaks in the spectra are weak phosphine absorption lines. (b) Curve fit of the peak obtained by subtracting the "dry" phosphine spectrum from the "wet" phosphine spectrum. The instrument used is the same as that used to acquire the data in Figure 7.16.

high density of overtones and broad backgrounds from the summation of the wings of multiple nearby spectral lines. These lines produce time-dependent background signals that can dominate the uncertainty in measurement of absolute water vapor concentration. An example for water vapor in phosphine gas is given in Figure 7.17, acquired in the same experimental setup as that used for the spectra of water vapor in nitrogen illustrated in Figure 7.16. The absolute background has shifted upward relative to data in Figure 7.16 from nearby phosphine peak wings. Peaks from weak phosphine absorption lines are also visible within the range of the scan. The shift in the background due to slight changes in cavity pressure and temperature is readily apparent in Figure 7.17a. Although this background can be subtracted from the "wet" phosphine spectrum after realigning the spectra for optimal overlap of phosphine features, the resulting curve in Figure 7.17b contains more noise than does its counterpart in nitrogen matrix gas. The effect of background absorption by matrix gases encourages exploration for spectral windows that contain strong water vapor absorption lines but minimal matrix gas absorption, as has been shown for additional water vapor absorption lines in phosphine [95]. As the spectra move lower in energy toward more fundamental transitions, the water absorption peaks increase in strength, and matrix background peaks tend to be more isolated. As near- and mid-infrared laser and detector technologies improve, these spectral regions are likely to become more widely used in CRDS technology.

Another consideration in working with industrial gases is that many of them are reactive and corrosive. The reactions can both change the composition of impurities over time, either generating new water vapor from reactions between hydrides and solid oxides in the gas cylinder wall, or consuming water and converting it to oxide particulates. Materials used in the construction of instrumentation must also be compatible with reactive gases. (Although a concern about mirror stability was raised early in the development of CRDS, modern dielectric mirrors have proven to be quite resistant to chemical attack by common semiconductor processing gases.) The uncertainties for water vapor concentration for corrosive gases are therefore higher than what can be obtained in inert matrices, and perhaps more significantly, depend on variable experimental conditions to a greater degree than identical measurements in inert matrices. The current state of the art for highly sensitive water vapor measurement in various industrial gases is summarized in Table 7.2. The column containing lower detectable limit (LDL) values should be interpreted somewhat loosely, as different authors applied different criteria for its determination. The units of this column are in parts per billion by volume ppbv, which is equivalent to $nmol\, mol^{-1}$ in SI units.

Table 7.2   Performance for water vapor detection in various gases using CRDS[a]

Matrix (Host Gas)	LDL (ppbv)	Comments	Ref.
Ammonia	10	Commercial instrument specification	[129]
Arsine	3.5		[127]
Chlorine	0.77	Assumes relative uncertainty in τ of 0.1 %	[90]
Hydrogen bromide	10	Estimated from stability of readings on gas with about 125 ppb concentration	[128]
	2.3	Assumes relative uncertainty in τ of 0.1 %	[90]
Hydrogen chloride	1.2	Assumes relative uncertainty in τ of 0.1 %	[90]
Nitrogen	2	Commercial instrument specification, instrument compatible with corrosives but no LDL specified	[130]
	0.44	Assumes relative uncertainty in τ of 0.1 %	[90]
	0.2	Commercial instrument specification	[131]
Oxygen	0.22	Assumes relative uncertainty in τ of 0.1 %	[90]
Phosphine	1.3	Absolute limit uncertain due to background absorption from phosphine	[126]
Silane	50	Corrected for dilution to 5 % silane in nitrogen to prevent particulate formation	[125]

[a] See the text for a discussion of lower detectable limit (LDL) values.

7.4.4   Direct Frequency-Comb Spectroscopy

Direct frequency-comb spectroscopy (DFCS) is an optical detection technique that has been developed in recent years to address sensing challenges that cannot be overcome by classical techniques such as TDLAS, CRDS, or FTIR spectroscopy. In particular, the combination of a broad spectral bandwidth with single-laser-line resolution and coherence is a unique detection capability for the DFCS technique, and it offers unprecedented potential for ultrasensitive, rapid multispecies detection for many experimental scenarios and applications [132–136].

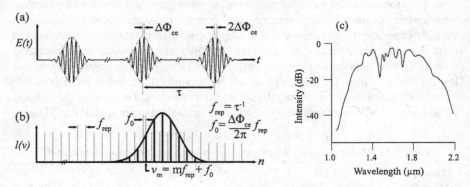

Figure 7.18   (a) Schematic time-domain picture of the pulse train of a mode-locked laser with the laser round-trip time τ and the carrier-envelope phase shift $\Delta\Phi_{ce}$. (b) Corresponding frequency-domain picture demonstrating the broadband comblike structure given by the laser repetition rate f_{rep} and the offset frequency f_0. (c) Typical spectrum of an Er-doped fiber frequency comb broadened with a short piece of highly nonlinear fiber.

Properties of Frequency Combs   Current frequency-comb sources are based almost entirely on mode-locked lasers. These systems were developed originally for their ultrafast pulse duration in the femtosecond or picosecond range. These pulses are emitted as an extremely regular train of pulses, resulting in a remarkably regular and stable frequency structure [137]. Figure 7.18a and b illustrate this connection: The short pulses in the time domain result in a very wide frequency spectrum according to the Fourier theorem; however, since the laser does not emit a single pulse but a train of pulses with extremely stable pulse-to-pulse period τ (given by the laser round-trip time), the spectrum becomes discretized and consists of a large number of narrow lines that are uniformly spaced by the laser repetition rate $f_{rep} = \tau^{-1}$. Resembling a comb in the frequency domain, this structure gives rise to the name "frequency comb." Considering the underlying electric field in each ultrashort pulse emitted and the effect of dispersion usually present in the laser cavity, pulses are not exact copies of each other, but may experience phase shifts with respect to their intensity envelopes, known as carrier-envelope-phase (CEP) shift. Taking this effect into account in the frequency domain, we find a constant offset f_0 of the frequency comb from the exact integer multiples of f_{rep}. The optical frequency associated with every comb line can be defined exactly by only three parameters via $\nu_m = f_0 + mf_{rep}$, where m is an integer.

For DFCS the most commonly used frequency-comb sources are currently based on Er-doped fiber lasers [138–140]. These systems are commercially available, compact, and easy to operate. Their near-infrared emission spectrum around 1.55 μm is a typical region for trace gas sensing and allows the use of many standard components from the telecommunications industry. The typical repetition rates range from 100 to 300 MHz. The direct spectrum emitted by Er-doped fiber mode-locked lasers covers the wavelength range from approximately 1.5 to 1.6 μm (or 6200 to 6700 cm^{-1} in wavenumbers), but may be extended further by using nonlinear optics for frequency conversion. Highly nonlinear fibers (HNFs) with extremely small core diameters allow a massive spectral broadening while preserving the comb's coherence properties. Figure 7.18c shows a typical spectrum of a Er-doped fiber laser broadened with a short piece (ca. 10 cm) of HNFs. At the −20-dB level relative to the spectral peak, the spectrum extends easily from 1.2 to 2.0 μm, a span of more than 3300 cm^{-1}. With a repetition rate of 250 MHz for the comb used here, the spectrum contains approximately 400,000 individual comb modes, each with a width of a few tens of kilohertz and an extremely well defined position ν_m given by $\nu_m = f_0 + m f_{rep}$. Furthermore, the position of each comb mode can be controlled precisely via f_0 and f_{rep}, and all lines can be used simultaneously, eliminating the need for wide and time-consuming frequency scans.

Coupling a Comb to an External Enhancement Cavity   One of the biggest advantages of frequency combs compared to incoherent broadband light sources such as thermal emitters or fiber-based white-light supercontinua is the regular frequency-mode structure, which allows coupling the comb efficiently to high-finesse external enhancement cavities such as those used in CRDS. By matching the comb's repetition rate with the cavity's free spectral range (FSR) and adjusting the absolute frequency positions (via f_0), every comb line can be coupled exactly to a cavity mode, thereby allowing maximum coupling efficiency. The spectral bandwidth that can be coupled to the cavity, however, is usually smaller than the total bandwidth of the frequency comb, due to dispersion in the enhancement cavity. Mirror coatings and dispersion of any sample gas inside the cavity cause the cavity modes to be nonuniformly spaced. As a result, the comb and cavity modes slowly walk off perfect alignment away from the frequency, where the coupling is optimized. In general, the transmission bandwidth is limited to several tens of nanometers in the near infrared, corresponding to approximately 100 to 200 cm^{-1}. Of course, this bandwidth limit scales inversely with the cavity finesse. This restriction on the bandwidth transmitted simultaneously may be overcome by changing the lock point (frequency where transmission is optimized) or by dithering f_{rep}, which allows all comb lines to come on resonance with a cavity mode during one sweep of f_{rep} and results in transmission of the full comb spectrum when averaged over multiple dither cycles. A more detailed overview of comb-cavity coupling and an outline of various coupling schemes are provided by Thorpe, Adler, and others [135,136].

Cavity-enhanced direct frequency-comb spectroscopy (CE–DFCS) is an extremely powerful technique combining the sensitivity and resolution advantages of CW–CRDS with the broadband coverage of FTIR. Although earlier experiments were limited primarily by technical noise due to the more complicated broadband laser-cavity coupling [122,141–143], quantum-limited detection sensitivity has been achieved recently for the first time in a CE–DFCS system [144].

Broadband, High-Resolution Detection   Ideally, one would like to detect every single mode of the frequency-comb individually, to take advantage of the high spectral resolution and still cover simultaneously the entire bandwidth of the broadband comb. The vast difference in a frequency comb's mode spacing (<0.01 cm^{-1}) and its simultaneous bandwidth (hundreds of wavenumbers) is a significant challenge for many commonly used spectrographic tools, such as simple grating spectrographs; however, several approaches exist that are able to fulfill these requirements.

Figure 7.19 illustrates three of the most widely used spectrograph designs used in CE–DFCS. Figure 7.19a shows the VIPA spectrometer [122,133,141,142]. It is based on a tilted étalon plate, called a virtually imaged phased array (VIPA), which is able to provide very high angular dispersion (usually, oriented vertically). Since the VIPA is per design an étalon, the dispersion pattern is folded in every free spectral range (typically, 1.5 to 3 cm^{-1}). Therefore, a diffraction grating is used to unfold the overlapping VIPA orders in the horizontal plane, resulting in a two-dimensional diffraction pattern. The grating only has to resolve the coarsely spaced VIPA mode orders and is also able to image a wide bandwidth onto the two-dimensional detection array (typically, 200 to 300 cm^{-1}). The spectral resolution of a VIPA spectrometer is typically on the order of 0.03 cm^{-1}. Although this resolution is not sufficient to resolve individual lines of most frequency combs, it is virtually ideal for most experiments operating close to room temperature and atmospheric pressure, owing to molecular absorption linewidths on the order of 1 GHz (0.03 cm^{-1}). The two-dimensional spectrum on the camera provides a simultaneous broadband snapshot of all imaged comb lines and is therefore very useful for measuring fast processes [142]. By reading out the VIPA mode orders line by line, the two-dimensional spectrum is converted to a classical one-dimensional spectrum. If more detection bandwidth is required or if the laser wavelength does not allow the use of a VIPA, a Fourier transform spectrometer (FTS) may be used to detect the spectrum of a frequency comb. Figure 7.19b shows a typical setup with a scanning interferometer behind the comb and cavity [143,145,146]. The advantages of an FTS are its high detection bandwidth (usually larger than the comb bandwidth) and the fact that the resolution can easily be scaled via the length of the delay line. Therefore, an FTS can be designed to resolve single comb modes if necessary. Compared to an incoherent FTIR system, the disadvantage of limited throughput (étendue) at high spectral resolution vanishes when using a coherent light source such as a frequency comb; therefore, comb-FTS systems allow much faster data acquisition for high-resolution scans because of enhanced light transmission. A recently demonstrated system performs one scan at less than 0.01 cm^{-1} resolution in only a few seconds [144,146]. The interferometer also provides two signals with a relative phase shift, π, which allow using balanced detection to subtract technical noise from the comb-cavity coupling, resulting in shot-noise-limited sensitivity [144]. Despite its performance advantages, a scanning FTS has two drawbacks: Since it requires a mechanical delay, its size is quite substantial if high resolution is required, and the spectral information is collected over a time frame of several seconds, ruling out the observation of dynamic processes. One method of speeding up the scan time of the FTS is dual-comb spectroscopy [147–149], which is illustrated in Figure 7.19c. Instead of a mechanical delay, a second frequency comb with a repetition rate that is slightly offset from the first comb is used to probe the light of the comb interacting with the sample. The small offset in f_{rep} mimics a fast scanning delay, as the pulse train from the second comb will evolve with respect to that of the first comb. A full scan (until the pulse from one laser overlaps again with a pulse from the other laser) can be performed in milliseconds or even microseconds, much faster than any mechanical delay; however, the spectral bandwidth, which can be scanned simultaneously without sample aliasing, decreases with faster sampling time [150]. Therefore, the right compromise will need to be found for a particular application. One further drawback of this method is also the need for two synchronized frequency combs. With compact and easily controllable Er:fiber systems, the additional laser is manageable; however, with more complex systems such as the ones used for mid-infrared spectroscopy [146], the dual-comb setup becomes rather complex.

Experimental Demonstration of Trace Water Detection in Arsine Recently, frequency-comb-based detection of impurities in specialty gases was demonstrated for the first time. In particular, the experiment showed ultrasensitive detection of trace water vapor in the semiconductor gas arsine [122]. The spectral band selected was a rarely explored window of transmission in arsine around 5500 res. The employed frequency comb was an Er:fiber system that was spectrally broadened with a piece of HNF, yielding the spectrum shown in Figure 7.18c. The remaining setup is similar to the one shown in Figure 7.19a, using an enhancement cavity with a finesse of 30,000 and a VIPA spectrometer with a resolution of $0.03\,cm^{-1}$. The high-finesse cavity contained the sample gas (arsine + trace water vapor). For data normalization, we also collected reference spectra when we replaced the sample gas with dry nitrogen. The sample absorption obtained is depicted in Figure 7.20 and clearly highlights the major advantage of the broadband and high-resolution CE–DFCS technique. Although the entire spectrum is dominated by densely spaced arsine absorption features, water absorption lines are clearly identified (using known positions from the HITRAN spectroscopic database [96]), such as those seen in the magnified inset. The spectrum collected contains dozens of water lines and hence enables a very precise and sensitive measurement of the water concentration. Here, the concentration of trace water was determined to be 310 ppbv with a minimum detectable concentration of 31 ppbv.

As a first demonstration of using CE–DFCS for this type of experiment, the results presented here still have a large room for improvement. For instance, the total covered spectral bandwidth of the CE–DFCS system was 5000 to $5800\,cm^{-1}$. Thus, the system was able to cover the absorption bands of multiple additional important impurity species, such as methane, hydrogen sulfide, carbon dioxide, or silane (demonstrated by using nitrogen as carrier gas instead of arsine). Only the limited transparency of arsine in the chosen spectral window restricted the trace detection to water. It is likely that different spectral windows and/or different specialty

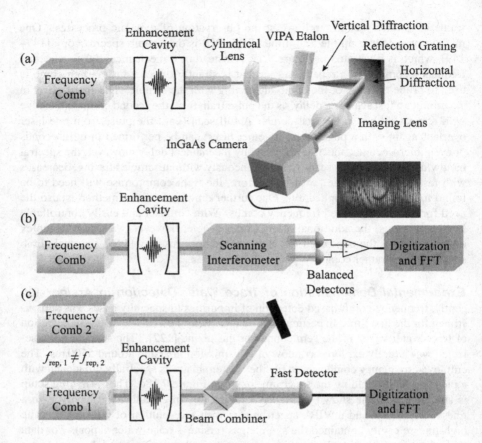

Figure 7.19 Detection schemes for cavity-enhanced direct frequency-comb spectroscopy: (a) VIPA spectrometer, (b) Fourier transform spectroscopy with a mechanically scanning interferometer, and (c) dual-comb spectroscopy using two synchronized mode-locked lasers.

gases (e.g., phosphine or silane) will allow the comb system to demonstrate its full potential more effectively. In addition, technical noise provided a major limitation to the sensitivity; recent efforts in overcoming technical noises have resulted in dramatic improvements of the sensitivity, reaching the shot-noise limit and promising much lower detection limits for future experiments [144]. An optimized CE–DFCS system is probably capable of achieving sub-ppbv detection sensitivity for trace water impurity.

Extension of Comb Spectroscopy into the Mid-infrared   One particular advantage of frequency combs, which is already used in the HNF-based spectral broadening, is the fact that nonlinear optics can be exploited to access spectral regions that are not covered directly by the laser system. One example is the extension of frequency-comb technology into the mid-infrared, where the fundamental vibra-

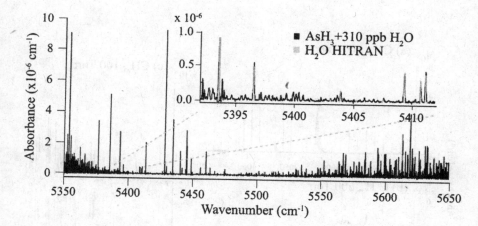

Figure 7.20 Spectrum of arsine gas around $5500\,cm^{-1}$ with trace quantities of water vapor (310 ppbv). The inset shows the positions of the water lines taken from the HITRAN database.

tional absorption bands of typical trace molecules are located. These bands can be significantly stronger than that of their near-infrared overtones. In the case of water, the fundamental O–H stretch vibration band around $3700\,cm^{-1}$ is approximately 10 times stronger than the band around $5350\,cm^{-1}$, which was used in the previously described experiment [96]. Other species offer even larger gains in line strength (e.g., 100 for methane and even 10^4 for carbon dioxide).

The first fully operable frequency-comb spectroscopy system in the molecular fingerprint region (frequencies below $4000\,cm^{-1}$) was demonstrated recently. Using a high-power Yb:fiber laser (fundamental operating wavelength at 1.05 μm) to pump an optical parametric oscillator synchronously [151], the frequency-comb source provides a tunable output ranging from center wavelengths of 2.8 to 4.8 μm (corresponding to approximately 3600 to $2100\,cm^{-1}$) with a simultaneous spectral bandwidth of up to $300\,cm^{-1}$ [146]. The entire spectroscopy setup is similar to Figure 7.19b, with a scanning Michelson interferometer for Fourier transform spectroscopy and a Herriott multipass cell instead of the enhancement cavity. Although the multipass cell provides only a total increased pathlength of 36.4 m compared to many kilometers of enhanced pathlength possible with an optical cavity, the access to strong absorption features in the mid-infrared still resulted in extremely low detection limits, to below the ppbv level. To implement cavity enhancement in this spectral region, high-quality low-dispersion mirrors will be needed; however, this requirement still presents a challenge for current optical coating technology. Figure 7.21 shows a collection of demonstration spectra collected with this mid-infrared comb Fourier transform spectroscopy system. Figures 7.21a and b show that the system's capability of covering a wide spectral window enables the detection and quantification of both species with a clear line spectrum such as methane and molecules with wide, continuous absorption features such as isoprene. At the same time, the system also offers high resolution as demonstrated in Figure 7.21c, which shows a magnified

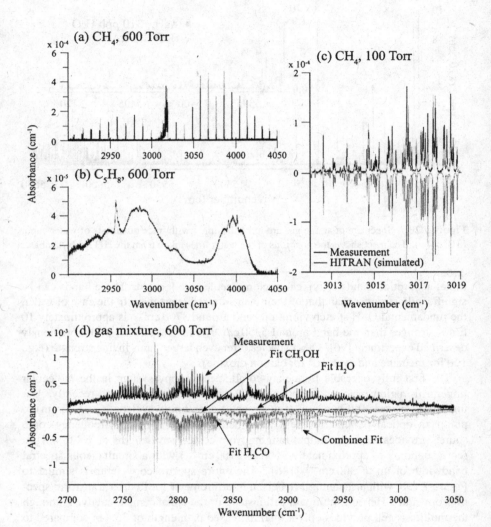

Figure 7.21 Typical spectra collected with a mid-infrared frequency-comb Fourier transform spectrometer: (a) spectrum of 10 ppmv methane in 600 torr (80 kPa) of nitrogen in the fundamental C–H-stretch spectral window around 3000 cm^{-1} at a resolution of 0.0056 cm^{-1}, (b) spectrum of 16 ppmv of isoprene in 600 torr (80 kPa) of nitrogen at a resolution of 0.058 cm^{-1}, (c) magnified view of the methane Q-branch to highlight the resolution of the mid-infrared comb-FTS (data in black above, HITRAN fit in gray below), and (d) spectrum of a mixture of formaldehyde, methanol, and water in 600 torr (80 kPa) of nitrogen (black) and the numerical concentration fits below based on spectral reference data for formaldehyde, methanol, water, and the combined fit.

view of the methane Q-branch. Finally, a combination of the comb-FTS's high resolution and wide bandwidth is utilized by quantifying a mixture of gases, as shown in Figure 7.21d. Here, the sample contains formaldehyde, methanol, and water. A simple fitting algorithm and a database with the molecules' reference spectra extract precise concentrations for the constituents, in this case, 47.1 ppmv of formaldehyde, 58.3 ppmv of methanol, and 813 ppmv of water in 600 torr (80 kPa) of nitrogen. The precision attained depends on the number of absorption features present for each molecule, but the concentration determined for methanol is particularly impressive. Whereas the continuous absorption background of methanol would impose a major challenge for any CW laser technique, the frequency-comb system is able to determine the concentration of methanol in this sample with a relative precision of 0.07 %.

Despite a variety of powerful demonstrations over the last few years, frequency-comb spectroscopy is still a technique under development. Current efforts focus on the generation of more stable, more broadband, and longer-wave mid-infrared combs, as well as novel ideas focusing on overcoming noise limitations to push for better sensitivity. Considering their unique capabilities, frequency combs will certainly play a major role in future trace detection applications.

7.5 Conclusions

Development of analytical methods for the measurement of water vapor impurity in inert and specialty gases is especially important in the manufacture of microelectronic devices that continue to decrease in size and become increasingly more complex. The need to control water vapor at trace levels in a range of gases used in critical process steps requires not only rapid, reliable, and selective gas-phase measurement but also high-sensitivity approaches. In the last decade many advances in the development and application of sensor and laser absorption spectroscopic techniques have taken place, and hence this chapter has focused primarily on these areas. Existing sensor technologies continue to be refined, and new sensor materials and approaches are being actively investigated. Even more significant is the rapid development of near-infrared laser spectroscopic methods using long-path cells and enhancement cavities, opening up new capabilities for the trace analytical chemist. Methods such as CRDS and TDLAS now enable high-sensitivity water vapor measurements down to 100 to 200 pptv in inert gas matrices in small, easy-to-use, relatively low-cost instrumentation. Just 10 years ago, such sensitivities were attainable only with much more expensive and complex research equipment (e.g., APIMS).

In the future we envisage advances in laser spectroscopic techniques continuing along the same path but extending into the mid-infrared region, as efficient and cost-effective mid-infrared laser sources become more widely available. These methods will offer enhanced sensitivity over current methods by making use of more intense fundamental absorption lines rather than less intense overtone or combination lines. Furthermore, in some cases the methods will widen the spectroscopic region available, enabling simultaneous detection of water vapor and other gas-phase impurities. CE–

DFCS is one such technique that shows considerable potential in this regard. Other emerging mid-infrared laser absorption spectroscopic techniques (some of which are described and discussed in Chapter 4) will probably also be developed specifically for moisture analysis. However, as already emphasized in this chapter, the successful application of any new spectroscopic method in multiple gas matrices will still depend on being able to locate a transmission window where strong analyte absorption lines can be accessed cleanly.

REFERENCES

1. Funke, H. H., Grissom, B. L., McGrew, C. E., & Raynor, M. W. (2003). Techniques for the measurement of trace moisture in high-purity electronic specialty gases. *Review of Scientific Instruments, 74*, 9, 3909–3933.

2. Monroe, S. (1998). Trace moisture in ammonia: Gas chromatography using calcium carbide. *Journal of the Institute of Environmental Sciences and Technology, 41*, 1, 21–25.

3. Siefering, K., Berger H., & Whitlock, W. (1993). Quantitative analysis of contaminants in ultrapure gases at the parts per trillion level using atmospheric pressure ionization mass spectroscopy. *Journal of Vacuum Science Technology A, 11*, 1593–1597.

4. Ketkar S., & Dheandhanoo, S. (2001). Use of ion mobility spectrometry to determine low levels of impurities in gases. *Analytical Chemistry, 73*, 2554–2557.

5. Ketkar, S., Dheandhanoo, S., & Scott, A. (2004). Comparision of APIMS and IMS for the analysis of trace impurities in inert gases. *SEMI Technology Symposium: Innovations in Semiconductor Manufacturing, Semicon West 2004, San Francisco, CA*, pp. 89–108.

6. Warrick, B. (2004). Detection of gas phase impurities: a comparison of API–MS and IMS technologies. *SEMI Technical Symposium (STS): Innovations in Semiconductor Manufacturing*. San Jose, CA: Semiconductor Equipment and Materials International, pp. 109–129

7. Bandy, A. R., Tu, F. H., Mitchell, G. M., & Ridgeway, R. G. (2002). Determination of moisture in anhydrous ammonia at the ppbv level using negative ion APIMS. *Proceedings of SEMICON West 2002, SEMI Technical Symposium: Innovations in Semiconductor Manufacturing, San Francisco, CA*, pp. 163–170.

8. Okhi A., Ohmi, T., Date, J., & Kijima T. (1998). Highly purified silane gas for silicon semiconductor devices. *Journal of the Electrochemical Society, 145*, 3560–3569.

9. Kitano, M., Shirai, Y., Ohki, A., Babasaki, S., & Ohmi, T. (2001). Impurity measurement in specialty gases using an atmospheric pressure ionization mass spectrometer with a two compartment ion source. *Japanese Journal of Applied Physics, 40*, 2688–2693.

10. Ma, C., Athalye, A., Fruhberger, B., & Ezell, E. (1998). Moisture dry-down in high purity hydrogen chloride. *Proceedings of Contamination Control, Phoenix, AZ*. Mount Prospect, IL: Institute of Environmental Sciences and Technology, pp. 285–289.

11. Meyer, C. W., Hodges, J. T., Hyland, R. W., Scace, G. E., Valencia-Rodriguez, J., & Whetstone, J. R. (2010). The second-generation NIST standard hygrometer. *Metrologia, 47*, 192–207.

12. Meyer, C. W., Hodges, J. T., Huang, P. H., Miller, W. W., Ripple, D. C., & Scace, G. E. (2008). *Calibration of Hygrometers with the Hybrid Humidity Generator.* NIST Special Publication 250-83. Washington, DC: USDOC and NIST.

13. Scace, G. E., Huang, P. H., Hodges, J. T., Olson, D. A., & Whetstone, J. R. (1997). The new NIST low frost-point humidity generator. *Proceedings of the 1997 National Conference of Standards Laboratories Workshop and Symposium.* Boulder, CO: NCSL International.

14. Scace, G. E., & Hodges, J.T. (2002). Uncertainty of the NIST low frost-point humidity generator. In B. Fellmuth, J. Seidel, and G. Scholz (Eds.), *Proceedings of TEMPMEKO 2001: Eighth International Symposium on Temperature and Thermal Measurements in Industry and Science.* Berlin: VDE Verlag, pp. 597–602.

15. Hodges, J. T., & Scace, G. E. (2006). Developing advanced humidity standards to measure trace water vapor in specialty gases. *MICRO, 24,* 59.

16. Ketkar, S. N, Scott, A. D., Jr., & Martinez de Pinillos, J. V. (1994). Dynamic dilution calibration system for calibrating analytical instruments used in gas analysis. *Journal of the Electrochemical Society, 141,* 1, 184–187.

17. McAndrew, J. J. F. (1997). Humidity measurement in gases for semiconductor processing. In J. D. Hogan (Ed.), *Specialty Gas Analysis: A Practical Guidebook.* New York: Wiley VCH, pp. 21–42.

18. Wiederhold, P. R. (2000). The principles of chilled mirror hygrometry. *Sensors, 17,* 7, 46–51.

19. Keidel, F. A. (1959). Determination of water by direct amperometric measurement. *Analytical Chemistry, 31,* 12, 2043–2048.

20. Zatko, D. A., & Maguire, J. F. (1989). *Low level moisture measurement system and method.* U.S. Patent No. 4,800,000. Washington, DC: U.S. Patent and Trademark Office.

21. Ma, C., Shadman, F., Mettes, J., & Silverman, L. (1995). Evaluating the trace-moisture measurement capability of coulometric hygrometry *MICRO, 13,* 4, 43–49.

22. Bell, S. A., Gardiner, T., Gee, R. M., Stevens, M., Waterfield, K., & Woolley, A. (2004). An evaluation of performance of trace moisture measurement methods. *Proceedings of 9th International Symposium on Temperature and Thermal Measurements in Industry and Science, 1,* 663–668.

23. Manickum, V., Prabhu, E., Jayaraman, V., Gnanasekar, K. I., Gnanasekaran, T., & Nagaraja, K. S. (2010). Electrolytic sensor for trace level determination of moisture in gas streams. *Measurement, 43,* 10, 1636–1643.

24. Wiederhold, P. R. (1997). *Water Vapor Measurement Methods and Instrumentation.* New York: Marcel Dekker.

25. Capuano, I. A., & Mathieu, R. J. (1999). A rapid response electrolytic amperometric moisture analyzer. *American Laboratory, 31,* 30–33.

26. Sauerbrey, G. Z. (1959). Use of quartz vibrator for weighing thin films. *Zeitschrift für Physik, 155,* 206–212.

27. O'Sullivan, C. K., & Guilbault, G. G. (1999). Commercial quartz crystal microbalances: theory and applications. *Biosensors and Bioelectronics, 14,* 8/9, 663–670.

28. Blakemore, C. B. (1985). Piezoelectric moisture analyzer for semiconductor gases. In K. Herbst (Ed.), *Proceedings of the 31st Annual ISA Analysis Division Symposium,* Vol. 21. Research Triangle Park, NC: Instrument Society of America, p. 125.

29. Kleinfeld, E. R., & Ferguson, G. S. (1995). Rapid, reversible sorption of water from the vapor by a multilayered composite file: a nanostructured humidity sensor. *Chemical Materials, 7*, 2327–2331.

30. Halper, S. R., & Villahermosa R. M. (2009). Cobalt-containing polyimides for moisture sensing and absorption. *ACS Applied Material Interfaces, 1*, 5, 1041–1044.

31. Bear, R. S., Jr., & Hauer, R. (2002). Designing a ppb class moisture analyzer: evolution and technical hurdles. *Standards Workshop on New Advances in Detection of Trace (< 1 ppb) Impurities in Bulk Inert Gases, SEMICON West 2002, July 23, San Francisco, CA.*

32. Feng, J., & Raynor, M. W. (2007). Trace water vapor detection in pure arsine gas using quartz crystal oscillator technology. *Pittsburgh Conference on Analytical Chemistry and Applied Spectroscopy (Pittcon), Chicago, February 25–March 2.*

33. Wei, J., Pillion, J. E., & Hoang, C. (1997). In-line moisture monitoring in semiconductor process gases by a reactive metal coated quartz crystal microbalance. *Journal of the Institute of Environmental Sciences and Technology, 40*, 43–48.

34. Wei, J., Pillion, J. E., King, M., & Verlinden, M. (1997). Using an in-line monitor to obtain real-time moisture measurements. *MICRO, 15*, 31–36.

35. Toth, M. I., Zatko, D. A., Yesenofski, D. F., Schneegans, M., & Helneder, J. (1997). Beta testing an in-line monitor for moisture measurement in inert gas lines. *MICRO, July/August,* 79–92.

36. Huang, P. H. (1998). Humidity standards for low level water vapor sensing and measurement. *Sensors and Actuators B, 53*, 1, 125–127.

37. Beaubien, D. J. (2005). The chilled mirror hygrometer: how it works, where it works and where it doesn't. *Sensors, 22*, 5, 30–34.

38. Flaherty, E., Herold, C., Murray, D., & Thompson, S. R. (1986). Determination of water in hydrogen chloride gas by condensation technique. *Analytical Chemistry, 58*, 1903–1904.

39. Chen, Z., & Lu, C. (2005). Humidity sensors: a review of materials and mechanisms. *Sensor Letters, 3*, 274–295.

40. Hasagawa, S. (1980). Performance characteristics of a thin film aluminum oxide humidity sensor. *Proceedings of the 30th Electronic Components Conference, San Francisco, CA.* New York: IEEE, pp. 386–391.

41. Mehrhoff, T. K. (1985). Comparison of continuous moisture monitors in the range 1 to 15 ppm. *Review of Scientific Instruments, 56*, 1930–1933.

42. Nahar, R. K. (2002). Physical understanding of moisture induced degradation of non-porous aluminum oxide thin films. *Journal of Vacuum Technology B, 20*, 382–385.

43. Clevett, K. J. (1986). *Process Analyzer Technology.* New York: John Wiley & Sons.

44. Kerney, J. (2007). New method for trace moisture measurement. *Advances in Wafer Processing Session,* SEMICON West, San Francisco, CA.

45. Feng, J., & Raynor, M. W. (2009). Trace water vapor analysis under 100 ppb using cavity ring-down, oscillating quartz crystal and electrical impedance technologies. *Specialty Gas Session,* Pittsburgh Conference on Analytical Chemistry and Applied Spectroscopy (Pittcon), Chicago, February 25–March 2.

46. Coates, J. (1998). Vibrational spectroscopy: instrumentation for infrared and Raman spectroscopy. *Applied Spectroscopy Reviews, 33*, 4, 267–425.

47. Griffith, P. R., & de Haseth, J. A. (2007). *Fourier Transform Infrared Spectroscopy.* Hoboken, NJ: John Wiley & Sons.

48. Smith, B. C. (2011). *Fundamentals of Fourier Transform Infrared Spectroscopy*, 2nd ed. Boca Raton, FL: CRC Press.

49. Saptari, V. (2004). Fourier-transform spectroscopy instrumentation engineering. In *SPIE Tutorial Texts in Optical Engineering*, Vol. TT61. Bellingham, WA: Society for Photo-optical Instrumentation Engineers.

50. Kinch, M. A. (2007). *Fundamentals of Infrared Detector Materials.* Bellingham, WA: SPIE, The Society for Photo-optical Instrumentation Engineers.

51. Rogalski, A. (2011). *Infrared Detectors*, 2nd ed. Boca Raton, FL: CRC Press.

52. Richardson, R. L., Yang, H., & Griffiths, P. R. (1998). Evaluation of a correction for photometric errors in FT-IR spectrometry introduced by a non linear detector response. *Applied Spectroscopy, 52*, 4, 565–571.

53. White, J. U. (1942). Long optical paths of large aperture. *Journal of the Optical Society of America, 32*, 5, 285–285.

54. Vahey, P. G., Meyer, R. T., & Perez, J. E. (2006). Overcoming hydrogen bonding in on-line FTIR spectroscopy of hydrofluoric acid. *Gases and Technology, 5*, 1, 18–22.

55. Millward, A., Davia, D., Wyse, C., Seymour, A., Vininski, J., Torres, R., & Raynor., M. (2009). *Anhydrous Hydrogen Fluoride for Selective Etch Processes: Measurement and Control of Water and Metallic Impurities.* Taiyo Nippon Sanso Corporation Technical Report 28, pp. 17–22.

56. Stallard, B. R., Rowe, R. K., Garcia, M. J., Haaland, D. M., Espinoza, L. H., & Niemczyk, T. M. (1993). *Trace Water Vapor Determination in Corrosive Gases by Infrared Spectroscopy.* Sandia National Laboratories Report SAND 93-4026 UC-411.

57. Espinoza, L. H., Niemczyk, T. M., Stallard, B. R., & Garcia, M. J. (1997). *Trace Water Vapor Determination in Nitrogen and Corrosive Gases using Infrared Spectroscopy.* Sandia National Laboratories Report SAND 97-1494 UC-411.

58. Stallard, B. R., Espinoza, L. H., & Niemczyk, T. M. (1995). Trace water vapor determination in gases by infrared spectroscopy. *Proceedings of the 41st Annual Technical Meeting of the Institute of Environmental Sciences, Anaheim, CA*, pp. 1–8.

59. Stallard, B. R., Espinoza, L. H., Rowe, R. K., Garcia, M. J., & Niemczyk, T. M. (1995). Trace water vapor detection in nitrogen and corrosive gases by FTIR spectroscopy. *Journal of the Electrochemical Society, 142*, 8, 2777–2782.

60. Espinoza, L. H., Niemczyk, T. M., & Stallard, B. R. (1998). Generation of synthetic background spectra by filtering the sample interferogram in FT-IR. *Applied Spectroscopy, 52*, 3, 375–379.

61. Nishikida, K., Nishio, E., & Hannah, R. W. (1995). *Selected Applications of Modern FTIR Techniques.* Tokyo: Kodansha.

62. Jaakkola, P., Tate, J. D., Paakkunainen, M., Kauppinen, J., & Saarinen, P. (1997). Instrumental resolution considerations for Fourier transform infrared gas-phase spectroscopy. *Applied Spectroscopy, 51*, 8, 1159–1169.

63. Anderson, R. J., & Griffiths, P. R. (1975). Errors in absorbance measurements in infrared transform spectrometry because of limited instrument resolution. *Analytical Chemistry, 47*, 14, 2339–2347.

64. Pivonka, D. E. (1991). The infrared spectroscopic determination of moisture in HCl for the characterization of HCl gas drying resin performance. *Applied Spectroscopy, 45*, 4, 597–603.

65. Salim, S., Litwin, M. M., & Natwora, J. P. J. (1998). FTIR spectroscopy for measurement of moisture levels in hydride and corrosive gases. Workshop on Gas Distribution Systems, *Proceedings of SEMICON West 1998, San Francisco, CA*, pp. D1–D18.

66. Miyazaki, K., & Kimura, T. (1993). Analysis of trace impurities in corrosive gases by gas-phase FTIR. *Bulletin of the Chemical Society of Japan, 66*, 11, 3508–3510.

67. Salim, S., & Gupta, A. (1996). Measurement of trace levels of moisture in UHP hydride gases by Fourier transform infrared spectroscopy. *Proceedings of CleanRooms 1996 West Session 602, Santa Clara, CA, 22*, 22–32.

68. Mitchell, G. M., Vorsa, V., Ryals, G. L., Milanowicz, J. A., Ragsdale, D. J. M., Marhefka, K. L., Wagner, M., & Ketkar, S. N. (2002). Trace impurity detection in ammonia for the compound semiconductor market. SEMI Technical Symposium: Innovations in Semiconductor Manufacturing. *Proceedings of SEMICON West 2002, San Francisco, CA*, p. 199.

69. Funke, H. H., Raynor, M. W., Yucelen, B., & Houlding, V. H. (2001). Impurities in hydride gases: 1. Investigation of trace moisture in the liquid and vapor phase of ultra-pure ammonia by FTIR spectroscopy. *Journal of Electronic Materials, 30*, 11, 1438–1447.

70. Yao, J., Funke, H. H., & Raynor, M. (2004). Spectroscopic methods for trace moisture detection in electronic specialty gases. *Nippon Sanso Corporation Engineering Report, 23*, 43–49.

71. Haaland, D. M. (1992). Multivariate calibration methods applied to quantitative analysis of infrared spectra. P. C. Jurs (Ed.), *Computer-Enhanced Analytical Spectroscopy*, Vol. 3. New York: Plenum Press, pp. 1–30.

72. Haaland, D. M., & Easterling, R. G. (1980). Improved sensitivity of infrared spectroscopy by the application of least squares methods. *Applied Spectroscopy, 34*, 5, 539–548.

73. Haaland, D. M., & Thomas, E. V. (1988). Partial least-square methods for spectral analysis. *Analytical Chemistry, 60*, 1193–1202.

74. Hovde, D. C., Hodges, J. T., Scace, G. E., & Silver, J. A. (2001). Wavelength-modulation laser hygrometry for ultrasensitive detection of water vapor in semiconductor gases. *Applied Optics, 40*, 6, 829–839.

75. Wu, S. Q., Morishita, J., Masusaki, H., & Kimishima, T. (1998). Quantitative analysis of trace moisture in N_2 and NH_3 gases with dual-cell near infrared diode laser absorption spectroscopy. *Analytical Chemistry, 70*, 15, 3315–3321.

76. Wu, S. Q., Masusaki, H., Ishihara, Y., Matsumoto, K., Kimishima, T., Morishita, J., Kuze, H., & Takeuchi, N. (1996). Trace moisture measurements with dual beam diode laser spectroscopy. *Proceedings of the 5th International Symposium on Semiconductor Manufacturing*. Piscataway, NJ: Institute of Electrical and Electronics Engineers, pp. 321–324.

77. Kermarric, O., Campidelli, Y., Bensahel, D., Ly, C. H., & Mauvais, P. (2002). The detrimental effect of moisture in SiGe epitaxy. *Solid State Technology, 45*, 55–60.

78. Wright, A. O., Wood, C. D., Reynolds K. J., & Malczewski, M. L. (2004). Detecting single digit part-per-billion levels of moisture in ammonia gas. *MICRO, 22*, 47–55.

79. Inman, R. S., & McAndrew, J. J. F. (1994). Application of tunable diode laser absorption spectroscopy to trace moisture measurements in gases. *Analytical Chemistry, 66*, 2471–2479.

80. Funke, H. H., Yao, J., Raynor M. W., & Wright A. O. (2004). Using tunable diode laser spectroscopy to detect trace moisture in ammonia. *Solid State Technology, 47*, 10, 49–54.

81. McAndrew, J., Bartolomey, M., Girard, J. M., Goltz, G., & Flan, J. M. (2000). Implementing on-line and in-situ moisture monitoring in reactive gas environments. *MICRO, 18*, 39–47.

82. Vorsa, V., Dheandhanoo, S., Yesenofski, D., & Wagner, M. (2003). Measuring moisture in corrosive gases using TDLAS and CRDS. *Workshop on Advances in Detection of Trace Moisture in Specialty Gases, Semicon West 2003, July 14, San Francisco, CA.*

83. McAndrew, J. (1998). Progress in in-situ contamination control. *Semiconductor International, 21*, 71–78.

84. Bomse, D. S., Stanton, A. C., & Silver, J. A. (1992). Frequency modulation and wavelength spectroscopies: comparison of experimental methods using lead-salt diode laser. *Applied Optics, 31*, 718–731.

85. Silver, J. A. (1992). Frequency modulation spectroscopy for trace species detection: theory and comparison of among experimental methods. *Applied Optics, 31*, 707–717.

86. Reid, J., & Labrie, D. (1981). Second harmonic detection with tunable diode lasers: comparison of experiment and theory. *Applied Physics B, 26*, 203–210.

87. Feher, M., & Martin, P. A. (1995). Review: tunable diode laser monitoring of atmospheric trace gas constituents. *Spectrochimica Acta A, 51*, 1579–1599.

88. Lisak, D., Hodges, J. T., & Ciurylo, R. (2006). Comparison of semiclassical line-shape models to rovibrational H_2O spectra measured by frequency-stabilized cavity ring-down spectroscopy. *Physical Review A, 73*, 012507.

89. Tran, H., Bermejo, D., Domenech, J. L., Joubert, P., Gamache, R. P., & Hartmann, J. M. (2007). Collisional parameters of H_2O lines: velocity effects on the line shape. *Journal of Quantitative Spectroscopy and Radiative Transfer, 108*, 1, 126–145.

90. Vorsa, V., Dheandhanoo, S., Ketkar, S. N., & Hodges, J. T. (2005). Quantitative absorption spectroscopy of residual water vapor in high-purity gases: pressure broadening of the 1.392,53 µm H_2O transition by N_2, HCl, HBr, Cl_2, and O_2. *Applied Optics, 44*, 4, 611–619.

91. Amerov, A., Maskas, M., Meyer, W., Fiore, R., & Tran, K. (2007). New process gas analyzer for the measurement of water vapor concentration. *Proceedings of Instrumentation, Systems and Automation Society 52nd Analysis Division Symposium, Houston, TX, April 15–10*, pp. 63–74.

92. Soleyn, K. (2009). Development of tunable diode laser absorption spectroscopy moisture analyzer for natural gas. *Proceedings of 5th International Gas Analysis Symposium and Exhibition, Rotterdam, The Netherlands, February 11–13.*

93. Dean, W. (2006). In-situ analysis of trace moisture in chlorine gas streams using a tunable diode laser. *Proceedings of the Instrumentation, Systems and Automation Society 51st Analysis Division Symposium, Anaheim, CA, April 2–6*, pp. 1–8.

94. Lackner, M. (2007). Tunable diode laser absorption spectroscopy (TDLAS) in the process industries: a review. *Chemical Engineering, 23*, 2, 65–147.

95. Lehman, S. Y., Bertness, K. A., & Hodges, J. T. (2004). Optimal spectral region for real-time monitoring of sub-ppm levels of water in phosphine by cavity ring-down spectroscopy. *Journal of Crystal Growth, 261*, 2/3, 225–230.

96. Rothman, L. S., Gordon, I. E., Barbe, A., Benner, D. C., Bernath, P. E., Birk, M., Boudon, V., et al. (2009). The HITRAN 2008 molecular spectroscopic database. *Journal of Quantitative Spectroscopy and Radiative Transfer, 110*, 533–572.

97. Lisak, D., Havey, D. K., & Hodges, J. T. (2009). Spectroscopic line parameters of water vapor for rotation–vibration transitions near $7180\,cm^{-1}$. *Physical Review A, 79*, 052507.

98. Lisak, D., & Hodges, J. T. (2008). Low-uncertainty H_2O line intensities for the 930-nm region. *Journal of Molecular Spectroscopy, 249*, 1, 6–13.

99. Aarts, I. M. P., Pipino, A. C. R., de Sanden, M., & Kessels, W. M. M. (2007). Absolute in-situ measurement of surface dangling bonds during a-Si:H growth. *Applied Physics Letters, 90*, 16.

100. Johnston, P. S., & Lehmann, K. K. (2008). Cavity enhanced absorption spectroscopy using a broadband prism cavity and a supercontinuum source. *Optics Express, 16*, 19, 15013–15023.

101. Busch, K. W., Hennequin, A., & Busch, M. A. (1999). Mode formation in optical cavities. In K. W. Busch and M. A. Busch (Eds.), *Cavity-Ringdown Spectroscopy an Ultratrace-Absorption Measurement Technique*. Washington, DC: American Chemical Society, pp. 7–48.

102. Anderson, D. Z., Frisch, J. C., & Masser, C. S. (1984). Mirror reflectivity based on optical cavity decay. *Applied Optics, 23*, 1238–1245.

103. Hodges, J. T., & Ciurylo, R. (2005). Automated high-resolution frequency-stabilized cavity ring-down absorption spectrometer. *Review of Scientific Instruments, 76*, 2, 023112.

104. Fox, R. W., & Hollberg, L. (2002). Role of spurious reflections in ring-down spectroscopy. *Optics Letters, 27*, 20, 1833–1835.

105. Paldus, B. A., Harb, C. C., Spence, T. G., Zare, R. N., Gmachl, C., Capasso, F., Sivco, D. L., et al. (2000). Cavity ringdown spectroscopy using mid-infrared quantum-cascade lasers. *Optics Letters, 25*, 9, 666–668.

106. Brumfield, B. E., Steward, J. T., Weaver, S. L. W., Escarra, M. D., Howard, S. S., Gmachl, C. F., & McCall, B. J. (2010). A quantum cascade laser CW cavity ringdown spectrometer coupled to a supersonic expansion source. *Review of Scientific Instruments, 81*, 6, 063102.

107. Tan, Z., Long, X., Yuan, J., Huang, Y., & Zhang, B. (2009). Precise wavelength calibration in continuous-wave cavity ringdown spectroscopy based on the HITRAN database. *Applied Optics, 48*, 12, 2344–2349.

108. van Zee, R. D., Hodges, J. T., & Looney, J. P. (1999). Pulsed, single-mode cavity ringdown spectroscopy. *Applied Optics, 38*, 18, 3951–3960.

109. Looney, J. P., Hodges, J. T., & van Zee, R. D. (1999). Quantitative absorption measurements using cavity ringdown spectroscopy with pulsed lasers. In K. W. Busch and M. A. Busch (Eds.), *Cavity-ringdown spectroscopy an ultratrace-absorption measurement technique*. Washington, DC: American Chemical Society, pp. 93–105.

110. Ye, J., Ma, L. S., & Hall, J.L. (1999). Using FM methods with molecules in a high finesse cavity: A demonstrated path to $<10^{-12}$ absorption sensitivity. In K. W. Busch and M. A. Busch (Eds.), *Cavity-Ringdown Spectroscopy an Ultratrace-Absorption Measurement Technique*. Washington, DC: American Chemical Society, pp. 233–257.

111. Huang, H., & Lehmann, K. K. (2008). CW cavity ring-down spectroscopy (CRDS) with a semiconductor optical amplifier as intensity modulator. *Chemical Physics Letters, 463*, 246–250.

112. Drever, R. W. P., Hall, J. L., Kowalski, F. V., Hough, J., Ford, G. M., Munley, A. J., & Ward, H. (1983). Laser phase and frequency stabilization using an optical-resonator. *Applied Physics B: Photophysics and Laser Chemistry, 31*, 2, 97–105.

113. Fox, R. W., Oates, C. W., & Hollberg, L. W. (2003). Stabilizing diode lasers for high-finesse cavities. In R. D. van Zee and J. P. Looney (Eds.), *Experimental Methods in the Physical Sciences: Cavity-Enhanced Spectroscopies*, Vol. 40. Amsterdam: Elsevier Science, pp. 1–46.

114. Hamilton, M. W. (1989). An introduction to stabilized lasers. *Contemporary Physics, 30*, 1, 21–33.

115. O'Keefe, A., & Deacon, D. A. G. (1988). Cavity ring-down optical spectrometer for absorption measurements using pulsed laser sources. *Review of Scientific Instruments, 59*, 12, 2544–2551.

116. Romanini, D., & Lehmann, K. K. (1993). Ring-down cavity absorption spectroscopy of the very weak HCN overtone bands with six, seven, and eight stretching quanta. *Journal of Chemical Physics, 99*, 9, 6287–6291

117. Hodges, J. T., Looney, J. P., & van Zee, R. D. (1996). Response of a ring-down cavity to an arbitrary excitation. *Journal of Chemical Physics, 105*, 23, 10278–10288.

118. Lehmann, K. K., & Romanini, D. (1996). The superposition principle and cavity ring-down spectroscopy. *Journal of Chemical Physics, 105*, 2, 10263–10277.

119. Hodges, J. T., Looney, J. P., & van Zee, R. D. (1996). Laser bandwidth effects in quantitative cavity ring-down spectroscopy. *Applied Optics, 35*, 21, 4112–4116.

120. Romanini, D., Kachonov, A. A., Sadeghi, N., & Stoeckel, F. (1997). CW cavity ring down spectroscopy. *Chemical Physics Letters, 264*, 316–322.

121. Ye, J., & Hall, J. L. (2000). Cavity ringdown heterodyne spectroscopy: high sensitivity with microwatt light power. *Physical Review A, 61*, 6, 061802.

122. Cossel, K. C., Adler, F., Bertness, K. A., Thorpe, M. J., Feng, J., Raynor, M. W., & Ye, J. (2010). Analysis of trace impurities in semiconductor gas via cavity-enhanced direct frequency comb spectroscopy. *Applied Physics B, 100*, 917–924.

123. Paldus, B. A., & Zare, R.N. (1999). Absorption spectroscopies: from early beginnings to cavity ring-down spectroscopy. In K. W. Busch and M. A. Busch (Eds.), *Cavity-Ringdown Spectroscopy an Ultratrace-Absorption Measurement Technique*. Washington, DC: American Chemical Society, pp. 49–70.

124. O'Keefe, A., Scherer, J. J., & Paul, J. B. (1999). Cavity-ringdown laser spectroscopy: History, development, and applications. In K. W. Busch and M. A. Busch (Eds.), *Cavity-Ringdown Spectroscopy an Ultratrace-Absorption Measurement Technique*. Washington, DC: American Chemical Society, pp. 71–92.

125. Ono, H. (2011). Moisture analysis in monosilane gas. *Pittsburgh Conference on Analytical Chemistry and Applied Spectroscopy (Pittcon), Atlanta, GA.*

126. Funke, H. H., Raynor, M. W., Bertness, K. A., & Chen, Y. (2007). Detection of trace water vapor in high-purity phosphine using cavity ring-down spectroscopy. *Applied Spectroscopy, 61*, 4, 419–423.

127. Feng, J., Clement, R., & Raynor, M. W. (2008). Characterization of high-purity arsine and gallium arsenide epilayers grown by MOCVD. *Journal of Crystal Growth, 310*, 23, 4780–4785.

128. Yao, J., Funke, H. H., & Raynor, M. W. (2004). Measurement and control of trace moisture in corrosive gases. In *SEMI Technical Symposium (STS): Innovations in Semiconductor Manufacturing*. San Jose, CA: Semiconductor Equipment and Materials International.

129. Tiger Optics (2011). Aloha-H_2O specification sheet. Warrington, PA: Tiger Optics.

130. Tiger Optics (2011). Halo–H_2O specification sheet. Warrington, PA: Tiger Optics.

131. Tiger Optics (2011). MTO-1000-H_2O specification sheet. Warrington, PA: Tiger Optics.

132. Thorpe, M. J., Moll, K. D., Jones, R. J., Safdi, B., & Ye, J. (2006). Broadband cavity ringdown spectroscopy for sensitive and rapid molecular detection. *Science, 311*, 5767, 1595–1599.

133. Diddams, S. A., Hollberg, L., & Mbele, V. (2007). Molecular fingerprinting with the resolved modes of a femtosecond laser frequency comb. *Nature, 445*, 7128, 627–630.

134. Gohle, C., Stein, B., Schliesser, A., Udem, T., & Hänsch, T. W. (2007). Frequency comb vernier spectroscopy for broadband, high-resolution, high-sensitivity absorption and dispersion spectra. *Physical Review Letters, 99*, 26, 263902.

135. Thorpe, M. J., & Ye, J. (2008). Cavity-enhanced direct frequency comb spectroscopy. *Applied Physics B, 91*, 3-4, 397–414.

136. Adler, F., Thorpe, M. J., Cossel, K. C., & Ye, J. (2010). Cavity-enhanced direct frequency comb spectroscopy: technology and applications. *Annual Review of Analytical Chemistry, 3*, 175–205.

137. Cundiff, S. T., & Ye, J. (2003). Colloquim: femtosecond optical frequency combs. *Reviews of Modern Physics, 75*, 1, 325–342.

138. Washburn, B. R., Diddams, S. A., Newbury, N. R., Nicholson, J. W., Yan, M. F., & Jørgensen, C. G. (2004). Phase-locked, erbium-fiber-laser-based frequency comb in the near infrared. *Optics Letters, 29*, 3, 250–252.

139. Schibli, T. R., Minoshima, K., Hong, F. L., Inaba, H., Onae, A., Matsumoto, H., Hartl, I., & Fermann, M. E. (2004). Frequency metrology with a turnkey all-fiber system. *Optics Letters, 29*, 21, 2467–2469.

140. Adler, F., Moutzouris, K., Leitenstorfer, A., Schnatz, H., Lipphardt, B., Grosche, G., & Tauser F. (2004). Phase-locked two-branch erbium-doped fiber laser system for long-term precision measurements of optical frequencies. *Optics Express, 12*, 24, 5872–5880.

141. Thorpe, M. J., Balslev-Clausen, D., Kirchner, M. S., & Ye, J. (2008). Cavity-enhanced optical frequency comb spectroscopy: applications to human breath analysis. *Optics Express, 16*, 4, 2387–2397.

142. Thorpe, M. J., Adler, F., Cossel, K. C., de Miranda, M. H. G., & Ye, J. (2009). Tomography of a supersonically cooled molecular jet using cavity-enhanced direct frequency comb spectroscopy. *Chemical Physics Letters, 468*, 1, 1–8.

143. Kassi, S., Didriche, K., Lauzin, C., de Ghellinck d'Elsenghem Vaernewijckb, X., Rizopoulos, A., & Herman, M. (2010). Demonstration of cavity enhanced FTIR spectroscopy using a femtosecond laser absorption source. *Spectrochimica Acta A, 75*, 1, 142–145.

144. Foltynowicz, A., Ban, T., Masłowski, P., Adler, F., & Ye, J. (2011). Quantum-noise-limited optical frequency comb spectroscopy. *Physical Review Letters, 107*, 233002-1–5.

145. Mandon, J., Guelachvili, G., & Picqué, N. (2009). Fourier transform spectroscopy with a laser frequency comb. *Nature Photonics, 3*, 2, 99–102.

146. Adler, F., Masłowski, P., Foltynowicz, A., Cossel, K. C., Briles, T. C., Hartl, I., & Ye, J. (2010). Mid-infrared Fourier transform spectroscopy with a broadband frequency comb. *Optics Express, 18*, 21, 21861–21872.

147. Keilmann, F., Gohle, C., & Holzwarth, R. (2004). Time-domain mid-infrared frequency-comb spectrometer. *Optics Letters, 29*, 13, 1542–1544.

148. Coddington, I., Swann, W. C., & Newbury, N. R. (2008). Coherent multiheterodyne spectroscopy using stabilized optical frequency combs. *Physical Review Letters, 100*, 1, 013902.

149. Bernhardt, B., Ozawa, A., Jacquet, P., Jacquey, M., Kobayashi, Y., Udem, T., Holzwarth, R., et al. (2010). Cavity-enhanced dual-comb spectroscopy. *Natural Photonics, 4*, 1, 55–57.

150. Coddington, I., Swann, W. C., & Newbury, N. R. (2010). Coherent dual-comb spectroscopy at high signal-to-noise. *Physical Review A, 82*, 4, 043817.

151. Adler, F., Cossel, K. C., Thorpe, M. J., Hartl, I., Fermann, M. E., & Ye, J. (2009). Phase-stabilized, 1.5 W frequency comb at 2.8–4.8 μm. *Optics Letters, 34*, 9, 1330–1332.

CHAPTER 8

GAS CHROMATOGRAPHIC COLUMN CONSIDERATIONS

Daron Decker[1] and Leonard M. Sidisky[2]

[1] Agilent Technologies, Pearland, Texas
[2] Supelco, Bellefonte, Pennsylvania

8.1 Introduction

When choosing a column for gas chromatographic (GC) analysis, the analyst must first decide on whether a packed column or capillary column is most appropriate. Packed columns have been the GC workhorse for many years, as they offer superior sample capacity and a wide variety of stationary phases, some of which are still unavailable in capillary dimensions. Packed and micro-packed columns do deliver lower efficiency or broader peak shapes, which can result in lower resolution and longer run times than in capillary columns, but the unique selectivity available based on the broad range of adsorbents that can be packed into a column maintains their status as a widely accepted and effective separation tool. For simple analyses of a few analytes of interest in simple matrices, packed columns may still be the most appropriate choice. If the best stationary phase for a desired separation cannot be manufactured into a capillary configuration for whatever reason, the packed configuration may be the only choice available to the analyst and employed for that reason.

The most common physical types of columns for gas analysis are packed, wall-coated open tubular (WCOT) columns and porous-layer open tubular (PLOT) columns. Figure 8.1 illustrates the general appearance and sizes of these three types of columns.

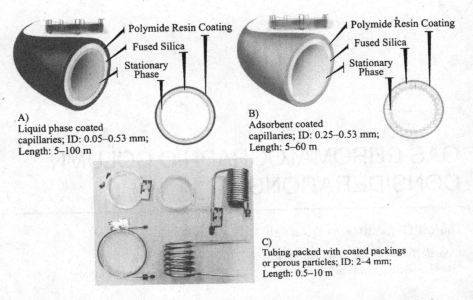

A)
Liquid phase coated
capillaries; ID: 0.05–0.53 mm;
Length: 5–100 m

B)
Adsorbent coated
capillaries; ID: 0.25–0.53 mm;
Length: 5–60 m

C)
Tubing packed with coated packings
or porous particles; ID: 2–4 mm;
Length: 0.5–10 m

Figure 8.1 General appearance and sizes of (A) WCOT, (B) PLOT, and (C) packed columns. Courtesy of Agilent Technologies, Inc.

The majority of packed column applications currently in service are used for relatively simple analysis of atmospheric constituents such as hydrogen, methane, nitrogen, argon, oxygen, and carbon dioxide in a product gas. They are generally operated isothermally since the measurement requirements are at low part per million (ppm) or part per billion (ppb) levels. Detectors for these gases at such levels would not perform particularly well with the flow and temperature changes attendant on temperature ramping. It is important to note that when making measurements of non-complex gases, packed columns will probably be used into the foreseeable future, due to their greater capacity.

8.2 Column Considerations with Packed Columns

Tubing, support, support particle mesh size, and stationary phase are four key variables the analyst evaluates when choosing a packed column for the analysis of a gas sample. The common tubing choices are glass, stainless steel (SS), glass-lined SS, aluminum, copper, nickel, and Teflon. Glass tubing is the choice when working with samples that are reactive. Glass is more inert than metal tubing and can also be deactivated to provide a more inert surface. Stainless steel is probably the most widely used metal column and is considered to be more inert than other metal tubing and also offers better packing efficiencies than that of the other metals. SS tubing coated with various proprietary silicon-based coatings are available from some manufacturers and provide enhanced inertness over uncoated SS tubing. Glass-lined SS is a combination of glass and SS tubing but the dimensions available are limited. Other metals, such as aluminum and copper, can be adsorptive and detrimental to chromatographic performance. Teflon also offers an inert surface for various analyses but has limited temperature stability.

A wide variety of chromatographic packing materials are available in the form of various supports and stationary phases, along with combinations of these two materials in a number of mesh sizes. In most cases, when we are analyzing gaseous samples we will be performing gas–solid chromatography (GSC) using an adsorbent material (support) to adsorb the compounds onto the packing surface or trap them within the pore structure of the adsorbents. These adsorbents are solid materials that contain a pore structure and, typically, a high surface area. The keys to GSC are a high surface area and defined pore structures to allow the adsorption process to occur. In comparison, gas–liquid chromatography (GLC) involves the sample components partitioning into the liquid stationary phase and retaining them by a differential migration through the stationary phase. Given the molecular size and volatility of the gaseous samples, packed columns performing GSC offer the best solution for analyzing these samples. There are also packed columns prepared using various adsorbent supports that are modified with a liquid phase to provide surface deactivation or to modify the selectivity of the adsorbent.

The mesh size of the packing will affect the overall efficiency and the back-pressure requirements of a chromatographic system. As is the case in most chromatographic systems, smaller dimensions provide higher efficiency. Therefore, as we decrease the mesh size and achieve the corresponding decrease in particle size, the efficiency of the chromatographic system is improved. There will also be a corresponding increase in the back-pressure required to operate the columns properly given the same column dimensions. Typical mesh sizes that are used in today's packed columns are 60/80 mesh, which equates to a particle size of 250 to 177 μm; 80/100 mesh with a particle size range from 177 to 149 μm, and 100/120 mesh with particles from 149 to 125 μm. The 80/100 is the most popular mesh size, but if increased efficiency is required, many analysts use 100/120 mesh.

Adsorption of the gases onto the solid surface or through the pore structure of the support is the key feature of GSC. Since a wide variety of materials can be used as adsorbents, we focus on the most widely used materials for an understanding of the performance of the adsorbents used to perform GSC.

8.2.1 Adsorbents

In evaluating the types of samples discussed in this chapter, a number of different adsorbent types have been utilized, depending on the type of sample and the carbon chain length if it is a hydrocarbon moiety. Both inorganic and organic adsorbents have been used to separate gaseous samples. The inorganic adsorbents include silica gel, alumina, molecular sieves, and carbon-based adsorbents. Each of these types has applicability to a limited range of target compounds. Silica gels and alumina, including phase-modified alumina, have been used to separate gaseous hydrocarbons from methane to butane and various isomers in the C_1 through C_4 carbon range. Zeolite-based molecular sieves are typically limited to the analysis of permanent gases such as hydrogen, oxygen, nitrogen, and carbon monoxide. Carbon-based molecular sieves can also be used to separate permanent gases, including carbon dioxide, but expand their applicable range up to the C_3 (propane) hydrocarbons. The organic adsorbents, or porous polymers, are all high-surface-area copolymers based on combinations of styrene and divinylbenzene. Various ratios of monomers used in polymer synthesis, as well as other additives, are used to modify the adsorbent selectivity. Most of the porous polymer applicability has been with light hydrocarbon analysis.

Silica gels have been used for a wide variety of gas–solid separations. The pore structure and surface area of the silica determines the analytical applications. Two silicas that have been used are Spherosil and Porasil. Both have been available with surface areas ranging from 5 to $500 \, m^2/g$. Correspondingly, as the surface area of the silica increases, there is a decrease in pore diameter. For example, Porasil B has a surface area of 125 to $250 \, m^2/g$ with an average pore diameter of 10 to 20 nm. Porasil C has a surface area of 50 to $100 \, m^2/g$ with a pore diameter ranging from 20 to 40 nm. Silica gels, however, are highly variable, depending on the amount of moisture that is adsorbed on the columns from the carrier gas, which will affect the retention of volatile compounds on the columns.

Alumina is another inorganic adsorbent that has been used for gas analysis. Alumina-base adsorbents provide similar types of separations as silica gel but with different selectivity, as their interactions with the solute molecules are dependent on the Lewis acid interaction of the aluminum ions. The aluminas are typically modified or deactivated with an inorganic salt such as potassium chloride or sodium sulfate. Similar to the silica gels, water retention will also affect their selectivity.

Molecular sieves have been used widely for the separation of permanent gases such as hydrogen, oxygen, argon, nitrogen, methane, and carbon monoxide. Carbon dioxide is typically adsorbed by molecular sieve columns. The most common molecular sieves are 3A, 5A, and 13X. These materials possess a relatively high surface area, and they resolve the permanent gases based on their pore size. The effective pore diameter for the 3A and 5A molecular sieve is 3 and 5 Å, and the 13X has a pore diameter of approximately 10 Å. To function properly as a molecular sieve, these materials need to be fully activated at high temperatures to remove water and other gases that may be trapped in the pore structure .

Carbon-based molecular sieves have been prepared with a wide variety of surface areas and pore sizes. These materials are typically hydrophobic and possess varying amounts of macro-, meso-, and micropores in their pore structure. Table 8.1 lists some of the carbon molecular sieves that have been used for GSC and includes the surface area, the density, the porosity, and the pore diameter of the carbons. Macropores

Table 8.1   Carbon molecular sieves for GSC[a]

Adsorbent[b]	BET Surface Area (mm^2/g)	Density (g/mL)	Porosity			Pore Diameter (Å)
			micro-	meso-	macro-	
Carboxen-563	510	0.53	0.24	0.15	0.24	7–10
Carboxen-564	400	0.60	0.24	0.13	0.14	6–9
Carboxen-569	485	0.58	0.20	0.14	0.10	5–8
Carboxen-572	1100	0.49	0.44	0.14	0.10	10–20
Carboxen-1000	1200	0.48	0.44	0.16	0.25	10–12
Carboxen-1001	500	0.61	0.22	0.13	0.11	5–8
Carboxen-1002	1100	0.43	0.36	0.28	0.3	10–12
Carboxen-1003	1000	0.46	0.38	0.26	0.28	5–8
Carboxen-1006	715	N/D	0.29	0.26	0.23	7–10
Carboxen-1010	675	0.60	0.35	N/A	N/A	6–8
Carboxen-1011	1100	0.48	0.41	0.19	0.24	10–12
Carboxen-1012	1500	0.50	N/A	0.66	N/A	19–21
Carboxen-1016[c]	75	0.40	N/A	0.34	N/A	260
Carboxen-1018	675	0.60	0.35	N/A	N/A	6–8
Carboxen-1021	600	0.62	0.3	N/A	N/A	5–7
Carboxen-1026	1300	0.46	0.82	0.13	N/A	5–20
Carboxen-1027[c]	130	N/A	N/A	0.5	N/A	173
Carbosieve S-III	820	0.61	0.35	0.04	N/A	4–11
Carbosieve S-II	1059	N/D	0.45	0.01	N/A	6–15
Carbosieve G	1160	N/D	0.49	0.02	N/A	6–15
SupelCarb	1150	0.46	0.47	0.26	N/A	5–8

[a] N/D indicates that the density has not been determined and N/A indicates that the carbon does not have any pores of that size.
[b] Carboxen is a trademark of Sigma-Aldrich Co.
[c] Graphitized polymer carbon.

are defined as being greater than 500 Å in size. Mesopores range from 20 to 500 Å and micropores are less than 20 Å in size. The distribution of the pores has a direct effect on the kinetics involved in resolving the gaseous samples. Carboxen-1000, with a very high surface area ($1200\,m^2/g$) and a large number of micropores, is one of the most widely used carbon sieves. Figure 8.2 illustrates the separation of permanent gases and C_2 hydrocarbons on this adsorbent. Note that the Carboxen-

1000 resolves the permanent gases and allows the elution of carbon dioxide and the C_2 hydrocarbons, which is not possible on a zeolite-based molecular sieve.

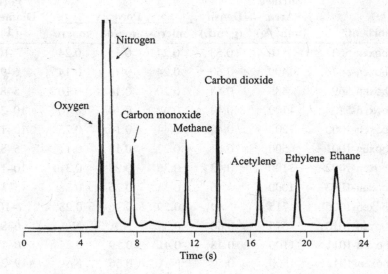

Figure 8.2 Separation of permanent gases and C_2 hydrocarbons on Carboxen-1000. Column: 15 ft × 0.125 in (2.1 mm) ID SS 60/80 Carboxen-1000; cat. no.: 12390-U (general configuration); oven temperature: 35 °C (5 min) to 225 °C at 20 °C/min; carrier: helium, 30 mL/min; detector: TCD; injection: 0.6 mL Scott gas mix (cat. no. 2-3437) with oxygen added, 1 % each analyte, 4 to 6 μg each analyte (except nitrogen on column).

Porous polymers are another group of adsorbents that are widely used for gas analysis. These adsorbents are large-surface-area resins that are made from core materials such as styrene and divinylbenzene, which are sometimes modified with other moieties to vary the selectivity of the polymer bead. The Chromosorb, Porapak, and HayeSep series are the most widely used porous polymers. These polymers, which have a high surface area as a result of micropores, are useful for the separation of gases and other volatile compounds. Chromosorb 102 and Porapak Q are two of the most used porous polymers because the surface area is greater than $300 \, m^2/g$, and their average pore diameter is in the microporous range (less than 20 Å).

The object of many gas analyses is not to measure the major constituent(s) but, rather, the impurities therein. If the impurities elute before the balance component, this is easily accomplished by a simple back-flush or two-column analysis with back-flush of the precolumn to vent, keeping the balance gas from damaging the analytical column or detector. However, if the component(s) of interest elute (closely) after the balance gas, the measurement becomes more challenging, especially at trace levels. A good example of a solution to this situation is a novel approach using three columns, two valves, and two different packing materials for low-level measurement of nitrogen and carbon monoxide in high-purity hydrogen.

Figure 8.3 represents the flowpath for this analysis. Two injections are made to obtain a single chromatogram. Initially, samples containing hydrogen, nitrogen, methane, and carbon monoxide are injected onto the silica gel (stripper) column and pass onto the 1.2-m MS 5A column. The 10-port valve is then reset to the load/back-flush position, removing any water, carbon dioxide, or heavies. Meanwhile, hydrogen, nitrogen, and methane have eluted to vent from the 1.2-m MS 5A column through the four-port valve vent line. Carbon monoxide is still in the 1.2-m MS 5A column. A second injection with the 10-port valve is now made, allowing most of the hydrogen to elute to vent. This is followed by bringing the 2-m MS 5A column in line with the 1.2-m MS 5A column, catching the "tail" of the hydrogen that contains nitrogen from the second injection and the carbon monoxide from the first injection. Once the carbon monoxide from the first injection and the nitrogen from the second injection have entered the analytical column, the four-port valve is returned to the vent position to minimize the size of the hydrogen tail. Figure 8.4 illustrates the resulting chromatogram. This application requires careful column tailoring and timing, but saves a great deal of time and duplication of effort in a process environment.

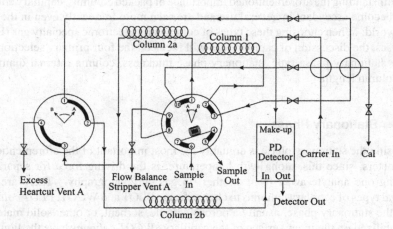

Figure 8.3 Flow path for analysis of trace nitrogen and carbon monoxide in high-purity hydrogen using a stripper/heart-cut column system.

Gas chromatography and, in particular, packed column gas chromatography has become a very mature technology over the past 60 years. Since much of the original column and liquid-phase technology was developed by instrument manufacturers or their employees in support of instrumentation, excellent basic references on liquid phases, packings, and applications have been published by these vendors [1–3]. Not current by any means, the technology and applications discussed are as relevant now as when originally published, and are valuable resources worth acquiring even if difficult to find.

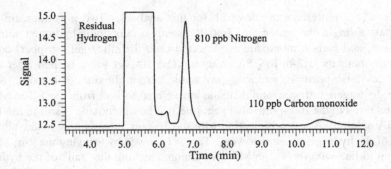

Figure 8.4 Chromatogram of trace nitrogen and carbon monoxide in high-purity hydrogen using a stripper/heart-cut column system.

8.3 Primary Selection Criteria for Capillary Columns

Notwithstanding the aforementioned importance of packed columns, capillary columns have become ascendant in general use and are seen more frequently even in the process world. When choosing these types of columns for electronic specialty gas (ESG) analyses, the discussion of capillary use will focus on the four primary selection criteria: stationary-phase type, stationary-phase thickness, column internal diameter, and column length.

8.3.1 Stationary Phase

Choosing the stationary phase is arguably the most important of the aforementioned parameters, since this, along with temperature, is the driving force for separation (getting one analyte away from another) in gas chromatography. There are two general types of capillary columns to choose from: PLOT and WCOT. *PLOT columns* have the stationary phase, usually a porous particle, sorbent, or other solid material, immobilized on the inner surface of the capillary. *WCOT columns* have the liquid or gum stationary phase coated and usually bonded to the inner surface of the capillary. With PLOT columns, analytes are separated by adsorption chromatography, which is a very strong mechanism of retention and is less efficient than the diffusion or dispersion chromatography that occurs with WCOT columns. However, this strong mechanism in PLOT columns allows retention of analytes that are difficult to retain, such as permanent gases or compounds that are gases at room temperature, an area of great interest to the types of analyses discussed here. One downside to PLOT columns is that large molecules and compounds with low vapor pressures and high retention are retained on the column permanently or are so well retained that by the time they elute, they are not discernible as peaks, therefore appearing as part of the background noise. Thus, PLOT columns are primarily, but not exclusively, preferred for the analysis of low-boiling-point solutes and gases.

Stationary phase affects compound solubility (or adsorptiveness in the case of PLOT columns), inertness, and net vapor pressure, which ultimately affects selectivity (separation) as well as peak shape, retention time, and sensitivity or compound response. A good rule of thumb when it comes to solubility is the old adage: "Like dissolves like, and oil and water don't mix." The more similar the compound is to the stationary phase, the more soluble it will be in that phase and will be able to interact more with the functional groups of that phase if present. Note in Figure 8.5 that there is a comparison of six analytes in two very different phases under the same chromatographic conditions. Only the stationary-phase type is different in this comparison. The nonpolar dimethylpolysiloxane, often referred to as a boiling-point column, has one order of elution, while poly(ethylene glycol) (PEG) of exactly the same dimensions and under the same conditions has an entirely different elution order.* It can be seen that the more soluble compounds have more retention in the phase in which they are more soluble and have less retention in the phase in which they are less soluble.

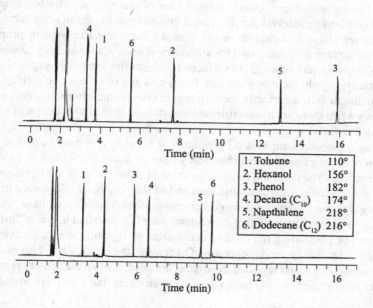

1. Toluene	110°
2. Hexanol	156°
3. Phenol	182°
4. Decane (C$_{10}$)	174°
5. Napthalene	218°
6. Dodecane (C$_{12}$)	216°

Figure 8.5 Comparison of elution order of six analytes in 100 % PEG phase (top) and 100 % methyl phase (bottom) under the same chromatographic conditions.

When comparing the dimethylpolysiloxane column on the bottom, one can see how the peaks shift due to solute/stationary-phase solubilities on the PEG phase even though the dimensions of the columns and chromatographic conditions are identical. Polar analytes have more retention in the polar PEG phase, and the nonpolar analytes

*Note that no column phase will elute all analytes in boiling-point order, including this one.

(i.e., decane and dodecane) have much less retention in the PEG phase than in the dimethylpolysiloxane phase.

Selectivity is very important when choosing a stationary phase for a particular analysis, but it isn't everything. Maximum temperature limits and column bleed can be critical factors and will play a role in the decision-making process. Analytes that have low vapor pressures or high boiling points require a stationary phase that can handle the temperatures required to elute these compounds without damaging the column or causing excessive stationary-phase bleed. For most applications covered in this book, low-vapor-pressure compounds are not an issue, so those challenges are not discussed here in detail.

Inertness is our last critical factor for phase selection. Stationary phases that are chromatographically active can reversibly and irreversibly absorb active analytes, which can cause peak shape distortion and/or loss of analyte, which can cause problems with reproducibility, sensitivity, and quantitation, especially at low concentrations. Typically, compounds that hydrogen bond or tend to be chemically reactive are said to be chromatographically active. One of the biggest culprits for activity from the columns themselves are the silanol groups from the fused silica tubing that the stationary phase is coated on or the silanol groups that can exist or propagate within the siloxane polymers used for stationary phases. Other activity comes from silanol groups on any other glass surfaces to which the sample is exposed in the sample analysis path, including the inlet liners and any other reactive entities in the sample path. In fact, a perfectly inert system is often compromised by the very act of analysis when there are nonvolatile residues present in the introduction of a liquid sample in gas chromatography.

Having a gaseous sample is no guarantee of keeping a system inert since there are often reactive gases that can chemically alter or attack the stationary phase and render it active. The common remedy for most of these issues is a diligent regimen of inlet maintenance, where the injection end of the column is adequately trimmed from time to time to remove damaged or compromised stationary phase. Another technique used for deactivation of the column after it is installed is one of "priming" the column or passivating the column active sites by injecting a high concentration of the active analyte or similar compound to "cover" the active sites. This is usually only a temporary fix, and repeated treatments are needed when the activity reappears. The only other remedy would be never to inject anything, but then there would be no analysis!

8.3.2   Stationary-Phase Film Thickness

The choice of stationary-phase film thickness when discussing WCOT columns can be tricky since many things besides retention characteristics can be affected. Not only will thicker films result in more retention for all analytes, they will also increase column bleed if or when the column is at high temperatures. Thicker films also give more sample capacity than do thinner films. All other things being equal a thick-film column tends to be more inert than a thinner film. This is probably due to better coverage of residual silanol groups on the surface of the fused silica tubing left, but there could be other reasons as well. When choosing film thickness, for most applications it usually comes down to retention needs. When analyzing volatiles, a thicker film is usually employed, and when analyzing high-boiling solutes, a thin film is used. If the sample contains analytes with a range of volatilites, then a compromise must be reached. For this type of sample, a standard film with a low starting temperature to enhance retention of the early eluters is the norm. A thick film would require too high a temperature to elute the high-boiling analytes and would also result in a lot of column bleed. A thin film would manage the high boilers well but would require very low starting temperatures to get adequate retention of the volatiles.

In the case of PLOT columns, phase thickness (rather than film) is discussed because there really is no film to speak of but, rather, a given thickness to the adsorbent layer. Since the chromatography is a surface phenomenon (adsorption) in these columns, the phase thickness tends to have less impact on the actual retention of the analytes but can, of course, affect the capacity of the column for an analyte. Phase thickness generally doesn't affect column bleed in most PLOT columns since the sorbents used as stationary phase are mostly nonvolatile.

8.3.3 Column Internal Diameter

Determining the internal diameter (ID) of the column is fairly straightforward but can be complicated by the number of choices available. The diameter can be decided simply by the capacity needed versus the efficiency (peak sharpness) desired. Smaller-diameter columns give sharper peaks than do larger-diameter columns unless they are overloaded. Hence, choosing a column that has sufficient diameter to handle the amount of sample needed to be injected to obtain the sensitivity requirements is the key and then not using a larger-than-necessary diameter will provide the most efficiency available to that analysis. Trace analyses typically use smaller-diameter (0.25 mm ID and smaller) columns, and wide-bore (0.32 mm ID or larger) columns are used for samples where capacity is an issue. This is again the one place that packed columns fare better, due to their high capacities. The one other consideration that tends to come up with diameter choice is that this determines the nominal carrier gas flows that will be used. The carrier gas velocity is chosen for reasons of chromatographic performance (the least amount of band broadening due to longitudinal diffusion but sufficient mass transport of the analyte molecules between mobile and stationary phases); this is then translated into a volumetric flow based on the diameter (volume) of the column. Volumetric flows also need to be considered when transferring the sample from some type of injecting or concentrating device, such as like a thermal desorption unit, static headspace, dynamic headspace (purge and trap device), or gas sampling valve. The faster the sample transfer to the column from these devices, the less band broadening (longitudinal diffusion again) that occurs and the easier the separation requirements. There is also the case of the mass spectrometer, where a vacuum is necessary and high flows from the column can compromise the performance of the detector. In that case, smaller-diameter (0.25 mm ID or smaller) columns are typically used.

8.3.4 Column Length

Length is the last column criterion to determine and generally the least impactful with regard to chromatographic resolution. Plain and simple, longer columns give more theoretical plates and therefore more resolution. The issue is that there is a square-root relationship between theoretical plates and chromatographic resolution, so to double the resolution one has to lengthen the column by a factor of 4! It takes a big change in length to enact a small change in resolution. Sadly, analysis times are almost proportional to length increases, and therefore long analysis times are often the result. One positive result of the relationship between resolution and column length is that often the use of a shorter column is possible without much loss in resolution, and hence shorter analysis times can be realized. When determining the proper column length for a given analysis, there are usually other, more compelling reasons than resolution, but it still plays a role. Longer length will mean better resolution but also longer analysis times [4].

8.3.5 Manufacturer

The last subject to touch upon briefly is that there will be differences between columns of the same phase type from one column manufacturer to another. This may not even be related to quality control issues. The analogy that applies is that of two cakes made in two different kitchens. Even if both kitchens have master chefs at the top of their game following the same recipes, there will be small variables that will make the two cakes different. To the unrefined palate there may be no discernible differences, but to the connoisseur, one cake may be a bit more moist than the other, while the other has a hint of a desired flavor (vanilla, mint, etc.). Every chef has his/her secrets and may make slight changes to recipes over time. Neither cake is bad, but one is more desirable than the other to one person and the other is more desirable to someone else. Columns can be very much like this. Many can use any brand of column because they all seem to do the job, but to some for a specific application they find that one works much better or even just slightly better. Even from the same manufacturer there can be different versions of the same phase, so the column manufacturer's technical support groups should be consulted when making a column selection to see what column phase best suits an application. Observe in Figure 8.6 that although both columns are sold as 5 % diphenyl–95 % dimethylpolysiloxane phases, there are slight selectivity differences when looking at the separation of *m*- and *p*-xylenes. The Agilent DB-5ms column has a siarylene phase and is a low-bleed version of the Agilent DB-5 column, which has a traditional diphenyl-substituted polymer phase.

8.4 Applications

It is now time to look at actual analyses and rationalize the columns chosen for those analyses. Not only the dimensions but also the stationary phase, PLOT or WCOT, is important. Some stationary phases are proprietary to the manufacturer, so not much is known about them other than they usually have some very desirable traits. This is

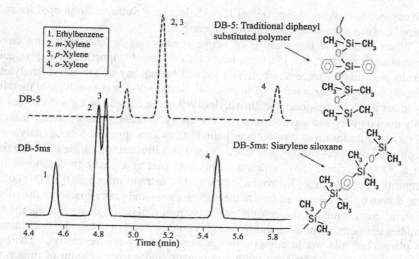

| 1. Ethylbenzene |
| 2. *m*-Xylene |
| 3. *p*-Xylene |
| 4. *o*-Xylene |

Figure 8.6 Comparison of two 5 % diphenyl–95 % dimethylpolysiloxane phase columns: Agilent DB-5 (top) and Agilent DB-5ms (bottom). Columns: 30 m × 0.25 mm ID × 0.25 μm; oven temperature: 60 °C isothermal; carrier: hydrogen at 40 cm/s.

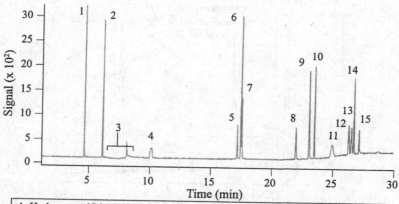

1. Hydrogen sulfide; 2. Carbonyl sulfide ; 3. Propylene; 4. Methyl mercaptan;
5. Ethyl mercaptan; 6. Carbon disulfide; 7. Dimethyl sulfide; 8. 2-Propanethiol;
9. Methyl ethyl sulfide; 10. Thiophene; 11. t-Butyl mercaptan; 12. 2-Butanethiol;
13. 2-Methyl-1-propanethiol; 14. Diethyl sulfide; 15. 1-Butanethiol

Figure 8.7 Agilent J&W Select Low Sulfur column used for sulfur compounds in propylene streams. Column: 60 m × 0.32 mm ID Agilent J&W Select Low Sulfur; cat. no: CP8575; GC conditions: carrier: helium, 2 mL/min; sample size: 1 mL; split ratio: 3 : 1; oven temperature: 40 °C; 9 °C/min, 120 °C; detector: PFPD; detector settings: hydrogen: 12 mL/min; air (1): 13 mL/min; air (2): 12 mL/min; PMT 610, trigger level: 500 mV. Copyright ©2011 Agilent Technologies, Inc. Reprinted with permission, courtesy of Agilent Technologies, Inc.

the case when looking at the Agilent J&W Select Low Sulfur column used for sulfur compounds in propylene streams (see Figure 8.7*).

When looking at the analysis of active compounds at low levels, not only is separation important, but inertness becomes of utmost importance. Reproducible results and finding trace amounts rely on not losing any of the active analytes of interest during the chromatographic process. That is why for the analysis of metalloid hydrides a thick-film nonpolar dimethylpolysiloxane column is a good choice. Not only does it give good separation of most of the hydride compounds, but the inertness of this column for these types of compounds allows no apparent loss of analyte even at ppb and part-per-trillion (ppt) levels. Figure 8.8 illustrates hydride retention times on a 300-m Agilent DB-1 column. This was part of a study to determine trace impurities in high-purity hydrogen selenide. The extraordinary length of this column was driven by the need to maximize theoretical plates and peak capacity of the column without compromising the column capacity from a mass-on-column standpoint. A smaller-diameter column of shorter length could be used to obtain the same number of plates, but that would compromise column capacity for the matrix. There is a massive amount of matrix, or main component, and a trace amount of impurity or analyte of interest. This separation is of particular importance when the analyte is on the tail of the bulk product.

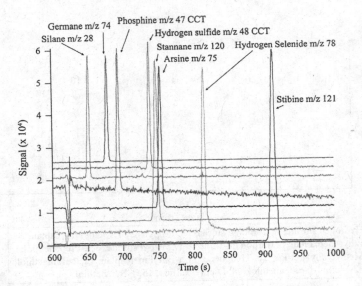

Figure 8.8 Hydride retention times on a 300-m Agilent DB-1 column. Column: 300 m × 0.53 mm ID × 5.0 μm Agilent DB-1; carrier: argon at about 6.5 mL/min; oven temperature: 35 °C isothermal; sample size: 400 μL.

*Carbonyl sulfide can be separated from propylene at an oven temperature of 40 °C, a high concentration of propylene will make a methyl mercaptan peak wider, and it is to be confirmed that 1-propanethiol may coelute with carbon disulfide [5].

When comparing the 300-m column performance to a 100-m column for the analysis of trace arsine in high-purity phosphine, somewhat better separation of analytes can be seen. This is without sacrificing significant resolution and hence sensitivity. Looking at Figure 8.9 we have about a 20-second time difference between phosphine and arsine with the 100-m column, but in Figure 8.8 this becomes 60 seconds with the 300-m column. Such enhancement lowers the practical detection limit by nearly an order of magnitude! Another reason for this particular configuration is for the multiple analytes that can be analyzed by this one setup. Sometimes a longer column is effective, but as discussed earlier, not without a cost in time. Luckily for this analysis, the analyte volatility minimizes this effect. Packed or PLOT columns were deemed unsuitable at these ultratrace levels, due to the poor peak shapes and partial adsorption losses of some of these analytes.

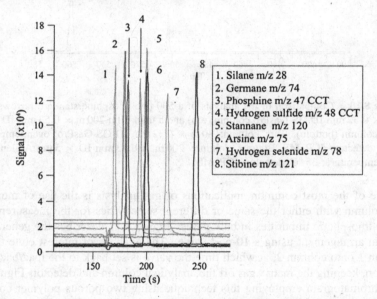

1. Silane m/z 28
2. Germane m/z 74
3. Phosphine m/z 47 CCT
4. Hydrogen sulfide m/z 48 CCT
5. Stannane m/z 120
6. Arsine m/z 75
7. Hydrogen selenide m/z 78
8. Stibine m/z 121

Figure 8.9 Hydride retention times on a 100-m Agilent DB-1 column. Column: 100 m × 0.53 mm ID × 5.0 μm Agilent DB-1; carrier: argon at about 8 mL/min; oven temperature: 35 °C isothermal; sample size: 400 μL.

Trace detection of one analyte in the presence of the percent level of another poses a unique kind of challenge. The column has to have the capacity to handle the high concentration of the one analyte and give extra separation of the trace analyte so as not to be swamped out in the chromatographic space occupied by the high-concentration component. This is often referred to as the *peak capacity* of the column [6–8]. Generally, the sample capacity problem (overload) is handled by using the larger-bore capillary columns with standard or even thicker phase film thickness. This is not always an option for many reasons, including matrix issues. A comparison of a 500-ppb phosphine standard on an 80 m × 0.32 mm ID GS-GasPro (proprietary) column analysis, and a thick film 200 m × 0.53 mm ID Agilent DB-1 column analysis, is shown in Figure 8.10. Even though the phosphine signal-to-noise, symmetry, resolution, and response are better on the higher-capacity Agilent DB-1, the GasPro column was chosen for the analysis since it provided far better separation from the matrix, which in this case was germane.

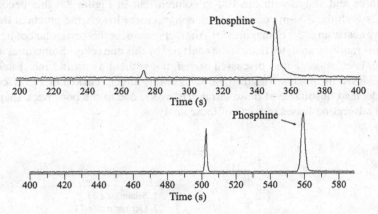

Figure 8.10 Comparison of the analysis of a 500-ppb phosphine standard (arrows) on an 80 m × 0.32 mm ID GS-GasPro column (top) and a thick-film 200 m × 0.53 mm ID Agilent DB-1 column (bottom). Top: column: 80 m × 0.32 mm ID GS-GasPro; oven temperature: 45 °C; detector: ICP–MS. Bottom: column: 200 m × 0.53 mm ID × 5 µm Agilent DB-1; oven temperature: 35 °C; detector: ICP–MS.

One of the most common applications of gas analysis is the use of more than one column with either the same or different selectivities for the measurement of trace atmospheric impurities in bulk gases. Figure 8.11 describes the generic two-column arrangement using a 10-port valve. The analytes of interest elute through column 1 onto column 2, at which time the valve is set back to the load/back-flush position, keeping the matrix gas off the analytical column and detector. Figure 8.12 is a chromatogram employing this technique using two porous polymer columns. The upper chromatogram is a 10-ppm standard of carbon monoxide, methane, and carbon dioxide in helium. The lower chromatogram describes the false-positive response generated by nitrogen when the sample is diluted to ppb levels. This interference could be avoided by the use of higher-capacity packed columns with a gain of separation and loss of resolution or the use of a molesieve (or molecular sieve) analytical column as in Figure 8.13. Note the significant response from oxygen and nitrogen even when using a reasonably specific AED. Careful consideration should be given to using packed columns for this application, as the increased surface area, use of steel columns, and the packing materials themselves are suspect when analyzing trace components and have a tendency to adsorb or react with trace analytes such as oxygen and carbon monoxide.

Another way of employing and taking advantage of two different column selectivities is with the use of a Deans switch. This is a technique first described by J. R. Deans in 1968 that has been revisited over the years [9]. It is now easily

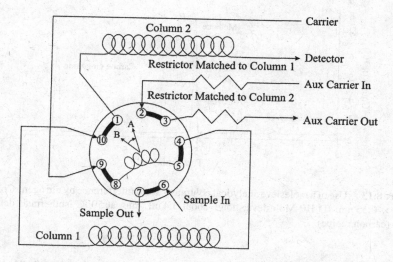

Figure 8.11   Generic two-column arrangement using a 10-port valve.

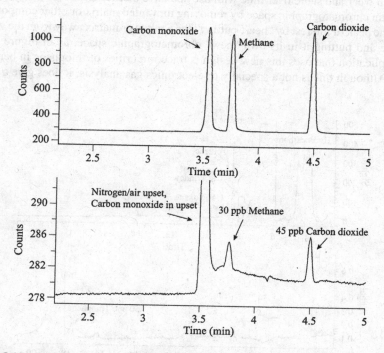

Figure 8.12   Measurement of trace atmospheric impurities in bulk gases using two porous polymer columns.   Top (10 ppm standard): 30 m × 0.53 mm ID GasPro GS-Q column; bottom: 30 m × 0.53 mm ID GasPro GS-Q column; flow: approximately 3.5 mL/min at 35 °C isothermal; detector: AED (carbon recipe).

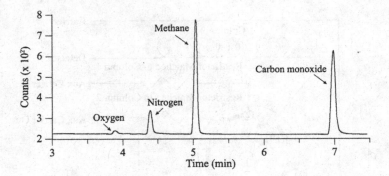

Figure 8.13   Use of a molesieve analytical column to avoid interference by nitrogen. Column: 30 m × 0.53 mm ID HP-Molesieve; flow: about 3.4 mL/min at 50 °C isothermal; detector: AED (carbon recipe).

automated and can use partial pressure flow switching instead of traditional valves that can wear and score over time with use and have thermal limits [10]. The idea is to obtain chromatographic space by removing unwanted matrix or other components from the peak of interest by "heart-cutting" the analyte or matrix away from the whole sample and putting it in its own "new" chromatographic space, as in Figure 8.14. An application that uses this new design is trace impurities of thiophene in benzene [11]. Although this is not a specialty or electronics gas analysis, it does give a good

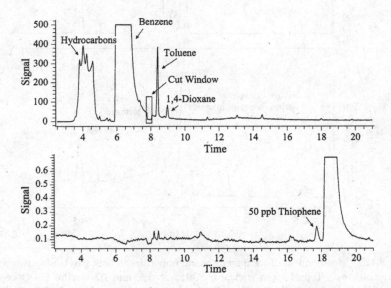

Figure 8.14   Heart-cut method. Column 1 (top): 60 m × 0.53 mm ID × 0.5 µm HP-Innowax; cut window: 7.74 to 8.02 min; column 2 (bottom): 15 m × 0.53 mm ID HP-PLOT Q.

example of looking at a trace component in a large matrix utilizing this technique. The traditional approach to this analysis requires a sulfur-selective detector such as a sulfur chemiluminescence detector (SCD) or AED that is not quenched by the benzene peak but still has good sensitivity for thiophene. Although these detectors are effective, they are costly, complex, and require more regular maintenance than does a standard GC detector. By using a standard flame ionization detector (FID) for detection and a Deans switch for heart-cutting, the HP-Innowax eluent from 7.74 to 8.02 minutes is cut to the HP-PLOT Q column. The eluent during this cut time contains both the thiophene and the benzene eluting in the tail of the main peak. The separation of thiophene from the benzene tail is shown on the chromatogram of the HP-PLOT Q column. Note that the HP-PLOT Q column selectivity is very different from the HP-Innowax since the thiophene now elutes before the benzene.

A good example of column optimization involving length, film thickness, and ID is demonstrated by the column used for the analysis of diborane (B_2H_6) in a 95 % nitrogen matrix (Figure 8.15). In this chromatogram the balance gas of nitrogen was separated sufficiently from the analytes of interest for good quantitative measurement. Although the product had a wide boiling range, there was minimal column bleed. A wider-bore thicker-film column could have been used, but a 0.32-mm ID column was required to keep the carrier gas volumetric flow rate down for GC/MS confirmation analysis.

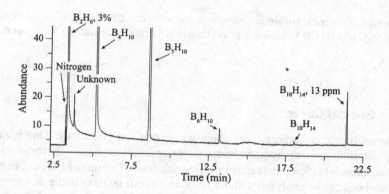

Figure 8.15 Chromatogram of column optimization involving length, film thickness, and ID for the analysis of diborane (B_2H_6) in 95 %+ nitrogen. Column: 60 m × 0.32 mm ID × 1.5 µm ID Restek RTX-Volatiles; carrier: helium at 20 psig; initial temperature: 35 °C; initial hold: 5 min; ramp: 15 °C/min to 240 °C; detector: AED (boron recipe).

A proprietary phase column that is quite useful for analyses of major and minor components in dimethylamine and trimethylamine is the Varian (now Agilent) CP-Volamine column (Figure 8.16). This column provides good resolution and inertness for ammonia even at very low concentrations, as can be seen in lower chromatogram. The asymmetric peak for water is due to its affinity for the gas sampling valve (GSV) rotor material and the amount of time the valve is left in the inject position.

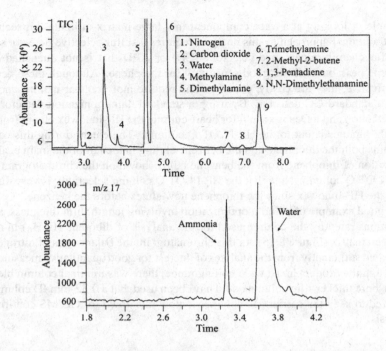

Figure 8.16 Separation of amines, ammonia, and water on a CP-Volamine column. Column: 60 m × 0.32 mm ID CP-Volamine; carrier: helium at 17 psig; temperature: 70 °C isothermal.

8.4.1 Special Cases

Occasionally, circumstances come about where one just gets lucky. One such case is the determination of hydrocarbon content in dichlorosilane. It turns out that dichlorosilane is sufficiently reactive with the substrate of alumina PLOT columns that hydrocarbons can be analyzed without elution or interference with the dichlorosilane matrix (see Figure 8.17). The one downside to this is that the hydrocarbon retention times are shifted slightly after each injection of product, probably due to modification of the porous layer of the column with the matrix. A replaceable guard column or precolumn might be the long-term solution to this problem.

A similar case is the use of the CP-SilicaPLOT column. It can be used for the speciation of sulfur compounds in silicon tetrafluoride (Figure 8.18). Silicon tetrafluoride also reacts irreversibly with the substrate while leaving the sulfur compounds unscathed. Further discussion of capillary columns, both theory and practice, can literally fill libraries. The intent here has been to provide a very brief introduction to their use and potential for application to ESGs. A more comprehensive discussion can be found in works that have been compiled by some of the groundbreakers in this field [12–15].

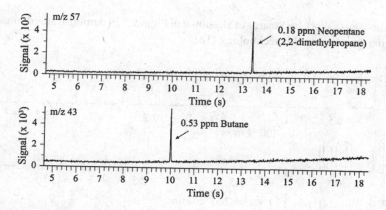

Figure 8.17   Determination of hydrocarbon content in dichlorosilane. Column: 50 m × 0.32 mm ID × 5 μm Al_2O_3/Na_2SO_4 Varian (now Agilent) CP-SilicaPLOT; carrier: helium at 14 psig; initial temperature: 45 °C; initial hold: 5 min; ramp: 10 °C/min to 195 °C; detector: mass spectrometer ($m/z = 57$ for neopentane and $m/z = 43$ for n-butane).

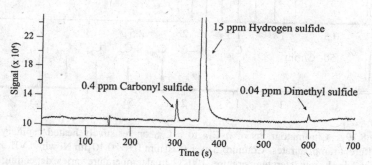

Figure 8.18   Speciation of sulfur compounds in silicon tetrafluoride. Column: 60 m × 0.53 mm ID × 6 μm Varian CP-SilicaPLOT; carrier: argon at 9 psig; initial temperature: 45 °C; initial hold: 4 min; ramp: 15 °C/min to 150 °C; detector: ICP–MS (CCT mode S = SO^+ $m/z = 48$).

8.5 The Future

In many endeavors, time is a critical factor. This is particularly true of chromatography. As analysts we strive for faster work turnaround as long as the quality is not compromised. Even experienced chromatographers can often be persuaded to abandon chromatography if a faster analytical solution such as FTIR is presented to them. However, this is not always possible. The answer has been to speed up the chromatography. The most dramatic example is the development of "fast GC," which has been in development since the 1960s. Today, several vendors offer add-on modules using small low-mass ovens or resistively heated short small-bore thin-film columns that can be heated and controlled much more rapidly than can conventional air bath ovens. Work by Stearns et al. shown in Figure 8.19 demonstrates dramatically the time saved using this technology [16].

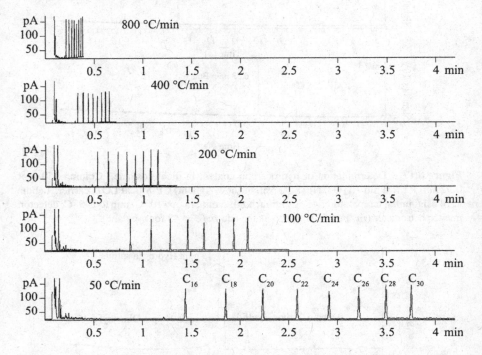

Figure 8.19   Chromatograms of n-C_{16} to n-C_{30} on resistively heated capillary column with different ramping rates. Column: 5 m × 0.10 mm ID × 0.10 μm Ni wired VB-5; carrier: helium at 3.1 mL/min; start temperature: 100 °C; final temperature varies, depending on when the last peak comes out: From 280 °C for 50 °C/min to 360 °C for 800 °C/min; injector: split, 50 psi, 40 : 1 split ratio, 280 °C; detector: FID, 350 °C. Reprinted from [16] with permission from Elsevier.

For high-purity gas analysis this is not a complete solution since there is still the issue of column capacity with the attendant loss of sensitivity for impurities. Nevertheless, it is a step in the right direction and is a technology that could be used for process control or other applications requiring only gross composition.

8.6 Conclusions

Gas chromatography is the predominant analytical tool for gas analysis and will continue to be in the foreseeable future. Regardless of advancements in detectors, or sophistication of chromatographic ovens, or data handling, it is the column that unarguably decides the quality of the analytical outcome. It is doubtful if the packed column will ever be replaced completely, due its capacity and robust nature, particularly in a production or process environment. Nevertheless, the new thick-film megabore and PLOT columns, along with modern techniques of passivating steel, are making inroads.

It has been fascinating to observe the rapid evolution of capillary column technology from crude steel coated by hand with liquid-phase solutions to today's bonded FS columns, which are made with such precision that their performance is 100 % reproducible. It is also remarkable, and even a little amazing, that column vendors are now able to manufacture columns of such quality and inertness that they can transmit analytes at ppt and even part per quadrillion (ppq) levels. We are hopeful that future developments in column technology will be as fruitful and exciting as they have been in the past.

REFERENCES

1. Lynn, T. R. (Ed.). (1975). *Guide to Stationary Phases for Gas Chromatography*, 10th ed. North Haven, CT: Analabs, Inc.

2. Thompson, B. (1977). *Fundamentals of Gas Analysis by Gas Chromatography*. Palo Alto, CA: Varian Instrument Division.

3. McNair, H. M., & Bonelli, E. J. (1969). *Basic Gas Chromatography*, 5th ed. Palo Alto, CA: Varian Instrument Division.

4. Jennings, W. (1987). *Analytical Gas Chromatography*. Orlando, FL: Academic Press.

5. Wang, C., Morales, M., & Firor, R. (2011). *Analysis of Low-Level Sulfur Compounds in Natural Gas and Propylene using a Pulsed Flame Photometric Detector*. Agilent Technologies Application Note 5990-9215EN.

6. Blumberg, L., & Klee, M. S. (2001). Metrics of separation in chromatography. *Journal of Chromatography A, 933*, 1–11.

7. Blumberg, L., David, F., Klee, M. S., & Sandra, P. (2008). Comparison of one-dimensional and comprehensive two-dimensional separations by gas chromatography. *Journal of Chromatography A, 1188*, 2–16.

8. Giddings, J. C. (1967). Maximum number of components resolvable by gel filtration and other elution chromatographic methods. *Analytical Chemistry, 39*, 8, 1027–1028.

9. Deans, D. R. (1968). A new technique for heart cutting in gas chromatography. *Chromatographia, 1*, 1/2, 18–22.

10. Zheng, R., Zhang, H., Zhao, J., Lei, M., & Huang, H. (2011). Direct and simultaneous determination of representative byproducts in a lignocellulosic hydrolysate of corn stover via gas chromatography–mass spectrometry with a Deans switch. *Journal of Chromatography A, 1218*, 31, 5319–5327.

11. McCurry, J. D., & Quimby, B. D. (2003). *Analysis of Trace (mg/kg) Thiopene in Benzene Using Two-Dimensional Gas Chromatography and Flame Ionization Detection*. Agilent Technologies Application Note 5988-9455EN.

12. Ettre, L. S. (1973). *Practical Gas Chromatography for the Users of Perkin-Elmer Gas Chromatographs*. Waltham, MA: Perkin-Elmer.

13. Barry, E. F., & Grob, R. L. (2007). *Columns for Gas Chromatography Performance and Selection*. Hoboken, NJ: John Wiley & Sons.

14. Lee, M. L., Yang, F. J., & Bartle, K. D. (1984). *Open Tubular Column Gas Chromatography*. Hoboken, NJ: John Wiley & Sons.

15. Rood, D. (1999). *A Practical Guide to the Care, Maintenance, and Troubleshooting of Capillary Gas Chromatographic Systems*, 3rd ed. New York: Wiley-VCH.

16. Stearns, S. D., Cai, H., Koehn, J. A., Brisbin, M., Cowles, C., Bishop, C., Puente, S., et al. (2010). A direct resistively heated gas chromatography column with heating and sensing on the same nickel element. *Journal of Chromatography A, 1217*, 27, 4629–4638.

CHAPTER 9

GAS MIXTURES AND STANDARDS

STEPHEN VAUGHAN

Custom Gas Solutions, Durham, North Carolina

9.1 Introduction

A necessary requirement for the successful performance of many specialty gas analytical techniques is the availability of one or more gas-phase calibration standards to
determine absolute concentrations. There are a few absolute techniques based on wet
chemical methods for quantitation; however, the vast majority of analyses [includ-
ing so-called absolute techniques, such as Fourier transform infrared spectroscopy
(FTIR) and mass spectrometry (MS)], are absolute only in the identification of analytes. While the need for calibration standards seems like a simplistic and obvious
requirement, it is an area fraught with difficulties, ranging from simple availability to
the specification of preparations and any additional requirements. Therefore, in this
chapter we discuss important aspects of gas mixtures and standards, including cylinder packages, preparation techniques, material compatibility considerations, stability,
and uses of alternative approaches.

Trace Analysis of Specialty and Electronic Gases.
By William M. Geiger and Mark W. Raynor. Copyright © 2013 John Wiley & Sons, Inc.

Specialty gas standards are available from most major cylinder gas suppliers and a multitude of smaller niche gas companies. Unfortunately, there are significant "equivalency" problems among suppliers. A cylinder gas standard procured from one supplier may differ significantly from that supplied by another company, even with similar purchase specifications [1].

9.2 Definition of Gas Standards

Each gas supplier names its standards differently, making it difficult to compare one with another; however, there are ways to tell them apart (at least on the face of manufacturers' claims). How would the perfect standard be specified? The perfect gas standard would match the exact request of the customer and be exactly the concentration specified; that is, the blend tolerance or how close the actual concentration matches the specification of the customer, would be zero and the analytical uncertainty would also be zero. Depending on preparation techniques, it is possible to achieve zero blend tolerance within the analytical uncertainty; however, this usually entails higher cost, due to the greater levels of labor involved in the standard preparation. Zero analytical uncertainty is a physical impossibility. In this chapter we define uncertainty, tolerance, and accuracy equivalently. It is also extremely useful to discuss concentration reporting conventions.

A number of different conventions are used to report the concentration of analyte material in compressed gas standards. Some of these conventions are specific to individual industry segments or uses. The typical statement of concentration, generally taken as a default in the industry at large, is concentration expressed as volume/volume, also expressed as concentration by volume, volume percent, ppmv (parts per million by volume), mole/mole, and molar concentration, to name a few. Concentrations expressed by volume can be, and generally are, made gravimetrically (by weight addition) or by volume addition (to pressure). However, volume additions by pressure readings must be monitored carefully for temperature increases and must allow for compressibility of materials involved in the mixtures. Thus, volume additions are normally reserved for gross or highly simplistic mixtures.

A secondary way to report concentration expresses the concentration as a weight-to-weight ratio. These are typically found in the hydrocarbon or oil and gas industries and are reported as weight percents or ppm w/w (parts per million by weight). An alternative reporting value, used in the environmental monitoring industry, is percent carbon or ppmc (parts per million by carbon). This value is based on a mole-per-mole concentration and is reported based on the number of carbon atoms in the molecular structure of the analyte of interest. An easy illustration of the determination of ppmc would be to compare the volume-per-volume concentration of 5 ppmv benzene to the analog ppmc concentration. To do this, the ppmv concentration is simply multiplied by the number of carbon atoms in the molecular structure of benzene. So, 5 ppmv benzene multiplied by 6 carbon atoms per benzene molecule equals 30 ppmc benzene. Regardless of the balance component (be it hydrogen, helium, nitrogen, etc.), a cylinder gas standard containing 5 ppmv benzene is equivalent to a 30 ppmc benzene standard.

Although there are many ways to express concentration, the only other typical concentration expression convention is the milligram per cubic meter (mg/m^3) or grams or micrograms per cubic meter. This is a combination of weight and volume reporting used in monitoring applications, toxicology measurements, or active bed absorption or catalysis applications. A comparison of the concentrations expressed in normal units is shown in Table 9.1 .

Table 9.1 Comparison of concentration units for a benzene/nitrogen gas standard

Compound	ppmv	ppmc	ppm w/w	mg/m^3	$\mu g/m^3$
Benzene	5	30	13.93	16.177	16177

The concentration of the analyte of interest in a gas standard can be certified in one of three ways. The first of these is to prepare the standard gravimetrically (discussed in detail in Section 9.4) with the materials weighed into the cylinder on a high-precision balance and the concentration calculated by added masses. Using this method, it is necessary to combine the analytical accuracy of the preparation technique with the analytical accuracy of the purity determined for the raw material used in the mixture. This method works particularly well for nonreactive analytes of interest. If there is any chance of concentration degradation in the cylinder after preparation, there is uncertainty unless the manufacturer uses flawless cylinder preparation and blending technologies and has a documented record of stability. The second way is to analyze the cylinder by a variety of analytical techniques against another standard of known concentration. In this instance, it is necessary to determine the analytical accuracy of the analysis and combine it with the analytical accuracy of the standard since these are cumulative uncertainties. In the specific case of EPA protocol standards preparation, the reference standard material uncertainty is defined as zero and not taken into account in the cumulative uncertainty of the final certified concentration [2]. The third method of certification is to verify the gravimetric concentration of the prepared standard by a confirming analytical measurement against a high-accuracy standard. If these methods agree, it is known as an *interlocking analysis*, so named by Scott Specialty Gases (now Air Liquide America Specialty Gases).

Now, with an understanding of the significance of blend tolerance and analytical uncertainty, it is possible to compare standards from different manufacturers. We begin the discussion by considering the highest-precision gas standards currently available. Three international standards organizations recognize each other's standards as equivalents: The U.S. National Institute of Standards and Technology (NIST) Standard Reference Materials (SRMs), the Dutch Metrology Institute (VSL) Primary Reference Standards (PRMs), and the U.K. National Physical Laboratory (NPL) Traceable Calibration Gas Standards. These standards are the highest accuracy available but are limited in production, materials, availability, and economy. Although these are generally recognized as the supreme standards available today, they are extremely expensive, they are typically produced at a few specific concentrations,

Gas and Liquid Mixtures Typical Specifications

Typical Specifications		Dual-Analyzed Standards	Dual-Certified Standards
		EPA Protocol	GRAVSTAT *
		RATA Class *	Compliance Class *	Reference Class *
Typical Applications		Calibration Standards to Meet Government Regulations		Master Calibration and Process Control
Concentration Range				
1 – 49 ppm	Blend Tolerance ±%	5	5	5
	Analytical Accuracy ±%	1	2	1
	Process Accuracy ±%	–	–	1
50 – 99 ppm	Blend Tolerance ±%	5	5	5
	Analytical Accuracy ±%	1	2	1
	Process Accuracy ±%	–	–	1
100 – 9999 ppm	Blend Tolerance ±%	5	5	2
	Analytical Accuracy ±%	1	2	1
	Process Accuracy ±%	–	–	1
1 – 1.9%	Blend Tolerance ±%	5	5	2
	Analytical Accuracy ±%	1	2	1
	Process Accuracy ±%	–	–	1
2 – 15.9%	Blend Tolerance ±%	5	5	1
	Analytical Accuracy ±%	1	2	1
	Process Accuracy ±%	–	–	1
16 – 49%	Blend Tolerance ±%	5	5	1
	Analytical Accuracy ±%	1	2	1
	Process Accuracy ±%	–	–	1
Reproducibility ±% for Concentration Range of 100 ppm – 1%		2	3	2
Analytical Traceability		NIST, VSL or Other Recognized Reference Standard		NIST, VSL or Other Recognized Reference Standard; or Air Liquide Reference Standard
Process Traceability		Not Applicable		NIST or VSL Weights
Component Verification		Laboratory Analysis of All Minor Components		Laboratory Analysis of All Minor Components
Critical Impurities Analysis		Not Applicable		Available

Minor Components Available (Balance)
  Inert Gas Phase
  Other Gas Phase
  Liquid Phase
		Inert and/or Reactive
		Not Applicable
		Not Applicable
				Inert and/or Reactive
				Inert and/or Reactive
				Inert and/or Reactive

Standard Industry Methods		EPA G1 or State Protocols Used	EPA G1 or EPA G2 Protocols Used	Not Applicable
Reference Standard Accuracy		Directly traceable to SRMs, NTRMs or PRMs Used	NTRMs or EPA GMISs Used	Meets or Exceeds Analytical Accuracy Specifications
Certificate of Accuracy		Protocol Concentration, Traceability and EPA Expiration Date Reported		Analyzed Concentration, Traceability and Shelf Life Reported; Requested and Blended Concentrations Shown
Physical Property Notifications		Mixture Property Warnings Available		Actual Mixture Properties Available (i.e. Dew Point, Temperature and Vapor Pressure Restriction)
Raw Material Quality Control		Verified for Specification Compliance		Complete Purity Analysis
Product Shelf Life		Meets Regulatory Specifications		ACULIFE ™ and Other Proprietary Processes Guarantee Stability and Shelf Life

Figure 9.1 Representative listing of standard names [3].

and they are sometimes difficult to obtain (unavailable with long production lead times). These standards are certified at less than ±1 % uncertainty and are required for some analyses. EPA protocol standards generation (typically, priority pollutants such as nitric oxide, sulfur dioxide, carbon dioxide, carbon monoxide, and nitrogen dioxide) require certification against SRMs or PRMs to qualify with less than ±1 % certified uncertainty [2]. Gas manufacturers have the option of creating a single lot of standards that may be analyzed by a number of samples from the lot sent to NIST to be qualified as NIST Traceable Reference Materials (NTRMs). There are addi-

Our seven product classes redefine mixture versatility. Each has unique specifications to meet different needs as summarized below. Refer to the following pages for additional information about each product class.

Single-Certified Standards			Custom Standards
ACUBLEND ™	Certified*		
Master Class ™	Master Class ™	Working Class ™	Custom Class ™
Calibration with Zero Blend Tolerance	Calibration and Process Control		Special Requirements
0	10	10	Select Blend Tolerance, Analytical and Process Accuracy by Component to Meet Your Specific Application Needs
2	2	5	
–	2	5	
0	5	10	
2	2	5	
–	2	5	
0	5	10	
1	2	5	
–	2	5	
0	5	10	
1	2	5	
–	2	5	
0	5	10	
1	2	5	
–	2	5	
0	2	5	
1	1	2	
–	1	2	
2	3	7	
Air Liquide or Other Reference Standards	Air Liquide or Other Reference Standards		Specify Traceability Source and Component Verification Level
Not Applicable	Weight, Pressure or Standards		
Analysis of All Minor Components	Confirmation of Minor Components		
Not Applicable	Not Applicable		Available
Inert and/or Reactive	Inert and/or Reactive		Inert and/or Reactive
Not Applicable	Inert and/or Reactive		Inert and/or Reactive
Not Applicable	Inert and/or Reactive		Inert and/or Reactive
Not Applicable	ISO Methods Available		ASTM, EPA, GPA, AQMD, ISO or Other Industry Methods Available
Achieves Analytical Accuracy Specifications	Achieves Analytical Accuracy Specifications		Customer Specified Accuracy
Analyzed Concentration, Traceability and Shelf Life Reported; Requested Concentrations Available	Certified* Concentration, Traceability and Shelf Life Reported; Requested Concentration Available		Custom Design Available
Actual Mixture Properties Available (i.e. Dew Point, Temperature and Vapor Pressure Restriction)	Mixture Property Warnings Provided		Actual Mixture Properties Available
Verified for Specification Compliance	Verified for Specification Compliance		Complete Purity Analysis
ACULIFE ™ and Other Proprietary Processes Guarantee Stability and Shelf Life	ACULIFE ™ and Other Proprietary Processes Guarantee Stability and Shelf Life		Component Dependent

* Special circumstances apply to certain low level reactive components (i.e. nitric oxide – please consult your Air Liquide representative).

tional requirements for in-house analysis as well as the NIST analyses; however, the quality of these standards is similar to the quality of the NIST-generated SRMs. Alternatively, gas manufacturers may generate a similar group of standards in a lot and perform only in-house analysis to qualify them as gas manufacturer internal standards (GMISs). These are normally used to qualify EPA protocol gas standards with less than ±2 % certified accuracy. There are a number of other projects that may require SRM or PRM reference certification, such as the qualification of Food and Drug Administration (FDA)–controlled current good manufacturing practices (cGMP) drug manufacturing; however, an analysis of standards cost versus requirements usually renders these standards "excessive."

The actual number of standards for which SRMs are available is fairly small and limits traceability to NIST standards by analysis to a small number of materials. VSL has proven itself to be more open to creating standards for other materials; however, this adds a significant time component to the acquisition process which may not be an affordable indulgence. There is another avenue to NIST traceability on mixtures, which is known as NIST traceable by weight. This type of traceability requires gravimetric standards preparation and calibration of the gravimetric weight scale by NIST traceable weights. The method works particularly well for non-reactive materials and allows a way to gain traceability for otherwise nontraceable materials. The negative aspect of this traceability lies in the dependence of the quality of traceability to the procedures and abilities of the manufacturer to control gravimetric addition processes precisely. In the correct instance this is effective traceability; however, in the wrong instance it could mean nothing.

Returning to the discussion of standards for generalized use, most gas manufacturers name their standards according to the accuracy defined in their list of blend and analytical tolerances. A representative listing of standard names is provided in Figure 9.1. Other manufacturers will have similar names, but careful attention must be paid to quotations regarding blend and analytical tolerances as well as traceability (if applicable).

Although there are many ways to specify standards, it is important to understand the variability and flexibility in requirements. The intricacies of gas standards selection may be reduced to a reasonable number of required pieces of information: materials in the standard, precision requirements of the standard desired, convention used in reporting the concentration, and the traceability desired, if any [1].

9.3 Cylinders and Valves: Sizes, Types, and Material Compositions

It is not within the scope of this chapter to present all possible types and combinations of cylinders and valves along with the compatibilities (incompatibilities); however, it is extremely useful to list cylinder sizes and designations, cross-correlating the major gas companies' nomenclature. Tables 9.2–9.5 and Figures 9.2 and 9.3 present a thorough description of various cylinder sizes and compositions.

Cylinder requirements for different materials become somewhat more complicated when a treatment (cylinder passivation technique) changes a material's compatibilities rather drastically [5]. Aluminum cylinders with correct passivation are the preferred standard for the majority of low-level standards, including reactive materials. Alternatively, nickel-coated steel cylinders are equally acceptable for low-level reactive standards that are inherently corrosive or otherwise reactive (either bonding to the cylinder surface or decomposing in a catalytic manner on the surface). Uncoated steel cylinders tend to accumulate rust layers (and even, in some instances, measurable levels of powdered rust in the bottom of the cylinder). Not only is rust (iron oxide) a reactive surface, but the nature of iron oxidation creates a highly porous layer that flakes away continually to allow more rust formation to occur and compro-

Table 9.2   Various sizes of standard high-pressure aluminum cylinders

Air Liquide	Scott	Custom Gas Solutions	Matheson	Linde	Airgas	Praxair
47AL	KAL	265AL				AT
30AL	AL	150AL	1R/1I	A31	150A	AS
16AL	BL	88AL	2R/2I	A16	80A	AQ
7AL	CL	33AL	3R/3I	A07	33A	AG
3AL	DL	15AL				A3
1AL		AL170	6R/6I			

Source: [3].

Table 9.3   Standard high-pressure aluminum cylinder specifications

Size	DOT Spec.	Svc. Press. (psig)	Approximate Capacity[a] ft³	Approximate Capacity[a] L	OD (in)	Height[b] (in)	Tare Wt.[c] (lb)	Internal Water Volume[d] in³	Internal Water Volume[d] L
47AL	3AL	2216	244	6909	9.8	51.9	90	2831	46.4
30AL	3AL	2015	141	3993	8	47.9	48	1800	29.5
16AL	3AL	2216	83	2350	7.25	33	30	958	15.7
7AL	3AL	2216	31	878	6.9	15.6	15	360	5.9
3AL[e]	3AL	2015	8	227	4.4	10.5	3.5	103	1.7
1AL[e]	3AL	2216	5	142	3.2	11.7	2.3	61	1

Source: [3].
[a] For N_2 at 70 °F, 1 atm. [b] Without valve. [c] With valve, nominal. [d] Nominal. [e] Resale cylinder only.

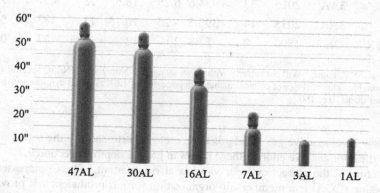

Figure 9.2   Standard high-pressure aluminum cylinder height and size comparison [3].

Table 9.4   Various sizes of standard high-pressure steel cylinders

Air Liquide	Scott	Custom Gas Solutions	Matheson	Linde	Airgas	Praxair
49	K	300ST(K)	1L	049(T)	300	T/UT
44	A	200ST(A)	1A	044(K)	200	K/UK
44H		3K	1H			3K
44hh		6K	1U	485	3HP	6K
16	B	80ST	2	016(Q)	80	Q/UQ
7	C	30ST	3	007(G)	35	G/UG
3	D	15ST	4	3		F
LB	LB	LB	LB	LBR(LB)	LB	LB/RB
LBX		LB-CGA	7X		LX	EB

Source: [3].

Table 9.5   Standard high-pressure steel cylinder specifications

Size	DOT Spec.	Svc. Press. (psig)	Approximate Capacity[a] ft^3	L	OD (in)	Height[b] (in)	Tare Wt.[c] (lb)	Internal Water Volume[d] in^3	L
50	9809-1[e]	2900	335	9373	9	58.2	130	3051	50
49	3AA	2400	277	7844	9.25	55	143	2990	49
44	3AA	2265	232	6570	9	51	133	2685	44
44H	3AA	3500	338	9571	10	51	189	2607	44
44HH	3AA	6000	433	12,261	10	51	303	2383	43
16	3AA	2015	76	2152	7	32.5	63	976	16
7	3AA	2015	33	934.6	6.25	18.5	28	427	7
3	3AA	2015	14	396.5	4.25	16.75	11	183	3
LB/LBX[f]	3E	1800	2	53.8	2	12	3.5	27	0.4

Source: [3].
[a] For N$_2$ at 70 °F, 1 atm. [b] Without valve. [c] With valve, nominal. [d] Nominal. [e] UN/ISO specification.
[f] Nonreturnable cylinder. Price of cylinder included in price of gas. LBX is an LB cylinder with a CGA valve other than 170 or 180.

mises low-level reactive gas standards that react proportionally to the actual surface area in the cylinder. We address these issues in greater depth in Section 9.7.

Selection of the correct cylinder valve is also important. The Compressed Gas Association (CGA) is a membership organization for gas producers that provides a number of services to the gas manufacturing community. In particular, the CGA provides guidelines that control the majority of cylinder gas connection selections

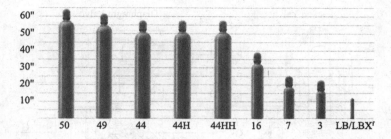

Figure 9.3   Standard high-pressure steel cylinder height and size comparison [3].

in the United States [4]. Although this organization holds no enforcement authority, their suggested usages are used as industry requirements and/or restrictions.

Valve selections (gas connection not withstanding) are typically based on both material compatibilities and purity specifications. Valves are usually constructed in two primary designs: packed stem and diaphragm pack-less (Figure 9.4). For lower-purity materials, the packed valve is normally chosen due to cost (lowest cost per unit available in cylinder valves). These packed valves have a stem that closes directly against the flow orifice. The stem is packed with a Teflon type of grease material for lower reactivity and is sealed by a compression nut that can be tightened if the valve shows signs of leaking. These valves are recognized as an industrial valve, and their use for high-purity and specialty gases and standards is extremely limited. Diaphragm pack-less valves use a two-part stem separated by multiple stainless steel spring diaphragms. The gas-wetted surfaces of these valves are normally stainless steel or a mixture of stainless steel and brass parts. This design is particularly easy to clean and takes passivation processes well, thereby becoming the industry standard for specialty gas standards. Some highly corrosive materials or materials that tend to deposit residual materials between the orifice and the diaphragms may cause the diaphragms to stick shut and render the valve inoperable. If this scenario occurs (and it can with a reasonable regularity for materials such as hydrogen chloride and other halogen acids), the remediation and hazardous materials disposal fees become excessive. There is a variation on this design that includes a tied diaphragm that minimizes this occurrence. In this design, the diaphragm is actually connected to the valve stem, and the hand wheel actually pulls the valve open.

Another major consideration with the valve selection for a particular application is the material of construction. Cylinder valves are manufactured from brass, stainless steel, and an alloy of aluminum, silicon, and bronze known as *ASB composition*. The valves constructed of brass are also available as chrome-plated brass. Again, it is beyond the scope of this discussion to give a detailed analysis of material compatibility for different gas products and mixtures.*

---

*Some information is available in the literature [4,7,8]. However, that discussion is best held with a knowledgeable representative of a gas standards supplier.

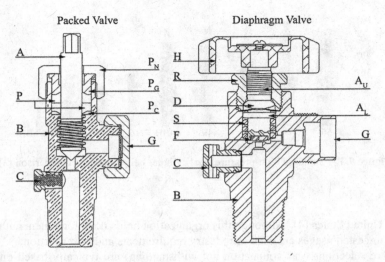

Figure 9.4 Valve diagrams for packed and diaphragm valves. A, stem; A_U, upper stem; A_L, lower stem; P, packing; P_N, packing nut; P_G, packing gland; P_C, packing collar; B, valve body; C, pressure relief device; G, outlet (with cap); H, handwheel; S, spring; F, valve seat; R, retainer; D, diaphragm.

9.4 Preparation Techniques for Gas Standards

As mentioned in Section 9.2, cylinder gas standards may be prepared in a number of different ways and in a number of different cylinder configurations. The typical preparation methods used in the industry today include gravimetric addition, volumetric addition, dynamic blending, and liquid injection. Each of these methods has benefits, and specific blend specifications usually dictate the use of one over another. It makes sense to address each one independently and then discuss relative accuracy.

9.4.1 Preparation of Gravimetric Standards

Gravimetric blending is defined as weighing controlled amounts of pure gases or gas blends into a cylinder to create a standard of known concentrations. The term *gravimetric* is taken loosely from the world of chemical analysis, where it can be defined as quantitative determination of an analyte based on the mass. In chemical analysis, there exist a set of methods whereby materials in a liquid solution are precipitated [either through chemical means, including solvent environment changes (polarity, pH, etc.), or through derivatization] and then weighed to determine the fraction of analyte originally present in the solution. It's not quite that simple, but the idea translates. In the case of gas standards, we know the initial weights of materials added to a cylinder, and we can therefore define the resulting concentrations.

The first step in manufacturing a gravimetric gas standard is a determination of the various component gram additions. As discussed earlier, the typical industry designation of concentration is a volume per volume quantity. The unit of material used is a mole. As an example, a typical large high-pressure aluminum cylinder (designated in the industry as a 1R, 1L, AL, 150AL, etc.) contains 165 mol of nitrogen at 2015 psia. The molar calculation is another unit taken from chemistry and is defined as the number of grams required to equal Avogadro's number of units (either molecules or atoms depending on the gas—molecules for nitrogen and atoms for monatomics such as helium or argon). One mole of material is defined as 6.023×10^{23} molecules. Direct calculation of the number of moles (n) is given by

$$n = \frac{m}{M} \tag{9.1}$$

where m is the mass in grams and M is the molecular weight.

For the purpose of this example we will calculate the gram additions of carbon dioxide and nitrogen required to make a standard at the concentration of 1 % carbon dioxide in nitrogen in a typical aluminum cylinder at full pressure. There are reasons to limit the final pressure for some mixtures, which is discussed in Section 9.5.

In this case we first calculate the required number of moles of carbon dioxide and then the required number of moles of nitrogen. Define the total number of moles as 165 for this example. The calculation to determine the number of moles of minor component, or carbon dioxide in our example, is

$$n_m = Cn_t \tag{9.2}$$
$$n_b = 1 - Cn_t \tag{9.3}$$

where n_m is the number of moles of minor component, C is the concentration of minor component expressed as a decimal, n_b is the number of moles of balance gas, and n_t is the total number of moles. The number of moles of carbon dioxide required to make a standard at the concentration of 1 % carbon dioxide in nitrogen in a typical aluminum cylinder at full pressure is 1.65.

When equation (9.1) is rearranged to $m = nM$, the number of moles required, along with the molecular weight of carbon dioxide, can be inserted to find the mass of carbon dioxide required in grams. In our example, 72.62 g of carbon dioxide is required. The same calculations can be performed for the major component or balance gas, which is nitrogen.

To manufacture this standard gravimetrically, we would first add 72.6 g of high-purity carbon dioxide and subsequently add 4575.4 g of nitrogen. Following manufacture, the cylinder is removed from the manifold, inspected for leaks, and transferred to a cylinder roller, where the cylinder is rolled for a minimum of 30 minutes to ensure homogeneity of the final product. Standards manufactured gravimetrically can be striated and require agitation (in this case in the form of the cylinder roller) to fully mix the gases together. Without this step, a sample analyzed from the newly manufactured cylinder would show a lower-than-expected concentration of carbon dioxide, with the concentration rising as a disproportionately larger quantity of nitrogen is removed initially. After the gases have mixed, in general, they will not separate. However, as discussed in Section 9.5, this is not always the case.

To manufacture lower concentration standards at commensurately high precision, this standard or gaseous premixture may be used as a component in the subsequent gravimetric addition. We continue this example by using this premixture or blending standard to create a lower concentration standard through dilution. This procedure can be performed repeatedly as *successive serial dilutions* to create extremely low concentration standards. In this example, a standard at 100 ppmv will be created. First, we calculate a mass composition of carbon dioxide in the pre-mixture as a mass ratio expressed as grams of minor component to total grams of standard known in the industry as a *gram per gram value* (g/g)*:

$$\frac{C_m}{T} = V \tag{9.4}$$

where C_m is the mass in grams of the minor component, T is the total mass in grams in the blend, and V is the gram per gram value.

In our example, the total mass of carbon dioxide in the blend is 0.0156 g/g of mixture. Knowing the target quantity of material needed to make a full pressure standard in an AL150 cylinder (165 mol), we use equation (9.1) to calculate the addition of pure carbon dioxide and pure nitrogen. The mass of carbon dioxide required is 0.726 g and the mass of nitrogen required is 4621.2 g. Since the amount of carbon dioxide is so small that it cannot be measured reliably, we use equation (9.4) to calculate a statistically precise gram-addition amount of material. In our example, that amount is 46.5 g of standard.

To determine the actual amount of the major component required to complete the dilution mixture, we must account for the amount of major component (nitrogen) in the pre-mixture. To do so, we must subtract the amount of minor component required from the gram-addition amount of material, then subtract that amount from the total amount of major component required. In our example, we find 4573.3 additional grams of nitrogen.

To complete the dilution of the standard through gravimetric addition to 100 ppmv, we add 46.5 g of standard (carbon dioxide in our example) followed by the addition of the balance gas (in this example, nitrogen) using the additional amount calculated above (4573.3 g), resulting in a final pressure of 2015 psia, followed again by the standard operations of leak check and cylinder roll for homogeneity.

This procedure can be repeated again and again to reach lower and lower concentrations with very reliable precision. A discussion of additive error will be included at the end of this section. In this section, a detailed discussion of gravimetric blending procedures has been given for three very important reasons. First, this method provides the backbone of high-precision standards manufactured in the specialty gas industry today. Second, many of the methods and terminology introduced in this section translate through all blending procedures. Third, gravimetric blending is easily misunderstood, misinterpreted, shortcut, or corrupted, leading to substantial errors in mixtures.

---

*Exercise care regarding carrying the correct number of significant figures in this calculation. It is easy to believe in greater precision because of the number of digits displayed by calculators.

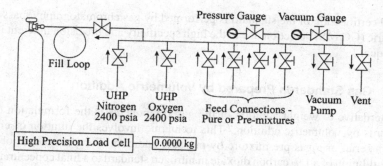

Figure 9.5 Typical gravimetric blending station.

A typical gravimetric blending station is illustrated in Figure 9.5. A gas cylinder to be used for standards preparation is situated on a high-precision floor balance and attached to a gas manifold by flexible connection. The components of a manifold usually include a vacuum pump, pressure gauges, and multiple connections to gas cylinders or high-pressure facility feed manifolds.

9.4.2 Gas Standards Prepared by Liquid Injection

A variation on the method presented in Section 9.4.1 involves the injection of a precisely measured amount of a liquid minor component directly into an evacuated gas cylinder followed by the addition of the balance gas according to equations (9.1)–(9.3). As the liquid is introduced into the vacuum, it volatilizes immediately in preparation of the standard. The discussion in Section 9.5 becomes much more relevant for materials that exist as liquids at or near room temperature, due to their lower vapor pressures.

After performing the standard preparation using liquid injection of the minor component, all remaining procedures are identical to the gravimetric blending procedures presented in Section 9.4.1. Again, to dilute these standards to lower concentrations, the standard prepared by liquid injection may be used for successive (or serial) dilutions.

The liquid injection technique may be used to prepare multicomponent standards in a similar manner. In many instances it is preferable to prepare a multiple component liquid pre-mixture prepared with the correct ratio of materials, followed by the injection of an aliquot of the liquid pre-mixture to the cylinder as a single component. Multicomponent mixtures such as EPA Methods TO-14 and TO-15 gas calibration standards, as well as an infinite variation with as many as 100 or more minor components, may be prepared with a high degree of accuracy [2]. The number of components is limited only by the imagination, creativity, and budget of the requesting analyst. This is limited, of course, by the requirement of noninteraction between components. Many of these mixtures contain components that are subject to loss during the injection and/or degradation in the cylinder. Therefore, it is always preferable to provide a final certification of the mixture by laboratory analysis. The typical certification of the standard is performed by gas chromatography/mass spectrometric (GC/MS) detection due to the high specificity and precision inherent in that technique.

9.4.3   Gas Standards Prepared by Volumetric Addition

An alternative to weight addition of the gas components is the formulation of gas standards by volumetric addition. This technique involves the dilution of either a pure material or a gas pre-mixture by pressure addition. For example, to obtain a 100 : 1 dilution of a 1 % carbon dioxide in nitrogen standard to a final concentration of 100 ppmv (as discussed in the gravimetric addition example in Section 9.4.1), 20 psia of the 1 % gas pre-mixture is added, followed by the addition of pure nitrogen to a final pressure of 2000 psia. There are several inherent weaknesses in the technique that must be understood to utilize the procedure effectively. First, to control and measure the addition of 20 psia of the pre-mixture, it is necessary to use a highly accurate and precise pressure gauge. Second, as gas is added to the receiving cylinder (i.e., the target gas standard), the compression of the gas to high pressure leads to heat evolution (the cylinder gets hot). Using the ideal gas law below, we can see a correlation between temperature and pressure:

$$PV = nRT \tag{9.5}$$

where P is the pressure, V is the volume, n is the number of moles, R is the ideal gas constant, and T is the temperature. The increased temperature of the gas in the cylinder will lead to an erroneous determination of pressure (higher temperature results in a higher pressure). It is necessary to prepare the mixture very slowly, allowing the heat to dissipate as it is generated, or allow the temperature of the cylinder to cool to room temperature and then adjust the mixture slightly to compensate for the artificially high pressure. Another option for mixtures prepared regularly (such as synthetic air blends performed on large multiposition manifolds) is to characterize the heating effect with sufficient accuracy to account for the effect as it occurs in mixture preparation.

The benefits of this technique are ease of use (for low-precision blends) and lower cost of equipment. The disadvantages of using this blending method include loss of accuracy and, usually, loss of any NIST traceability in the blend preparation, and the loss of time required to prepare the mixture with any reasonable accuracy or precision at all. In this instance, the only true traceability of the mixture would be by analytical comparison in the laboratory with recognized gas standards (NIST or VSL).

9.4.4   Gas Standards Prepared by Dynamic Addition

An important preparation technique used for the preparation of gas standards is known as *dynamic blending*. This method was designed to prepare a larger quantity of standards simultaneously with identical results for each cylinder. In this method, the gaseous materials (pure or pre-mixtures) are diluted with the balance gas at a low pressure, analyzed in situ, and referenced to a known standard (locally prepared or NIST certified) and then compressed to high pressure to fill the final products. This method could be considered a large-scale version of volumetric blending. The benefit of this process is the simultaneous fill of multiple cylinders with identical components and concentrations. Methods similar to this are normally used to manufacture batch lots of mixtures that ultimately become NIST or VSL primary standards.

9.4.5   Notes on Additive Uncertainty

The specific uncertainty in the certification depends on a number of factors [11–13]. For gravimetric mixtures these can be expressed in an equation form as follows: For the preparation of an initial mixture from pure materials by gravimetric addition;

$$(U_T)^2 = (U_B)^2 + \Sigma(U_M)^2 \tag{9.6}$$

where U_T is the total uncertainty expressed in grams, U_B is the uncertainty of the balance measurement expressed in grams, and U_M is the material purity uncertainty expressed in grams.

For example, can a gravimetric blend of $\pm 1\%$ uncertainty be made using the following parameters: ethylene purity of greater than 99% (indicates a relative uncertainty of $\pm 1\%$) with a weight addition of 150 g and gravimetric balance accuracy of ± 0.5 g absolute?

$$U_T = [(1.5)^2 + (0.5)^2]^{1/2} = (2.25 + 0.25)^{1/2} = (2.50)^{1/2} = 1.58$$

Therefore, the absolute or total uncertainty (U_T), divided by the addition, then multiplied by 100, will give the percent uncertainty, which when rounded down equals 1% in our example:

$$\frac{1.58}{150} \cdot 100 = 1.05$$

So the answer is yes, with an uncertainty of $\pm 1\%$, a gravimetric mixture can be manufactured using these parameters.

For the next mixture in a serial dilution preparation of a standard requiring multiple dilutions, the total uncertainty of the first mix then becomes a component of the equation for the second mixture. This progression continues and it can be seen that the ultimate uncertainty of the final product becomes greater as well. So, for mixture 2;

$$(U_{Tn})^2 = (U_B)^2 + \Sigma[(U_{D1})^2 \cdots (U_{D(n-1)})^2] \tag{9.7}$$

where U_{Tn} is the total uncertainty of the nth serial gravimetric dilution expressed in grams, U_B is the uncertainty of the balance measurement expressed in grams, and $\Sigma[(U_{D1})^2 \cdots (U_{D(n-1)})^2]$ is the sum of uncertainties for all previous dilutions in grams.

These equations may be manipulated to show that for a standard floor balance used in gravimetric blending, the lowest amount of material allowable to achieve a $\pm 1\%$ or better uncertainty in the blend would be 40 g. So-called *mass comparators* have higher precision, readability, and repeatability; however, more extreme measures must be taken to isolate these weighing devices from vibrations, air movement, and changes in humidity and temperature.

For the analytical certification of standards, the cumulative uncertainty would look more like

$$(U_T)^2 = (U_S)^2 + (U_C)^2 + (U_P)^2 \tag{9.8}$$

where U_T is the total uncertainty express as a percentage, U_S is the standard uncertainty expressed as a percentage, U_C is the calibration residual uncertainty expressed as a percentage, and U_P is the measurement precision (or coefficient of the variability) expressed as a percentage.

Based on this information, we may characterize standard gas certification uncertainties. For a certification to be appropriate for $\pm 1\,\%$ certification, the following must be true: U_S is below $1\,\%$, U_C is below $0.8\,\%$, and U_P is below $0.4\,\%$, yielding the following calculation:

$$U_T = (1.0^2 + 0.8^2 + 0.4^2)^{1/2} = 1.34$$

which rounds to $1\,\%$ uncertainty. For a certification to be appropriate for $\pm 2\,\%$ certification, the following must be true: U_S is below $1.5\,\%$, U_C is below $1\,\%$, and U_P is below $0.8\,\%$, yielding the following calculation:

$$U_T = (1.5^2 + 1.0^2 + 0.8^2)^{1/2} = 1.97$$

which rounds to $2\,\%$ uncertainty. In addition, for a certification to be appropriate for $\pm 5\,\%$ certification, the following must be true: U_S is below $3\,\%$, U_C is below $3\,\%$, and U_P is below $3.04\,\%$, yielding the following calculation:

$$U_T = (3.0^2 + 3.0^2 + 3.0^2)^{1/2} = 4.7$$

which rounds to $5\,\%$ uncertainty. In this manner a true determination (sometimes based on informed estimates) of the standard uncertainty may be made.

9.5   Pressure Restrictions and Compressibility Considerations

The actual physical contents of a cylinder of gas depends on pressure and temperature, as we discussed in Section 9.4.3 in relation to equation (9.5), known as the *ideal gas law*. The ideal gas law is an approximation for gases that behave in a certain manner (ideal interaction). Most real gases even obey this law near ambient conditions. In Section 9.5.1 we discuss the restrictions placed on the pressure of a gas-phase standard, and in Section 9.5.2 we introduce the compressibility factor, which predicts how real gases behave under high pressure [4].

9.5.1   Vapor Pressure Restrictions for Gas-Phase Mixtures

An important consideration in the manufacture of gas standards using liquid or solid components is the adjustment of the final pressure of the mixture to take into account the vapor pressure of the component material at a given temperature (usually taken to be 0 °C by North American gas mixture manufacturers). If the final pressure of the mixture exceeds the allowable vapor pressure of the component, the minor component in the mixture will condense out of the gas mixture onto the walls of the cylinder. There are differing opinions about whether the material can be returned to the gas phase and recreate the accuracy and dependability of the original formulation. Some practitioners in the gas business believe that heating the cylinder (carefully, in a controlled manner), followed by rolling, can revaporize the minor component.* The author has enough doubts regarding this practice as not to utilize it or recommend it to customers and partners.

For simple two-component mixtures, calculation of the vapor restriction is a fairly simple process. Equation (9.9) allows a simple and quick evaluation of the allowable final pressure of the mixture, based only on the concentration required for the minor component and the vapor pressure of that minor component at some given temperature. Most manufacturers choose to restrict gas standards to withstand temperatures down to 0 °C. As an example, we will calculate the final pressure restriction for a 100 ppm mixture of ethanol in nitrogen. The equation for the determination of the pressure restriction is

$$P_f = \frac{P_m}{C} \tag{9.9}$$

where P_f is the final pressure, P_m is the vapor pressure of the minor component at 0 °C, and C is the concentration expressed as a decimal.

For example, if 100 ppm expressed as a decimal is 0.0001, and the vapor pressure of ethanol at 0 °C is 0.229 psia, the final pressure will be 2290 psia. In this instance, the standard can be manufactured to full pressure in any high-pressure aluminum cylinder with a maximum pressure rating of 2216 psia or less. If the parameters of the blend requirement are changed to 1000 ppm, the final pressure becomes 229 psia.

As shown by the calculation, scaling occurs in a linear manner, and standard requirements can be adjusted accordingly, depending on use requirements. Higher concentrations may be chosen as an alternative to higher mix pressures (corresponding to a lower gas content in the cylinder), or a higher pressure may be chosen for the mixture, resulting in a lower concentration to maximize the amount of standard gas in the mixture. These requirements depend on the intended use of the standard.

9.5.2 The Compressibility Factor

The physical amount of gas that can be put into a cylinder at high pressure can be expressed as (or rather derived from) a variation of the ideal gas law [equation (9.5)]:

$$PV = ZnRT \tag{9.10}$$

---

*Note that aluminum high-pressure gas cylinders should never be heated, due to the potential for temper destruction in the metal, which could ultimately cause catastrophic cylinder failure, such as cylinder rupture and/or explosion.

where P is the pressure, V is the volume, Z is the compressibility factor, n is the number of moles, R is the ideal gas constant, and T is the temperature. For gases that deviate significantly from ideal behavior at high pressures, such as xenon, the compressibility factor, Z, is significantly less than 1. In this instance, there is significantly more gas in the cylinder at high pressure than would be predicted by the simplistic version of the ideal gas law [equation (9.5)]. Gases that have a compressibility factor closer to 1 will have real contents in the cylinder much closer to that predicted by the simplistic ideal gas law expression. An example of this is the amount of gas contained in a normal high-pressure aluminum cylinder such as the AL150 cylinder filled to 2015 psia. The cylinder filled to this pressure will contain 162 mol of helium. The same cylinder filled to the same pressure contains 463 mol of xenon. In this example, the compressibility factor of xenon for this fill is 0.362, while the compressibility factor of helium in this fill is 1.035.

Although this concept is important to the understanding of the amount of standard contained in a cylinder, its primary importance lies in the preparation of the mixture itself. The gas manufacturer must understand and use this knowledge in determining actual weights of material to add to the cylinder.

9.6 Multicomponent Standards: General Considerations

Most of the discussion regarding gas standard preparation in earlier sections used simplistic examples of two-component mixtures consisting of a primary component and a balance gas. These same concepts can be extended to gas standards containing almost as many components as one can imagine. Although conceptually and theoretically possible and reasonable, there are a few caveats surrounding multicomponent standards preparations that must be addressed prior to manufacturing the mixture. These are additive vapor pressure effects and material compatibility within the gas standard. Even with these additional concerns, gas manufacturers regularly produce reliable low-level gas standards according to EPA guidelines TO-14 and TO-15, and ozone precursors requiring exact blending of 39, 57, 69, or more components in a single gas mixture [2]. The prudent purchase of standards of this complexity requires a knowledgeable grasp of requirements as well as knowledge of the true capabilities of a potential gas standard vendor.

9.6.1 Additive Vapor Pressure Restriction Effects

The simplest explanation of additive vapor pressure effects is to list the steps in determining the ultimate restriction on a gas standard pressure:

Step 1: List components and concentrations by vapor pressure in ascending order.

Step 2: Using equation (9.9), calculate the mixture restriction for the first component.

Step 3: Add the concentrations of first and second components and use equation (9.9) to calculate the restriction using the sum of concentrations and the vapor pressure for the second component.

Step 4: Repeat step 3 through all components (creating the additive concentration for all materials with lower vapor pressures).

Step 5: Choose the lowest pressure allowable under all previous calculations and set that as the maximum pressure for the mixture.

Step 6: Calculate additions according to Section 9.4.1 and manufacture the standard.

Step 7: Test and certify the concentrations.

This is a critical calculation that must be performed to determine allowable final pressures of multicomponent mixtures. Without this consideration there is always a danger of minor component condensation and subsequent degradation of standard concentrations and dependability.

9.6.2 Material Compatibility Considerations

Another important consideration in the manufacture of multicomponent standards is the internal compatibility of the minor components. This is a separate issue from cylinder and valve compatibilities and passivation (discussed in Section 9.3) and should not be confused with degradation reactions requiring cylinder treatment or passivation. Every multicomponent mixture should be reviewed carefully by a person trained in material properties prior to manufacture. This can be a chemist, materials engineer, or experienced gas manufacturer. Typically, reactions occurring in a cylinder will lead to the formation of materials that are unwanted or the disappearance of desired components. Reactions of this type include, but are not limited to, acid-base reactions, oxidation-reduction reactions, and combustion. In most cases the degradation reaction are benign, leading only to the unreliability of the standard; however, in extreme cases and higher concentrations, these reactions could potentially generate enough heat and/or pressure to rupture the cylinder, causing any number of safety concerns [9,10]. Reputable gas manufacturers will recognize these dangers and recommend safe alternatives.

9.6.3 Additional Considerations

A number of methods are used to manufacture multicomponent standards, and most are specific to the actual gas standard mixture or to the manufacturers themselves. It is beyond the scope of this section to explain all of these methods. Suffice it to say that there are any number of ways to produce these standards precisely and just as many (or more) ways to make them badly. Again, the educated end user asking the gas supplier the correct questions will help mitigate problems.

When deciding on a gas standard, a realistic assessment of the lower limit of detection and quantitations for the method under study should be taken into account. If a specific method or instrument suggests the detection range to be 0.1 to 100 ppm, a gas standard at 0.1 ppm should not be ordered unless the method is established and optimized and ready to quantify the extreme lower end of the detection range. A newly installed instrument will probably not be able to detect or quantify the extreme lower limit of the published specification without considerable optimization.

A final concern to the end user is the cost of purchasing complex gas standards. Standards used for air monitoring (a normal product such as TO-14 or TO-15) will typically cost approximately $50 to $150 per component. Complex mixtures with 78 components might cost anywhere from $3900 to $11,700 for the standard. Complex mixtures with more exotic components can cost even more. It is advisable to assess carefully the number of components required in the gas standards as well as approaching the ordering process with a realistic assessment of the limit of quantification (LOQ) for the technique requiring the gas standard.

9.7 Cylinder Standard Stability Consideration

An important characteristic of any gas standard is the stability of the standard. One specification not discussed in Section 9.2 is shelf life or expiration period. Although most quality systems require a published expiration date, many gas standards can remain stable for a much longer period of time than documented. This is one reason that many gas manufacturers offer recertification services, albeit reluctantly.

The single most important factor in the continuing stability of a gas standard is the absence of contaminants in the standard at the time of blending as well as the absence of reacting contaminants in the pure materials used to make the mixture. The most common contaminants present in all cylinders to some degree are moisture, atmospheric oxygen, and rust. Varying levels of cylinder preparation are used to remove these contaminants and prepare a cylinder for a specific service—some of these were mentioned in Section 9.3. Many low-purity bulk gases, such as nitrogen, oxygen, argon, helium, and air, are packaged in cylinders that have virtually no preparation. They may be subjected to a forced-air drying cycle prior to valve installation, or they may have been subjected to purging or a rough vacuum. Higher-purity bulk gases and simple gas standards require the additional steps of a heated evacuation in a cylinder treatment oven. These devices are well-controlled heated vacuum systems in which readiness is characterized by the achievement of a specific vacuum reading. Gas standards at percentage levels or nonreactive materials such as hydrocarbons are suitable to be filled in either steel or aluminum cylinders prepared in this manner.

A higher level of cylinder preparation is required for corrosive gas standard manufacture. At concentrations below 1 % it is advisable to use a steel cylinder that has been coated with nickel internally (normally, an electroless nickel-coating process). The coating is usually applied to a newer clean cylinder with minimal or no rust, or to a cylinder that has been bead- or sand-blasted to clean off surface rust. This final cleanliness verification is performed by visual inspection. Although rust is not necessarily a reactive material, the exponential increase in surface area created by rust/flake cycles is a huge contribution to degradation. As the surface area increases, the potentially reactive metallurgical components are exposed, thereby increasing chances and rates of standard degradation. In processes where particle count is important, rust is a particularly onerous contaminant.

For standards where the highest level of purity and nonreactivity is required, there are a number of proprietary methods of cylinder passivation and preparation. Most of the true intellectual properties of gas manufacturers reside in this part of the process. Therefore, it falls loosely into the category of "magic." Realistically, this involves the baseline heated evacuation cycle mentioned earlier, followed by a passivation process that could be as simple as exposing the cylinder to the reactive material, which will ultimately be the minor component of the final mix, at some higher level than the final blend (often called the *pickling process*). The belief is that if all the active sites in a cylinder have been bound previously to the material in the mixture, there will be no further degradation of the standard when it is manufactured in the following step. If the degradation process is a simple adsorption process, this would potentially passivate the cylinder. The danger is that when the cylinder standard pressure drops below a certain level, the material will begin to migrate out of the cylinder walls, causing the measured concentration of the standard inexplicably to increase over time.

Several more advanced treatments exist that are proprietary and based essentially on silanization processes. The processes succeed to a greater or lesser extent depending on the exact treatment used and the control of the process, in creating a "glass-like" coating on the inside of the cylinder, thereby decreasing reactivity. Most of these processes have been around for more than 30 years and constitute "public secrets." The majority of these methods were modified versions of treatment technologies used in the manufacture of gas chromatography columns (mostly of capillary gas column manufacture). The development of more advanced cylinder passivation technologies representing the first true advances in cylinder passivation, derived from actual cylinder/molecule interaction models based on chemical interaction and standards observation/measurements, are in progress in some laboratories. These developments are driven by the requirements of the end user and regulatory agencies for lower concentrations of more reactive minor components [14,15].

While the cylinder passivation is of inestimable importance, the purity of the raw materials in the mixture also contributes to the accuracy and stability of the packaged gas standard. A perfect example of this type of degradation is the manufacture of low-level nitric oxide standards [5]. Nitric oxide degrades by the following process:

$$2NO + O_2 \rightarrow 2NO_2$$

This equation shows that a single molecule of oxygen reacts with two molecules of nitric oxide to form two molecules of nitrogen dioxide (actually, a dynamic equilibrium exists where the arrow goes both ways and the balance of the equation depends on temperature and pressure conditions). In this example, it requires only 0.5 ppm of oxygen in the nitrogen balance gas to degrade a 1 ppm nitric oxide standard completely, even when the best cylinder passivation technology is used. However, in that case, the end user would read the standard as 1 ppm nitrogen dioxide with little if any nitric oxide present. That would make this a "bad" gas standard. Analogous scenarios exist for other gases.

It is beyond our scope in this chapter to discuss cylinder passivation and standards degradation in more depth. A knowledgeable and experienced manufacturer of gas standards should be able to guide users through this maze to order the gas standards that best fit their needs. In some instances it becomes necessary to question the certified concentrations and/or stability of a particular gas standard. Most gas manufacturers are pleased to have technical support services assisting users through the learning process involved.

9.8 Liquefied Compressed Gas Standards: Preparation Differences and Uses

An alternative formulation of standards used in specific industries (primarily the petrochemical industries) involves the use of liquefied gas standards for certain calibration requirements. These can be produced in normal cylinders with the use of dip tubes or educator tubes with a pressurized head gas used to force liquid out of the cylinder, or they may be manufactured into piston cylinders (Figure 9.6).

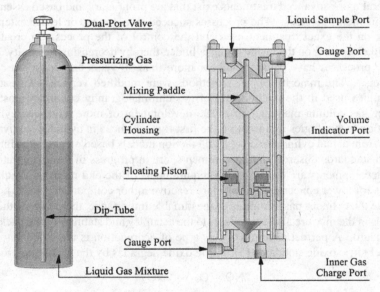

Figure 9.6 Dip-tube and piston cylinder diagrams.

The simplest version of a liquid standard uses a standard gas cylinder, either low pressure or high pressure, dictated by the mixture and the use proposed. The components may be added before the valve is inserted into the cylinder if the materials have sufficiently low vapor pressures. Additional components may be added after the valve has been inserted and the headspace removed. Depending on the specific components in the mixture, the pressure of the head gas can be adjusted to keep the majority of the components in the liquid phase. Note the use of the word *majority*.

There still exists a measurable vapor pressure even under significant head pressure. Therefore, even with extreme care and master blending techniques, there can still exist an additional uncertainty in the standards because of the physical properties of the materials themselves.

A more accurate method for the preparation of liquid standards is through the use of piston cylinder technology. In this case, liquids (primarily liquefied compressed gases) are added to the piston cylinder under pressure and in the liquid state. This is achieved because a piston cylinder has a floating piston that separates the pressurizing head gas from the standard material. During and after addition to the piston cylinder, the pressure behind the piston is adjusted to maintain all components in the standard in the liquid state. In many of these cylinders, there also exists a floating mixer that allows a final homogeneity to be accomplished prior to use. Carefully prepared standards in these cylinders are extremely accurate and reproducible from first to last use. Both styles of liquid mixture preparations are equally valid; however, use of the piston cylinder technology guarantees the consistency of samples from first to last (including the continual delivery of liquid sample).

9.9 Cylinder Standard Alternatives

Cylinder gas standards are normally an excellent calibration choice; however, some applications are better served with dynamic calibration generators. These devices generate highly characterized standard concentrations by passing a diluent gas stream across a semipermeable membrane. The diffusion of the target compound across the membrane occurs in a reproducible manner dependent primarily on temperature and membrane surface area [6]. The flow of diluent gas across the membrane can be changed to vary the final concentration of the standard generated. The generator tubes are certified as producing some specific quantity of material per unit time (e.g., nanograms per minute at a specified temperature). These devices are known as permeation tube devices and are recognized by the NIST as reliable standards generators (on a case-by-case basis). The primary manufacturers of permeation tube calibration generators are KIN–TEK Laboratories and VICI Metronics. Figure 9.7 illustrates the device.

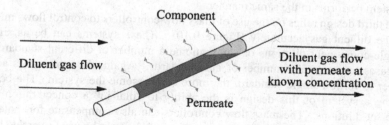

Component

Diluent gas flow

Diluent gas flow with permeate at known concentration

Permeate

Figure 9.7 Permeation tube principle.

Devices of this design are quite effective for generating gas standards with a small number of components (usually in the range of one to four different materials) and within given concentration ranges; however, the requirement of many components in a gas standard can make the hardware quite cumbersome. They are also not particularly effective with higher concentrations.

9.10 Dilution Devices and Calibration Uses

Another useful strategy of the gas standard user is the gas diluter. Dilution devices combine a gas standard or multiple gas standards with a dilution gas that is also known as the *diluent*, to allow the accurate and reproducible generation of multiple gas standards from a single cylinder. In this way a full calibration curve of multiple points may be created with one gas cylinder and tested or verified with one or more check standards.

The first design uses the pressure drop across glass capillary tubes of varying lengths to create varying flows that are combined with the diluent gas (Figure 9.8). The caveat associated with this design is that the physical characteristics of the gas standard and diluent gases must not be too different, and the pressures of the diluent gas and the gas standard delivered to the diluter must be set carefully and controlled per the manufacturer's specifications. This type of diluter can only be used to dilute a single standard gas into a number of preset dilutions (usually, 0 to 100 % in 10 % change increments).

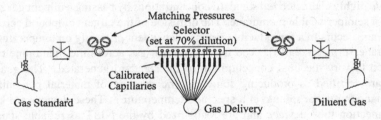

Figure 9.8 Capillary dilution device.

The second design is similar except that the pressure drops occur through fixed orifices instead of capillary tubes (Figure 9.9). The concept is the same, and functionally the system performs in the same manner.

The third design relies on the use of mass flow controllers to control flows mixing with the diluent gas accurately (Figure 9.10). These systems can be as simple as single-component dilutions or can combine a number of different standards or pure gases to create any number of gas standards. Again, the more gases added together to form a single standard, the more cumbersome the system. The benefit of using a system of this design is the ability to "dial in" a concentration with no pre-set dilutions. The mass flow controllers can also compensate for materials with significantly differing physical properties, and they can be used to create high-concentration calibration standards as easily as low concentrations.

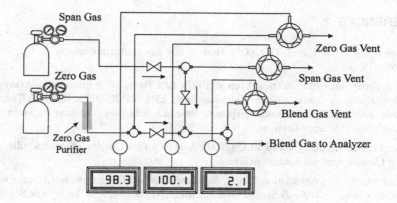

Figure 9.9   Pressure drop dilution device.

The caveat associated with using a system of this design is the regular requirement of calibration. Mass flow controllers are notorious for losing calibration or "drifting" over long periods of time. Although it has not been my experience, it is strongly recommended that the complete dilution device be returned to the manufacturer every year for calibration and certification of accuracy. This maintains the peace of mind of the user as well as providing an important supporting document for quality programs such as ISO 9002 systems.

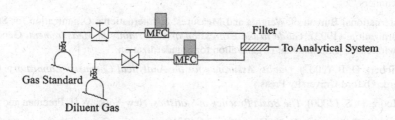

Figure 9.10   Mass–flow configured dilution device.

Gas diluters are available from a number of vendors, but they are based on only two designs. Neither design is more reliable or accurate than the other for any given application; however, the mass flow design requires regular calibrations. In either case, the diluter allows the gas standard user a way to reduce a large number of cylinder gas standards to a few critical standards without losing calibration accuracy or capabilities while significantly reducing cylinder rental fees and storage space requirements.

REFERENCES

1. Scheuring, S., & Bartel, D. (2005). How to buy gas calibration mixes. *LCGC, 23*, 7, 668–675.

2. U.S. Environmental Protection Agency (1997). *EPA Traceability Protocol for Assay and Certification of Gaseous Calibration Standards.* EPA-600/R97/121. Research Triangle Park, NC: U.S. EPA, National Exposure Research Laboratory, Human Exposure and Atmospheric Science Division.

3. Air Liquide Specialty Gases LLC (2009). *Specialty Gas Catalog.* Plumsteadville, PA: Air Liquide America Specialty Gases.

4. Compressed Gas Association (2009). *Standard Method of Determining Cylinder Valve Outlet Connections for Industrial Gas Mixtures, V-7,* 5th ed. New York: Van Nostrand Reinhold.

5. Vaughan, S. (2007). Low level nitric oxide and nitrogen dioxide cylinder gas stabilization and analysis by FTIR. *Gases and Instrumentation, 1,* 3, 26–29.

6. KIN–TEK. *How permeation tubes work.* Retrieved from http://www.kintek.com/how_permeat.html.

7. Yaws, C. L. & Matheson Company (2001). *Matheson Gas Data Book,* 7th ed. Parsippany, NJ: McGraw-Hill.

8. Braker, W., Mossman, A. L., & Matheson Company (1980). *Matheson Gas Data Book,* 6th ed. Lyndhurst, NJ: Matheson Company.

9. Sax, I. (1984). *Dangerous Properties of Industrial Materials,* 6th ed. New York: Van Nostrand Reinhold Company.

10. Masterton, W. L., & Slowinski, E. J. (1969). *Chemical Principles.* Philadelphia: W.B. Saunders Co.

11. International Bureau of Weights and Measures, & International Organization for Standardization. (1993). *Guide to the Expression of Uncertainty in Measurement.* Genéve, Switzerland: International Organization for Standardization.

12. Hibbert, D. B. (2007). *Quality Assurance for the Analytical Chemistry Laboratory.* Oxford: Oxford University Press.

13. Moore, D. S. (2000). *The Basic Practice of Statistics.* New York: W.H. Freeman and Co.

14. Jacksier, T., Benesch, R., & Haouchine, M. (2010). *Reactive Gases with Concentrations of Increased Stability and Processes for Manufacturing Same.* US Patent No. 7,850,790 BE. Washington, DC: U.S. Patent and Trademark Office.

15. Benesch, R., Haouchine, M., & Jacksier, T. (2004). The stability of 100 ppb hydrogen sulfide standards. *Analytical Chemistry, 76,* 24, 7396–7399.

APPENDIX A

CYLINDER AND SPECIALIZED FITTINGS

A.1 Cylinder Fittings

The fittings specified in Table A.1 may be dependent on the pressure and makeup of a gas. Always confirm the chemical, physical, and pressure characteristics with the gas manufacturer to ensure proper fitting strength and suitability. Unless otherwise noted, the same connection is used for both liquid and gas withdrawal. Never rely on the valve connection as the sole sampling method of cylinder contents. Proper paperwork should be provided by the manufacturer. All CGA fittings with a left-handed thread will be marked by a V-groove in the hexagon nut.

More detailed information may be found in Compressed Gas Association (2005). *Standard for Compressed Gas Cylinder Valve Outlet and Inlet Connections*. Chantilly, VA: Compressed Gas Association, or Matheson Tri-Gas (2011). *Semiconductor Products Catalog*. Basking Ridge, NJ: Matheson Tri-Gas.

A.2 Specialized Fittings

Table A.2 describes a few specialized fittings that we have found especially useful in assembling manifold systems, chromatographic systems, and other plumbing applications. Some are particularly useful in minimizing extra fittings that require transitions from different plumbing categories. It is not comprehensive but serves as a list of recommended solutions to plumbing problems. We highly recommend reference material from vendors such as Valco Instruments Co., Inc., Swagelok, and Evans, to name a few, for a more detailed listing of options. All figures courtesy of Valco Instruments Co., Inc. and Swagelok.

Table A.1: Cylinder fittings

| Gas | U.S. | | U.K. | Germany | France | Japan |
	CGA	DISS	BSI, BSIEN	DIN	NF	JIN
Acetylene	300, 510, 520, 200, 410[a]		2/4, 18	3, 4	A, H	
Air	599, 347, 702, 346[b], 590[b], 950[c]		3[b], 3[b]		B[b]	22–R[b], 23–R[b]
Air (breathing)	850[c], 855[cd]		Fig. 14[c]	16.9[c]	D	9[c]
Ammonia	705[be], 240[b], 660[bd], 800[ce], 845[ce]	720	10[b]	6[b]	C[b]	22–R[b]
Argon	580, 680, 677, 295	718	3	6	C	22–R, 23–R
Arsine	350, 660[e]	632[f]	4	1	E	22–L
Boron trichloride	660[e]	634[f]	6	8	K	
Boron trifluoride	330[e]	642[f]	6	8	P	22–L
Bromine pentafluoride	670[e]		6	8		
Bromine trifluoride	670[e]		6	8		
Bromoacetone	660[e]					
Bromochlorodifluoromethane	165, 182, 660[e]		6	6	C	
Bromochloromethane	165, 182, 660[e]		6			
1,3-Butadiene	510		2/4, 15	1	E	23–L

Continued. . .

Gas	U.S. CGA	U.S. DISS	U.K. BSI, BSIEN	Germany DIN	France NF	Japan JIN
Butane	510, 555		4		E, (GPL)	23-L
Butylenes	510		4		E	23-L
Carbon dioxide	320[be], 940[c], 323[f], 295, 622	716[f]	8[b]	6[b]	C[b]	22-R[b], 8[c]
Carbon monoxide	350	724	4	5	E	22-L
Carbonyl sulfide	330[e]		15		E	22-L
Chlorine	660[be], 820[ce], 820[ca]	728	6[b], 14[b]	8[b]	J[b]	26-R[b]
Chlorine pentafluoride	670[e]		6, 14	8		
Chlorine trifluoride	670[e]	728	6, 14	8	P	26-L
Chlorodifluoromethane	165, 182, 660[e]		6	6	C	
Chloroethane	300		7, 17	1	E	26-L
Chlorofluoromethane	510		4			
Chlorotrifluoroethylene	510		6	1	E	26-L
Chlorotrifluoromethane	165, 182, 320[e], 660[e]	716[f]	6	6	C	26-R
Cyanogen	660[e]		4	5	E	
Cyclobutane	510		4			
Cyclopropane	510[b], 920[c]		2/4[b], Fig. 11[c]	1[b], 16[c], 8[c]	E[b]	22-L[b], 6[c]
Deuterium	350	724	2/4	1	E	22-L
Diborane	350	632[f]	4	1	E	22-L

Continued...

Gas	U.S.		U.K.	Germany	France	Japan
	CGA	DISS	BSI, BSIEN	DIN	NF	JIN
Dibromodifluoroethane	165, 182, 660[e]		6			
Dibromodifluoromethane	165, 182, 660[e]		6			
Dichlorodifluoromethane	165, 182, 660[e]	716[f]	6	6	C	
1,2-Dichloroethylene	165, 182, 660[e]		4			
Dichlorofluoromethane	165, 182, 660[e]		6	6	C	
Dichlorosilane	678[e]	636[f]	15	5		
1,1-Dichlorotetrafluoroethane	165, 182, 660[e]		6			
1,2-Dichlorotetrafluoroethane	165, 182, 660[e]		6	6	C	
Diethyltelluride		726				
Dimethylamine	705[e]		11	1	E	22-L; 26-L
2,2-Dimethylpropane	510	632[f]	4	1	E	26-L
Disilane		632[f]	2	1		
Ethane	350	724	4	1	E	22-L
Ethene	350[b], 900[c]	724	4[b], Fig. 9[c]	1[ab], 16[c]	E[b]	4[c], 22-L[b]
Ethylacetylene	510		4			
Ethylamine	705[e]		11	1	E	22-L, 26-L
Ethylene oxide	510		7, 15	1	E	
Ethyl ether	510		4			
Ethyl fluoride	660[e]		7			

Continued...

Gas	U.S.		U.K.	Germany	France	Japan
	CGA	DISS	BSI, BSIEN	DIN	NF	JIN
Ethylidene fluoride	510		7	1	E	26–L
Fluorine	679[e]	728	6, 14	8	P	26–R
Germane	350, 660[e]	632[f]	2	1	E	
Germanium tetrafluoride	330[e]	642[f]	6, 14	8	E	
Helium	680, 677, 580[b], 930[c], 792, 295	718	3[b]	6[b]	C[b]	22–R[b], 23–R[b]
Hexafluorocyclobutene	660[e]					
Hexafluoroethane	165, 182, 660[e], 320[e]	716[f]	3	6		
Hexafluoro-1,3-butadiene	350	724				
Hexafluoropropylene	165, 182, 660[e]			6		
Hydrogen	350, 695, 703, 795, 350	724	4	1	E	22–L
Hydrogen bromide	330[e]	634[f]	6, 14	8	K	26–R
Hydrogen chloride	330[e]	634[f]	6, 14	8	K	26–R
Hydrogen cyanide	660[e]		15	5		
Hydrogen fluoride	660[e], 670[e]	638[f]	6	8	K	26–R
Hydrogen iodide	330[e]		6, 14	8	K	
Hydrogen selenide	350, 660[e]	632[f]	15	1	E	
Hydrogen sulfide	330[e]	722	15	5	E	
Isobutene	510		4	1	E	23–L

Continued...

Gas	U.S. CGA	U.S. DISS	U.K. BSI, BSIEN	Germany DIN	France NF	Japan JIN
Krypton	580, 680, 677	718	3	6	C	22–R, 23–R
Methane	510, 350, 695, 703, 450	724	2/4	1	E	22–L
Methylacetylene	510		4			26–L
Methylamine	705^e		11	1	E	22–L, 26–L
Methyl Bromide	320^e, 330^e		7	8		26–R
3-Methyl-1-butene	510		4		E	22–L, 26–L
Methyl chloride	510, 660^e		7, 17	1		26–R
Methylene fluoride	350	724				
Methyl ether	510		4	1	E	26–L
Methyl ethyl Ether	510		4		E	26–L
Methyl fluoride	350	724	4			
Methyl iodide	660^e		7			
Methyl mercaptan	330^e		7	5	E	22–L
2-Methylpropene	510		4	1	E	23–L
Natural gas	510, 350, 695, 703, 450	2/4			E	
Neon	580, 680, 677, 792	718	3	6	C	22–R, 23–R
Nickel carbonyl	660^e		15			
Nitric oxide	660^e	728	14			22–R

Continued...

Gas	U.S. CGA	U.S. DISS	U.K. BSI, BSIEN	Germany DIN	France NF	Japan JIN
Nitrogen	621, 680, 677, 580[b], 555[b], 590[b], 960[c], 295	718	3[b]	10[b]	C[b]	22–R[b], 23–R[b], 10[c]
Nitrogen dioxide	660[e]		14	8	P	A
Nitrogen trifluoride	330[e], 670[e]	640[f]	14	8	K	
Nitrous oxide	326[b], 910[b], 6245	712[f]	13[b], Fig. 10[c]	8[b]	G[b]	5[c]
Octafluorocyclobutane	165, 182, 660[e]	716[f]		6	C	
Octafluorocyclopentene	660[e]	716[f]				
Octafluoropropane	165, 182, 660[e]	716[f]	6			
Oxygen	601, 540, 870, 577, 701, 440	714[f]	3, Fig. 6	9, 16.4	F	22–R, 23–R, 1
Pentachlorofluoroethane	165, 182, 660[e]		6			
Pentafluoroethane	165, 182, 660[e]	716[f]	6			
Perfluorobutane	165, 182, 660[e]		6			
Perfluoro-2-butene	165, 182, 660[e]	716[f]	6			
Phosgene	160, 660[e]	632[f]	6, 14	8	K	26–R
Phosphine	350, 660[e]	642[f]	4	1	E	
Phosphorus pentafluoride	330[e], 660[e]		6	8		
Phosphorus trifluoride	330[e], 660[e]		6	8		
Propadiene	510		2/4, 15			

Continued...

Gas	U.S.		U.K.	Germany	France	Japan
	CGA	DISS	BSI, BSIEN	DIN	NF	JIN
Propane	510, 600, 791, 810^{ed}, 555, 790^{ed}		4	1,2,4	E, (GPL)	23–L
Propene	510, 600, 791		4	1	E	23–L
Silane	350, 510	632^f	4	1	E	
Silicon tetrachloride		636^f	14			
Silicon tetrafluoride	330^e	642^f	6, 14	8		22–L
Stibine	350		4	1		
Sulfur dioxide	660^e		10, 16	7		26–R
Sulfur hexafluoride	590	716^f	6	6	C	26–R
Sulfur tetrafluoride	330^e		6	8		
Sulfuryl fluoride	660^e		6	8		
1,1,1,2-Tetrafluoroethane	165, 167^e, 660^e	716^f	6			
Tetrafluoroethylene	165, 182, 350		6	6		22–L
Tetrafluoromethane	320^e, 580	716^f	3			
Trichlorofluoromethane	660^e		6	6	C	
Trichlorosilane		636^f	15	5		
Triethylaluminum	510	726				
Triethylborane	660^e			8		
Trifluoroacetyl chloride	330^e		8			

Continued...

Gas	U.S. CGA	DISS	U.K. BSI, BSIEN	Germany DIN	France NF	Japan JIN
Trifluorobromomethane	165, 182, 660[e]		6	6	C	
1,1,1-Trifluoroethane	510		4	1		
Trifluoroethylene	510		4			
Trifluoromethane	165, 182, 660[e], 320[e]	716[f]	6	6	C	
Trimethylamine	705[e]		11	1	E	26–L
Trimethyl silane		632[f]				
Tungsten hexafluoride	670[e]	638[f]	0.25", VCR3	8		
Uranium hexafluoride	330[e]					
Vinyl bromide	290, 510		7	1	E	26–L
Vinyl chloride	290, 510		7	1	E	26–L
Vinyl fluoride	350		7	1	E	22–L
Vinyl methyl ether	290, 510		4	1	E	26–L
Xenon	580, 660[e], 677	718	3	6	C	22–R

[a] Canada; [b] Threaded; [c] Yoked; [d] O-ring required; [e] Washer required; [f] Gasket required.

Table A.2: Specialized fittings

Fitting	Description	Purpose
Fused silica (FS) adapter	Types: one piece and two piece (*note*: nut and ferrule sold separately); materials: Valcon polyimide, PEEK, virgin polyimide (one piece only); the determining factor in selecting the adapter size is the FS tubing's OD; OD sizes: less than 0.25 to 1 mm; product no.: FS–varies by size (Valco).	The one-piece FS adapter, essentially a reducing ferrule, is recommended for use in fittings where the polyimide ferrule will not be removed; connections are made and disconnected by loosening the fitting nut and sliding the tube out; PEEK recommended for temperatures up to 175 °C; Valcon and virgin polyimide recommended for temperatures up to 350 °C.
Zero dead volume (ZDV) reducing ferrule	Types: standard, internal and external (*note*: nut and ferrule sold separately); materials: PEEK, CTFE, PTFE-glass filled, Valcon polyimide, virgin polyimide; sizes: $^1/_{16}$ to $^1/_8$ in; product no.: ZRF (Valco).	Adapts internal or external fitting details to use smaller tubing; inexpensive way to connect small temporary transfer lines to valves or fittings designed for larger tubing.
Internal zero-volume reducer	Types: standard and with frit for filtering; materials: stainless steel (SS) body, PEEK or SS nut and ferrule; bore sizes: 380 µm to 2 mm; product no.: IZR (Valco).	Allows smaller tubing to be used in valves with fitting details for larger tubing, forming a positive leak-free seal with ZDV.

Continued. . .

Fitting	Description	Purpose
ZDV tube adapter	Material: 300 series SS; bore: $1/16$-in; lengths: 0.7, 1.8 and 2.8 in (*note*: $1/16$ in nut and ferrule included; $1/4$ in nut and ferrule not included); product no.: ZTA41, ZLTA41, ZXLTA41 (Valco).	Ideal for connecting $1/16$-in tubing to a detector or injector with a $1/4$-in fitting; adapts $1/4$-in Swagelok tube fitting to $1/16$-in Valco tube fitting.
Aerosol adapter bulkhead union	Material: 300 series SS; product no.: ZBAA1 (Valco).	Provides an easy, direct method of connecting the nozzle of a standard aerosol can to a $1/16$-in Valco ZDV fitting.
FS makeup adapter	Lengths: 1.5 and 3.5 in; bore sizes: 0.5 to 1 mm (*note*: $1/32$-in FS adapter must be ordered separately); product no.: FSMUA(S) (Valco).	Connects a FS column to a valve or detector while adding a makeup gas; in the reverse mode it works like a splitter, without the uneven or erratic split seen with basic tees; also as a gas diverter, as in Figure A.1.

Continued...

Fitting	Description	Purpose
Internal/external reducing union 	Size: $1/16$ in internal to $1/32$ in external; bore: 0.25, 0.50 and $1/32$ in; materials: 300 series SS; the $1/16$-in tubing is made up with an internal fitting and the FS tubing is made up with an external fitting; in the bulkhead version, the bulkhead nut is on the side with the internal fitting (*note*: use only the one-piece FS adapters in the $1/32$-in external detail of this union, as metal ferrules will distort it); includes a $1/16$-in SS ferrule for SS tube, $1/32$-in FS adapter ordered separately; product no.: EZRU.51 (Valco).	Connects a standard ZDV fitting to FS tubing; bulkhead versions designed to be mounted on an instrument panel also available.
External/internal reducing union 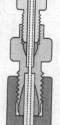	Materials: 300 series SS, Hastelloy C, gold-plated SS, titanium; ODs: $1/32$, $1/16$ and $1/8$ in; bore sizes: 0.25 mm to $1/8$ in; male threads on one end and female threads on the other. (*Note*: $1/16$-in external fittings have very thin, easily distorted walls; this style of union recommended only when connecting to an installed external nut; select the union with the bore that matches the ID of the tubing; if the IDs are different, choose the union with a bore that matches the smaller tube bore; product no.: EZRU (Valco).	Connects a standard ZDV fitting to FS tubing; also good for a needle adapter; bulkhead versions designed to be mounted on an instrument panel also available.

Continued...

Fitting	Description	Purpose
Column-to-valve connector	Also known as an internal to external reducer/adapter; sizes: $1/16$ to $1/32$ in; bore sizes: 0.25 to 1 mm; unique internal to external reducer/adapter permits the $1/32$-in nut to be tightened without affecting the $1/16$-in connection (*note*: $1/32$-in FS adapters must be ordered separately); product no.: IZERA1.5 (Valco).	Connects a FS capillary to a valve.
Male national pipe thread (NPT) to Valco internal adapter	Materials: 300 series SS, Hastelloy C, titanium; NPT male: $1/8$, $1/4$ and $1/2$ in to ZDV: $1/16$ to $1/4$ in; bore sizes: 1 to 4.6 mm; product no.: PZA (Valco).	Makes a minimum volume connection from female pipe fittings on pressure gauges and regulators to Valco ZDV internal fittings.
Female NPT to Valco internal adapter	Materials: 300 series SS, Hastelloy C, titanium; NPT female: $1/8$, $1/4$ and $1/2$ in to ZDV: $1/16$ to $1/4$ in; bore sizes: 1 to 4.6 mm; product no.: FPZA (Valco).	Makes a minimum volume connection from the male pipe fittings typically found in gas distribution plumbing to Valco ZDV internal fittings.

Continued...

Fitting	Description	Purpose
Controlled radius internal nut	Materials: SS, PEEK; sizes: $1/16$ and $1/8$ in [note: the short version (ZSN1R) can only be used in certain applications]; product no.: ZN1R, ZN2R (Valco).	These special-purpose nuts facilitate a tight bend as the $1/16$-in tube exits the fitting and can also prevent kinks in very thin wall tubing.
Needle adapter	—	Used to adapt $1/16$-in zero-volume fitting to needle.
Ferrule removal kit	Product no.: FRK1 (Valco).	These tapered tools have teeth designed to grip and remove FS adapters if they get stuck in a fitting detail; each kit has two sizes of tools, so they can retrieve $1/32$-in and $1/16$-in adapters.
Swagelok VCR to $1/16$-in Valco adapter	Product no.: VCRZ41 (Valco).	Adapts female $1/4$-in Swagelok VCR fitting to $1/16$-in Valco fitting.

Continued...

Fitting	Description	Purpose
Swagelok VCR gland to $1/16$-in Valco adapter	Product no.: VCR4ZA1 (Valco).	Adapts $1/4$-in male or female Swagelok VCR gland to $1/16$-in Valco fitting; appropriate gasket required.
Female Luer adapter	Product no.: ZLA (Valco.)	Adapts Luer to male Valco fitting for direct connection to ZDV fittings and valves.
Female Swagelok VCR split-nut assembly	Materials: SS type 316; product no.: SS-4-VCR-1-SN (Swagelok).	Adapts $1/4$-in male Swagelok VCR nut to female Swagelok VCR fitting allowing for installation of nut after a close-fitting weld.
Male Swagelok VCR split-nut assembly	Materials: SS type 316; product no.[a]: SS-x-VCR-4-xx (Swagelok).	Adapts $1/4$-in female Swagelok VCR nut to male Swagelok VCR fitting allowing for installation of nut after a close-fitting weld.

Continued. . .

Fitting	Description	Purpose
Swagelok VCR metal gasket face seal to male NPT connector	Materials: SS type 316; product no.[a]: xx-x-WVCR-1-xx (Swagelok).	Adapts male NPT to Swagelok VCR with female nut; appropriate gasket required.
Swagelok VCR metal gasket face seal to tube fitting connector	Materials: SS type 316; product no.[a]: xx-x-WVCR-6-xx (Swagelok).	Union/reducing union; use to adapt male Swagelok VCR to tube fittings; appropriate gasket required.
Swagelok VCR metal gasket face seal rotating female union	Materials: SS type 316; product no.[a]: xx-x-WVCR-6-DF (Swagelok).	Connects male Swagelok VCR nut to male Swagelok VCR nut; appropriate gasket required.
Swagelok VCR metal gasket face seal female NPT connector	Materials: SS type 316; product no.[a]: xx-x-WVCR-7-xx (Swagelok).	Adapts male NPT to female Swagelok VCR; appropriate gasket required.

Continued...

Fitting	Description	Purpose
Swagelok VCR metal gasket face seal reducing adapter	Materials: SS type 316; sizes: $1/8$ to $1/4$ in, $1/4$ to $1/2$ in; product no.[a]: SS-x-VCR-7-xVCRF (Swagelok).	Adapts larger male Swagelok VCR fitting to smaller female Swagelok VCR fitting; appropriate gasket required.
AN thread tube adapter	Materials: SS type 316, aluminum, brass, carbon steel, nylon, PFA, PTFE, titanium; sizes: various tube OD, AN tube flare, and thread; product no.[a]: -x-TA-1-xAN (Swagelok).	Adapts AN flare fitting to tube fitting.
Reducing union tube fitting (straight fitting)	Materials: SS type 316, aluminum, brass, carbon steel, nylon, PFA, PTFE, titanium; sizes: various in both imperial and metric; product no.[a]: xx-xxx-6-xxx (Swagelock).	Adapts various tubing sizes to male tube.

Continued...

Fitting	Description	Purpose
Reducer	Materials: SS type 316, aluminum, brass, carbon steel, nylon, PFA, PTFE, titanium; sizes: various in both imperial and metric; product no.[a]: xx-xxx-R-xxx (Swagelok).	Adapts various tubing sizes to male tube fitting.
AN thread female tube adapter	Materials: SS type 316, aluminum, brass, carbon steel, nylon, PFA, PTFE, titanium; sizes: various in both imperial and metric; product no.[a]: xx-xxx-A-xANF (Swagelok).	Adapts tube to AN flare.

[a] Specific product number varies by material and size specifications.

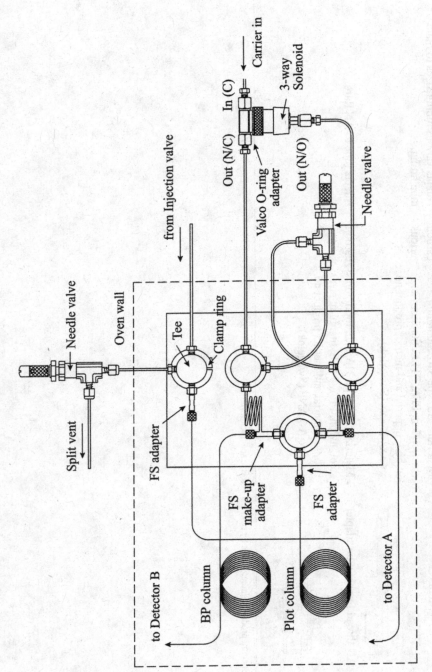

Figure A.1 Deans switch fabricated using off-the-shelf Valco Instruments Company, Inc. fittings.

APPENDIX B

MATERIALS OF CONSTRUCTION

This appendix includes material selections that can be used to mitigate handling problems in analysis and help to prevent equipment failure due to reactions caused by sample and material incompatibilities. Since most specialty and electronic gases are categorized as hazardous materials, it is important to know the sample composition not only for potential reactions, but for safety purposes as well. We will outline the crucial parts of analytical instrumentation based on their materials of construction. We address primarily materials that are associated with sample presentation to either a gas chromatograph (GC) or a Fourier transform infrared (FTIR) instrument. Obviously, many other instruments are used in the laboratory, but GC and FTIR are representative of most instrumentation in the sense that either valving is involved in sample introduction or some sort of cell or chamber will be used, possibly with mirrors and windows that will come in contact with the sample.

B.1 Tubing, Transfer Lines, and Other Hardware

There are numerous combinations of materials that can be used for an array of experiments. Generally, the ideal transfer lines will stand up to the chemical environment

Trace Analysis of Specialty and Electronic Gases.
By William M. Geiger and Mark W. Raynor. Copyright © 2013 John Wiley & Sons, Inc.

provided by the matrix gas while also being able to withstand the temperature and pressure that will be encountered. The analyst will be limited in hardware selection based on the matrix gas and expected impurities. Aside from fittings, the analyst should consider the tubing used for transfer lines. Table B.1 is a summary of the properties of various tubing available.

Table B.1: Metal compatibilities[a]

	Hazard			Material Type					
Gas	Toxic	Flammable	Corrosive	Aluminum	Carbon Steel	Stainless Steel	Monel	Kel-F	Teflon
Acetylene		◆		S	S	S	S	S	S
Ammonia	◆	◆	◆	S	S	S	S	S	S
Arsine	◆	◆		U	S	S		S	S
Boron trichloride	◆		◆	U	S	S	C	S	S
Boron trifluoride	◆		◆	U	S	S	S	S	S
Carbon dioxide				S	S	S		S	S
Carbon monoxide	◆	◆		S	S	S		S	S
Carbonyl sulfide	◆	◆		S	S	S		S	S
Chlorine	◆		◆	U	S	S	C	S	S
Chlorine trifluoride[b]	◆		◆	U	S	S	S	C	C
Deuterium		◆		S	S	S		S	S
Dichlorosilane	◆	◆	◆	U	S	S	S	S	S
Dimethylamine	◆	◆	◆	U	S	S		S	S
Dimethyl ether		◆		S	S	S		C	S
Disilane		◆		S	S	S	S	S	S
Ethyl chloride	◆	◆		U	C	C		S	S
Ethylene oxide	◆	◆		U	S	S		S	S
Fluorine[b]	◆		◆	C	C	C	S	U	C
Germane	◆	◆		S	S	S	S	S	S
Hydrogen		◆		S	S	S	S	S	S
Hydrogen bromide	◆		◆	U	C	C	C	S	S
Hydrogen chloride	◆		◆	U	C	C	C	S	S
Hydrogen fluoride	◆		◆	U	S	S		S	S

Continued. . .

Gas	Hazard			Material Type					
	Toxic	Flammable	Corrosive	Aluminum	Carbon Steel	Stainless Steel	Monel	Kel-F	Teflon
Hydrogen selenide	◆	◆		U	S	S		S	S
Hydrogen sulfide	◆	◆		C	C	S		S	S
Methyl acetylene		◆		S	S	S		S	S
Methyl bromide	◆	◆		U	S	S		S	S
Methyl chloride	◆	◆		U	S	S		S	S
Methyl mercaptan	◆	◆		U	S	S		S	S
Monoethylamine	◆	◆		U	S	S		C	S
Nitric oxide	◆	◆	◆	S	S	S	U	S	S
Nitrogen dioxide	◆		◆	S	S	S	U	S	S
Nitrosyl chloride	◆		◆	U	U	U	S	S	S
Nitrous oxide	◆			C	S	S		S	S
Oxygen				C	C	C	S	S	S
Phosgene	◆		◆	U	C	C		S	S
Phosphine	◆	◆		S	S	S		S	S
Silane	◆	◆		S	S	S	S	S	S
Silicon tetrachloride	◆		◆	U	S	S		S	S
Silicon tetrafluoride	◆		◆	U	C	C		S	S
Sulfur dioxide	◆		◆	C	S	S		S	S
Sulfur hexafluoride	◆			S	S	S		S	S
Sulfur tetrafluoride	◆		◆	C	S	S		S	S
Trimethylamine	◆	◆		U	S	S		S	S
Tungsten hexafluoride	◆		◆	U	S	S	S		S

[a] ◆, primary hazard; U, unsatisfactory; S, satisfactory; C, conditional use.
[b] Extremely reactive and requires surface passivation.

Brass Used where a soft metal ferrule is desirable but no corrosive materials are present.

Electroformed nickel (EFNI) Electroplated pure nickel over a diamond-drawn mandrel in a continuous process, then carefully separated and removed from the mandrel. The result is an extremely inert and smooth interior surface (1 to 2 μin finish). It is widely used for transfer lines, since it minimizes the potential for carryover or cross-contamination often found with rougher-surfaced mill-drawn Nickel 200. Unlike glass- or silica-lined stainless, electroformed nickel can easily accept tight bends and cutting without heating, and does not release damaging glass fragments or silica particles. Electroformed nickel has more in common with fused silica than does drawn nickel tubing in terms of surface inertness and smoothness.

Hastelloy C series The material most often recommended for corrosion resistance — it works when nothing else will. This versatile nickel–chromium molybdenum alloy has excellent resistance to most acids, including strong oxidizers such as ferric acid; cupric chlorides; nitric, formic, and acetic acids; wet chlorine; seawater; and brine solutions. Hastelloy C has excellent resistance to pitting, stress corrosion cracking, and oxidizing atmospheres up to temperatures well beyond any other standard components of a chromatographic system. See Table B.2 for the approximate chemical composition of Hastelloy C and Figure B.1 for a comparison of the corrosion rate of Hastelloy C and other alloys in hydrogen fluoride vapor and liquid.

Inconel 600 One of the few metals that can be used with hot, strong solutions of magnesium chloride. Good for most severely corrosive environments at elevated temperatures. Resistant to sulfuric and hydrofluoric acid and to all concentrations of phosphoric acid at room temperature. Poor resistance to nitric acid. See Table B.2 for the approximate chemical composition of Inconel 600, Figure B.1 for a comparison of the corrosion rate of Inconel 600 and other alloys in hydrogen fluoride vapor and liquid, and Figure B.2 for the percent weight change in corrosion testing of new (all-Inconel) unplated, old (Inconel-stainless) gold-plated, and old unplated UG bellows.

Monel 400 High resistance to hydrochloric, hydrofluoric, and sulfuric acid under reducing conditions. Attacked by oxidizing acid salts and hypochlorites. High resistance to chlorinated solvents and nearly all alkalis. See Table B.2 for the approximate chemical composition of Monel 400 and Figure B.1 for a comparison of the corrosion rate of Monel 400 and other alloys in hydrogen fluoride vapor and liquid.

Nickel 200 Excellent resistance to caustics, high-temperature halogens and hydrogen halides, and salts other than oxidizing halides. Good resistance to caustic soda and other alkalis except ammonium hydroxide. The industry standard in nickel alloy tubing, containing trace amounts of copper, carbon, silicon, and other elements that impart certain mechanical characteristics. Like stainless steel type 316, this tubing is cold-drawn to close ID and OD specifications, and is suitable for many applications where a relatively inert and low-cost nickel is required. Although more inert than 316 SS in most applications, it is still absorptive and has a relatively rough interior. Use electroformed nickel tubing for applications requiring a high level of inertness or finish.

Nitronic 50 Good resistance to chlorides, sulfuric acid, and seawater. Resistant to sulfur gases such as hydrogen sulfide and sulfur dioxide.

Nitronic 60 Chemical resistance similar to that of stainless steel type 316, but its resistance to galling and oxidation make it superior to stainless steel types 316 and 303 in the majority of applications.

Silcosteel General-purpose silicon passivation layer applied to stainless steel as described in U.S. patent 6,511,760. The passivation layer is applied via chemical vapor deposition, which enables the entire surface of tubing or components to be covered. Hence, it is effective for minimizing loss of sensitive analytes. The coating is compatible with most organic solvents except amines. It also significantly reduces water vapor adsorption and corrosion of the metal surface in acid environments such as hydrogen chloride. It is, however, attacked by aggressive fluorinated gases such as hydrogen fluoride and tungsten hexafluoride.

Stainless steel, gold-plated Improved inertness and high-integrity sealing for applications such as high-purity gas analysis.

Stainless steel, type 303 Recommended for GC use and general-purpose connections, good resistance to corrosion and high-temperature oxidation. Susceptible to attack by chlorides, iodides, and bromides.

Stainless steel, type 316 Standard tubing material used for chromatography, suitable for a wide variety of applications. It is cold-drawn seamless, not welded, with close tolerances held on both ID and OD. Austenitic stainless steels may be used for most chromatographic applications. Type 316 is most commonly used for high-performance liquid chromatography (HPLC) because of its superior chloride ion resistance. See Table B.2 for the approximate chemical composition of stainless steel, type 316.

Sulfinert Proprietary treatment similar to that of Silcosteel in providing metal surface isolation. This passivation layer as described in U.S. patent 6,444,326 is particularly effective in preventing loss of sulfur-containing gases such as hydrogen sulfide. It is also effective with organosulfur compounds. Chemical compatibilities similar to those of Silcosteel can be expected.

Titanium Good for organic and inorganic salts except aluminum and calcium chlorides, and all alkalies except boiling concentrated potassium hydroxide. Good with dilute low-temperature formic, lactic, sulfuric, hydrochloric, and phosphoric acids, but attacked rapidly by hydrofluoric acid. Good with dilute nitric acid at low temperatures; corrodes at high concentrations and temperatures. Can ignite with fuming nitric acid. Attacked by oxalic acid, concentrated phosphoric acid, hot trichloroacetic acid, and zinc chloride.

Zirconium Excellent resistance to hydrochloric acid, good with hot sulfuric acid at concentrations up to 70% and boiling nitric acid at up to 90%. Attacked by hydrofluoric acid.

Table B.2 Approximate chemical compositions of selected alloys

Alloy	Composition (wt%)						
	Ni	**Cu**	**Fe**	**Cr**	**Mo**	**Mn**	**W**
Monel 400	67	31.5	1.2	–	–	–	–
Inconel 600	76	–	8	15.5	–	–	–
Hastelloy C–276	57	–	–	16	16	–	4
Stainless steel 316	12	65	17	2.5	1–2	–	–

Source: Rebak, R. B., Dillman, J. R., Crook, P., & Shawber, C. V. V. (2001). Corrosion behavior of nickel alloys in wet hydrofluoric acid. *Materials and Corrosion, 52*, 4, 289–297.

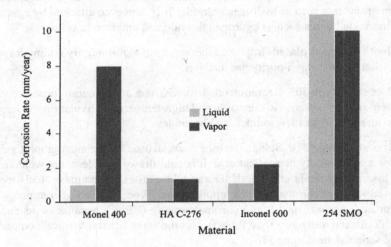

Figure B.1 Comparison of the corrosion rate of alloys in hydrogen fluoride vapor and liquid. Adapted from Rebak, R. B., Dillman, J. R., Crook, P., & Shawber, C. V. V. (2001). Corrosion behavior of nickel alloys in wet hydrofluoric acid. *Materials and Corrosion, 52*, 4, 289–297.

There are two types of transfer lines, metallic and polymer tubing, the industry standard being stainless steel type 316. There are, of course, alternative types of metallic tubing such as Nickel 200 alloy and electroformed nickel tubing. Electroformed nickel tubing, although more expensive and not as readily available, has a pristine interior surface, which will limit internal memory due to its inactive surface. The rough internal surface of 316 and Nickel 200 alloy leaves the analysis vulnerable to carryover and cross-contamination. There is an electropolished stainless steel type 316 option which provides a smoother interior surface but is more expensive. Type 316 is composed of a mixture of nickel, chromium, manganese, and a small amount of carbon, which allows for high performance against corrosion. However, the resistance to corrosion will be compromised if the tubing is to be heated to 500 to 800 °C. This will cause the chromium to react with any carbon in the steel to create chromium

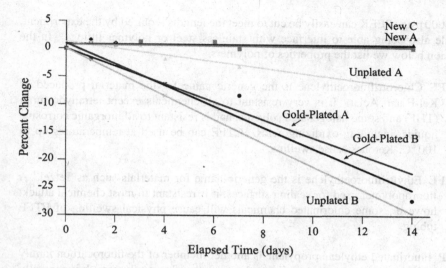

Figure B.2   Percent weight change in corrosion testing of new (all-Inconel) unplated, old (Inconel-stainless) gold-plated, and old unplated UG bellows. *Source:* Osborne, P. E., Icenhour, A. S., Del Cul, G. D. (2002). *Corrosion test results for Inconel 600 vs Inconel-stainless UG bellows.* Oak Ridge, TN: Oak Ridge National Laboratory. Retrieved from http://www.ornl.gov/~/cppr/y2002/rpt/114428.pdf.

carbides. In aggressive-material and high-temperature applications, Inconel tubing is recommended. When used with fluoride, Inconel/nickel will provide a good barrier for passivation.

Moisture can be problematic, due to its tendency to initiate a reaction. Purging strategies will reduce excess moisture contained in the transfer lines and cylinder fittings. Pulse or cycle purging is a quick way to reduce contamination by effectively diluting the concentration. This includes pressurizing transfer lines to 20 psig with an inert gas, usually helium. The type of inert gas used to purge will be defined by the experiment. Moisture absorbed on the internal surface of the transfer lines can be reduced by heat tracing the lines. Heat tracing, used in combination with pulse purging, has proven to be an effective method. A vacuum source to evacuate the hardware will improve this procedure.

Electropolished 316 stainless steel tubing and components are often used for sampling manifolds due to material compatibility issues and because the surface area can be minimized through electropolishing. However, studies in hydrogen and inert gases have shown that Silcosteel passivated stainless steel adsorbs less water vapor than does the 316 stainless steel. Therefore, Silcosteel may be preferable in certain applications.

When analyzing for metal impurities, a metallic transfer line can compromise the content measured within the sample. For these particular analyses, the use of polymeric tubing is suggested (see Table B.3), specifically poly(ether ether ketone) (PEEK). PEEK remains chemically inert to most solvents and is pressure rated to up to 5000 psig. PEEK can easily be cut to meet the lengths required by the experiment, while also being able to interface with stainless steel or polymer fittings. In the section below we list the properties of polymers.

CTFE Chlorotrifluoroethylene is the generic name for the material produced as Kel-F and Aclar. It is very resistant to all chemicals except tetrahydrofuran (THF) and some halogenated solvents, and is resistant to all inorganic corrosive liquids, including oxidizing acids. CTFE can be used at temperatures up to 100 °C. Ketones cause swelling.

ETFE Ethyltrifluoroethylene is the generic name for materials such as Tefzel. A fluoropolymer used for sealing surfaces, it is resistant to most chemical attack; however, some chlorinated chemicals will cause physical swelling of ETFE tubing.

FEP Fluorinated ethylene propylene is another member of the fluorocarbon family with chemical properties similar to those of other fluoropolymers. It is generally more rigid that PTFE, with somewhat increased tensile strength. It is typically more transparent that PTFE, slightly less porous, and less permeable to oxygen. FEP is not as subject to compressive creep at room temperature as is PTFE, and because of its slightly higher coefficient of friction, it is easier to retain in a compression fitting.

PAEK Poly(aryether ketone) is the generic name for the family of polyketone compounds. PAEK includes poly ether ketone (PEK), poly(ether ether ketone) (PEEK), poly(ether ketone ketone) (PEKK), and poly(ether ketone ether ketone ketone) (PEKEKK), which differ in physical properties and, to a lesser degree, in inertness. These composites resist all common high-performance liquid chromatography solvents and dilute acids and bases. However, concentrated or prolonged use of halogenated solvents may cause the polymer to swell. Avoid concentrated sulfuric or nitric acids (over 10 %).

PEEK Considered relatively inert and biocompatible, poly(ether ether ketone) tubing can withstand temperatures up to 100 °C. Under the right circumstances, 0.005- to 0.020-in ID tubing can be used up to 5000 psi for a limited time, and 0.030 in to 3000 psi. Larger IDs are typically good to 500 psi. Their limits are reduced substantially at elevated temperatures and in contact with some solvents or acids. Its mechanical properties allow PEEK to replace stainless in many situations and in some environments where stainless would be too reactive. However, PEEK can be somewhat adsorptive of solvents and analytes, notably methylene chloride, dimethyl sulfoxide (DMSO), tetrahydrofuran, and high concentrations of sulfuric and nitric acid.

PEEK, glass-filled This form of PEEK has better mechanical properties than those of natural PEEK, and performs extremely well in products such as ferrules.

PFA Perfluoroalkoxy is a fluorocarbon with chemical and mechanical properties similar to those of FEP. More rigid than either PTFE or FEP. Commonly used for injection-molded parts.

PPS Poly(phenylene sulfide) is the generic name for the material produced as Fortron, Ryton, and others. It is very resistant to all solvents, acids, and bases.

PTFE Poly(tetrafluoroethylene) is the generic name for the class of materials such as Teflon. It offers superior chemical resistance but is limited in pressure and temperature capabilities. Because it's so easy to handle, it is often used in low-pressure situations where stainless steel might cause adsorption. PTFE tubing is relatively porous, and compounds of low molecular weight can diffuse through the tubing wall.

PTFE, glass-filled Form of PTFE nearly as inert as the virgin PTFE but is much more mechanically stable.

Polyimide, graphite Graphite-filled polyimide. Due to its brittle nature, it is generally used for ferrules.

Polyimide, Valcon High-temperature (350 °C) graphite-reinforced polyimide composite used for all FS and FSR ferrules (fused silica adapters with and without liner) and many standard ferrules. Valcon polyimide is specially prepared by a process known as hot isostatic pressing (HIP) prior to being machined into individual adapters. This two-step process yields a product with high-temperature stability far exceeding that of parts produced by molding. It cannot be used with steam or with bases such as strong alkali and aqueous ammonia solutions.

Polyimide, virgin Not recommended for general use, due to its tendency to be sticky and brittle at high temperatures. Often used as a high-temperature electrical insulator.

Polypropylene Widely used polymer for nonwetted parts. Attacked by strong oxidizers, and aromatic and chlorinated hydrocarbons.

PVDF Poly(vinylidene fluoride) has excellent resistance to most mineral and organic acids, aliphatic and aromatic hydrocarbons, and halogenated solvents; poor resistance to acetone, methyl ethyl ketone (MEK), tetrahydrofuran, and potassium and sodium hydroxide. Often supplied as Kynar.

Table B.3: Polymer compatibilities (resistance at 20 °C)[a]

Compound	PEEK	PPS	ETFE	PTFE	FEP
Acetaldehyde	A				A
Acetic acid (80 %)	A	A		A	A

Continued...

Compound	PEEK	PPS	ETFE	PTFE	FEP
Acetone	A	A	A	A	A
Ammonia (anhydrous)	A	A			A
Ammonium hydroxide	A	A	A	A	A
Benzene	A	A	A	A	A
Bromine	C	C			A
Chlorine (gas)	A	C			A
Cyclohexane	A	A	A	A	A
Diethylamine	A		A	A	A
Diethylether	A	A			A
Ethanol	A	A	A	A	A
Ethyl acetate	A	A	A	A	A
Ethylene chloride		A	A	A	A
Hexane	A	A	A	A	A
Hydrobromic acid (100%)	C	A		A	A
Hydrochloric acid (100%)	A	C	A	A	A
Hydrofluoric acid (100%)	C	C	A	A	A
Hydrogen peroxide (100%)	A	C		A	A
Ketones	A	A		A	A
Methanol	A	A	A	A	A
Nitric acid (100%)	C	C	A	A	A
Perchloric acid	A				B
Phosphoric acid (100%)	A	A	A	A	A
Potassium hydroxide (70%)	A	A			A
Sulfuric acid (100%)	C	A	A	A	A
Toluene	A	A	A	A	A
Trichloroacetic acid	A	A	A	A	A
Triethylamine			A	A	A
Water	A	A	A	A	A

[a] A, acceptable; B, marginally acceptable/conditional use; C, unsuitable.

B.1.1 Valves

Valve bodies are available in metals or polymers and are generally constructed from the same materials as transfer tubing (i.e., stainless steel type 316, Hastelloy, Nickel 200, and Inconel 600). Polymer valve material is also similar to its transfer line counterpart.

The valve body contains a rotor, which comes into contact with the sample to direct the flow. Rotors are available in a variety of polymer compositions to prevent corrosion. In the section below we outline available rotor materials.

Valcon E A poly(arylether ketone)/PTFE composite, the Valcon E material receives wide GC use in what had previously been a problematic gap between the optimum temperature ranges of pressure and temperature, and in high-performance liquid chromatography applications where the temperature requirement is higher than what can be handled by the Valcon H material and where a lower pressure limit can be tolerated. Standard specifications are 400 psi at 225 °C, but higher pressure ratings are possible at reduced temperatures. However, this polymer cannot be used in prolonged contact with high concentrations of sulfuric and nitric acids, dimethyl sulfoxide, tetrahydrofuran, or liquid methylene chloride.

Valcon E2 A proprietary reinforced TFE composite, Valcon E2 works well at lower pressures and is suitable for temperatures up to 75 °C. This material is resistant to most chemicals but should not be used in prolonged contact with high concentrations of sulfuric and nitric acids, dimethyl sulfoxide, or liquid methylene chloride.

Valcon E3 This designation indicates a proprietary polyimide blend with chemical properties similar to Valcon T, but with higher compressive strength.

Valcon H Carbon fiber–reinforced PTFE-lubricated inert engineering polymer, long used as the standard for typical high-performance liquid chromatography applications in which pressures are around 5000 psi and temperatures are not more than 75 °C. It is not unusual for such valves to be ordered for use at 7000 psi, or even 10,000 psi. However, at 10,000 psi the lifetime may be shortened by as much as 50 %.

Valcon M This material, basically a hydrocarbon in structure, is the most impermeable to light gases of all the rotor materials currently available, with wide acceptance in low-temperature (50 °C maximum) trace gas applications. Avoid use with aromatic hydrocarbons.

Valcon P This composite, the majority of which is PTFE and carbon, was the standard choice for most GC applications before the development of Valcon E. Standard specifications are 400 psi at 175 °C. Routinely used at 1000 psi, 75 °C, it can also be used at temperatures approaching 200 °C with decreased sealing tension; however, at that point Valcon E is probably a better choice from a lifetime standpoint. Valcon E can replace Valcon P in most applications.

Valcon R Although rarely used today, Valcon R (a PTFE composite) finds use in low-temperature/pressure situations that require its nearly universal chemical inertness. Of the chemicals encountered in commercial practice, only molten sodium and fluorine at elevated temperatures and pressures produce any detrimental effects. Its most severe limitation is that it cannot go over 75 °C, even at only at 400 psi.

Valcon T   This polyimide/PTFE/carbon composite has been used successfully for many years, and still cannot be surpassed when applications demand operating temperatures in the range 250 to 350 °C. Standard spec for most series is 300 psi at 330 °C. However, at temperatures below 150 °C there is a tendency for the seal material to stick to the valve body, making the valve difficult to turn and causing the rotor to crack in extreme cases. Literature provided at the time of purchase contains instructions for reconditioning the material if this condition should arise. The Valcon T material is susceptible to attack from steam, ammonia, hydrazines (anhydrous liquids or vapor), primary and secondary amines, and solutions having a pH of 10 or more. Chemical reagents that act as a powerful oxidizing agents (nitric acid, nitrogen tetroxide, etc.) must also be avoided. Valcon T can be used in "hot" GPC/SEC applications with *o*-dichlorobenzene as a solvent.

Valcon TF   This is the series designation for a valve with a virgin PTFE seal. Its mechanical characteristics are poor compared to the other choices, but occasionally its use is dictated by the presence of oxidizing agents too strong for even Valcon R material.

B.2   FTIR Materials

The flow path of the sample in an FTIR system is rather simple; however, there are vulnerabilities where the sampling gas comes into contact with vital parts of the hardware. This includes the O-rings, gas cell, optical windows, and mirrors. From this point, the gas material and contaminants should be identified to make a material selection. Since the optical windows are in the sample flowpath, careful consideration is needed to select the proper material fpr the windows (see Table B.4). The detector output depends on the transmission range of the windows, and the material for the windows depends on the sample gas. A proper window must be able to withstand the chemical environment produced by the sample while having a wide-enough transmission range to "see" the contaminant in question. In the section below we outline material composition and transmission ranges for FTIR windows.

Let us then consider an extreme case for which hydrogen fluoride is the matrix gas and tetrafluorosilane is an impurity. Although calcium fluoride will provide sufficient protection against hydrogen bonding on the surface of the window, it will not provide the wave range to detect tetrafluorosilane. An effective alternative option will be barium fluoride windows.

B.2.1   O-Rings/Gaskets

O-rings are included in the flow path in two crucial areas. A seal is typically required to connect the mirrors to the gas cell and to connect the windows within the mirror. Before beginning an experiment, pressure checking the system could prove to be a necessary addition to the overall experiment method. For if there is a leak, the experiment, or even the health of the analyst, could be compromised. Just as the windows were selected, the O-ring selection will depend greatly on the environment in which it is used. O-ring composition varies greatly; there are, however, a few types of O-rings that will cover most experiment conditions. Viton O-rings are an inexpensive solution for general analysis (i.e., a nonacidic matrix gas). Teflon encapsulated with Viton will add the corrosion benefits of Teflon, together with the compression rating of a Viton O-ring. Kalrez is the more expensive option, for good reason. Kalrez can withstand virtually any matrix composition while providing thermal stability up to 323 °C. This means that Viton O-rings can be a cost-effective solution for most experiments, while Kalrez O-rings will provide the extra stability required for highly acidic and reactive gases. O-ring materials are described briefly in Table B.5.

In addition to the more common polymeric materials listed, metallic and non-metallic gaskets that are harder than O-rings are used in many cases. The most common is copper or nickel in VCR fittings. Historically, copper, gold, and even lead have been used in vacuum systems requiring minimal outgassing. Even indium has been used in high-integrity FTIR systems. The choices of these materials are driven by their ductility versus sealing surfaces, temperature requirements, and of course, reactivity.

B.2.2   Cells/Mirrors

Gas cells can range from glass to aluminum to nickel to stainless steel and even to very exotic materials and, of course, are highly dependent on intended use. These options have been discussed earlier, but it is worth noting that vendors can prepare many of these with proprietary coatings to enhance their resistance to attack from the matrix gas.

Mirrors, which are a vital part of the measurement, are also available in a variety of materials, which may influence their light transmission or precision of analysis. Again, vendors can provide specialty coatings or platings that offer some extended protection to the surfaces.

Table B.4: FTIR window materials and properties

Materials		Transmission Range (cm^{-1})	Refraction Index	Description
AMTIR-1	Ge-As-Se	13,333–714	2.606	Similar to germanium; not soluble in water; expensive
Barium fluoride	BaF_2	50,000–869	1.46	Sensitive to thermal shock; insoluble in water; resists fluorine and fluorides
Calcium fluoride	CaF_2	66,667–1,111	1.40	Low solubility; insoluble in water, resists most acids and alkalides; do not use with solutions of NH_3 salts
Cesium iodide	CsI	6,667–200	1.74	Hygroscopic; easily scratched
Germanium	Ge	5,000–869	4.0	Chemically inert, hard and brittle; tends to fracture
Magnesium fluoride	MgF_2	90,909–1,333	1.37–1.38	More soluble than CaF_2; sensitive to thermal shock
Potassium bromide	KBr	40,000–400	1.53	Low cost, extended transmission range; withstands thermal and mechanical shock; hygroscopic
Potassium chloride	KCl	55,555–500	1.46	Less hygroscopic and more resistant than NaCl to thermal shock
Silicon dioxide	SiO_2	25,000–2,500	1.5[a]	Insoluble in water; birefringent
Silver bromide	AgBr	20,000–300	2.2	Insoluble; will cold flow; darkens less than AgCl in UV light
Silver chloride	AgCl	25,000–434	2.0	Soluble in water; corrosive to metals
Sodium chloride	NaCl	40,000–667	1.52	Low cost; rugged; hygroscopic
Thallium bromoiodide (KRS–5)		20,000–285	2.37	Soluble in bases, insoluble in acids; slightly water soluble

Continued...

Materials		Transmission Range (cm^{-1})	Refraction Index	Description
Zinc selenide	ZnSe	10,000–556	2.40	Insoluble in water, resistant to most solvents; brittle, handle with caution
Zinc sulfide	ZnS	10,000–714	2.20	Insoluble in water, slightly soluble in acids HNO_3, H_2SO_4, and KOH

[a] At 3333 cm^{-1}.

Table B.5: O-ring materials

O-ring Type	Properties	Temperature Range (°C)	Compatible	Not Compatibile
Viton (fluorocarbon type A)	Good wear resistance; moderate short term resilience; excellent permeation resistance	−26–200	Most acids; halogenated hydrocarbons; petroleum fuels	ketones; amines; acetate; acetone
Kalrez/Chemraz (perfluoroelastomers)	Good wear resistance; excellent chemical resistance; high temperature resistance	−23–323	Excellent resistance to nearly all chemicals with the exception of fluorinated solvents	Gaseous alkali metals; uranium hexafluoride
Teflon encapsulated with Viton	Chemical resistance; heat resistance	−26–200	Properties of Viton	Properties of Viton

Index

Trace Analysis of Specialty and Electronic Gases.
By William M. Geiger and Mark W. Raynor. Copyright © 2013 John Wiley & Sons, Inc.

337